Methods in Molecular Biology

For further volumes:
http://www.springer.com/series/7651

For over 35 years, biological scientists have come to rely on the research protocols and methodologies in the critically acclaimed *Methods in Molecular Biology* series. The series was the first to introduce the step-by-step protocols approach that has become the standard in all biomedical protocol publishing. Each protocol is provided in readily-reproducible step-by step fashion, opening with an introductory overview, a list of the materials and reagents needed to complete the experiment, and followed by a detailed procedure that is supported with a helpful notes section offering tips and tricks of the trade as well as troubleshooting advice. These hallmark features were introduced by series editor Dr. John Walker and constitute the key ingredient in each and every volume of the *Methods in Molecular Biology* series. Tested and trusted, comprehensive and reliable, all protocols from the series are indexed in PubMed.

Stem Cell-Based Neural Model Systems for Brain Disorders

Edited by

Yu-Wen Alvin Huang

Department of Molecular Biology, Cell Biology, and Biochemistry, Center for Translational Neuroscience, Carney Institute for Brain Science and Brown Institute of Translational Science, Brown University, Providence, RI, USA

ChangHui Pak

Department of Biochemistry and Molecular Biology, University of Massachusetts Amherst, Amherst, MA, USA

Humana Press

Editors
Yu-Wen Alvin Huang
Department of Molecular Biology
Cell Biology, and Biochemistry
Center for Translational Neuroscience
Carney Institute for Brain Science and
Brown Institute of Translational Science
Brown University
Providence, RI, USA

ChangHui Pak
Department of Biochemistry and Molecular Biology
University of Massachusetts Amherst
Amherst, MA, USA

ISSN 1064-3745 ISSN 1940-6029 (electronic)
Methods in Molecular Biology
ISBN 978-1-0716-3286-4 ISBN 978-1-0716-3287-1 (eBook)
https://doi.org/10.1007/978-1-0716-3287-1

This Humana imprint is published by the registered company Springer Science+Business Media, LLC, part of Springer Nature.
The registered company address is: 1 New York Plaza, New York, NY 10004, U.S.A.

Preface

Studying the complex and lengthy pathogenesis of brain disorders, spanning across neurodevelopment, neuropsychiatric, and neurodegenerative diseases, is challenging at all levels, and this is exacerbated in particular by the lack of reliable and robust models that can capture human-specific disease characteristics. Human brain cells are markedly different from rodent neurons, astrocytes, and microglia, and such differences between model organisms are well reflected by the low success rate of clinical trials that were based on initial tests in mice. In addition, the inherent complexity inside the brain remains a principal hurdle to disease modeling, as standard approaches are unable to replicate the intricate interplay among different cell types. All of the technical concerns discussed here highlight the utility of human-induced pluripotent stem cell (iPSC)-based models, in which major brain cell types—neurons, astrocytes, microglia, oligodendrocytes, and endothelial cells—can be generated to harbor a desired genotype precisely and robustly. Considering the individual brain cell types have been documented to exhibit different and even conflicting effects in the face of the same stimulus or genetic manipulation that promotes neurological pathology, the established reductionist approach of pure or mixed neural cell cultures, in a 2D or 3D platform, uniquely offers cell type-specific information that is obscured with the established models in vivo or in vitro that provide only aggregate readout from a collection of brain cell types. As a result, the stem cell-based culture systems have become a valuable resource for researchers studying a wide range of neurophysiological conditions in various experimental settings, including disease modeling, drug discovery, and regenerative medicine.

The scope of this method collection is to present validated and well-adapted procedures involving humanized and/or stem cell-based neural model systems that have been proven helpful in better understanding the essential brain functions involved in the pathogenesis of brain disorders. These chapters are intended to guide readers with simple step-by-step approaches such that experiments can be readily adopted in a laboratory setting. Protocols entailing the generation of multiple neural cell types in 2D and 3D (Chapters 1, 2, 3, 4, 5, 6, 7, 8, 9, 10, 11, 12, 13, and 14) as well as cutting-edge techniques to assay neural function (Chapters 15, 16, 17, 18, 19, 20, 21, and 22) are described. With contributions from leading experts in the field who are actively generating and utilizing these techniques, we hope that this book will be an essential resource for researchers and students in neuroscience, stem cell biology, and related fields. The diverse content of these chapters provides a broad view of the current state of the field and highlights the enormous potential of iPSCs for advancing our knowledge and ultimately for the development of much-needed therapeutics.

Providence, RI, USA *Yu-Wen Alvin Huang*
Amherst, MA, USA *ChangHui Pak*

Contents

Contributors

BENEDETTA ASSETTA • *Department of Molecular Biology, Cell Biology and Biochemistry, Center for Translational Neuroscience, Carney Institute for Brain Science and Brown Institute of Translational Science, Brown University, Providence, RI, USA*

ELAINA ATHERTON • *Department of Molecular Pharmacology, Physiology, and Biotechnology, Brown University, Providence, RI, USA*

ANNE G. BANG • *Conrad Prebys Center for Chemical Genomics, Sanford Burnham Prebys Medical Discovery Institute, La Jolla, CA, USA*

NATHANIEL BARTON • *Program in Bioinformatics and Integrative Biology, University of Massachusetts Chan Medical School, Worcester, MA, USA; Department of Neurology, University of Massachusetts Chan Medical School, Worcester, MA, USA; NeuroNexus Institute, University of Massachusetts Chan Medical School, Worcester, MA, USA; Department of Molecular, Cell and Cancer Biology, University of Massachusetts Chan Medical School, Worcester, MA, USA*

JOEL W. BLANCHARD • *Nash Family Department of Neuroscience, Ronald M. Loeb Center for Alzheimer's Disease, and Friedman Brain Institute, Icahn School of Medicine at Mount Sinai, New York, NY, USA; Black Family Stem Cell Institute, Icahn School of Medicine at Mount Sinai, New York, NY, USA*

ANDREW J. BORELAND • *Child Health Institute of New Jersey, Robert Wood Johnson Medical School, New Brunswick, NJ, USA; Department of Neuroscience and Cell Biology, Robert Wood Johnson Medical School, Piscataway, NJ, USA*

DAVID A. BORTON • *Center for Biomedical Engineering, Brown University, Providence, RI, USA; Department of Veteran Affairs, Providence Medical Center, Center for Neurorestoration and Neurotechnology, Providence, RI, USA; Department of Neurosurgery, Rhode Island Hospital, Providence, RI, USA; School of Engineering, Brown University, Providence, RI, USA; Carney Institute for Brain Science, Brown University, Providence, RI, USA*

DIEDE BROEKAART • *Nash Family Department of Neuroscience, Ronald M. Loeb Center for Alzheimer's Disease, and Friedman Brain Institute, Icahn School of Medicine at Mount Sinai, New York, NY, USA; Department of Neurology Icahn School of Medicine at Mount Sinai, New York, NY, USA*

SOPHIE BROWN • *Center for Biomedical Engineering, Brown University, Providence, RI, USA; School of Engineering, Brown University, Providence, RI, USA*

MICHAEL BULA • *The Scripps Research Institute, La Jolla, CA, USA*

YINGLEONG CHAN • *Program in Bioinformatics and Integrative Biology, University of Massachusetts Chan Medical School, Worcester, MA, USA; Department of Neurology, University of Massachusetts Chan Medical School, Worcester, MA, USA; NeuroNexus Institute, University of Massachusetts Chan Medical School, Worcester, MA, USA*

SHIH-WEI CHEN • *Department of Life Sciences and Institute of Genome Sciences, College of Life Sciences, National Yang Ming Chiao Tung University, Taipei, Taiwan*

SUNDARI CHETTY • *Center for Regenerative Medicine, Massachusetts General Hospital, Boston, MA, USA; Department of Psychiatry, Massachusetts General Hospital, Harvard Medical School, Boston, MA, USA; Harvard Stem Cell Institute, Cambridge, MA, USA*

SANGMI CHUNG • *Department of Cell Biology and Anatomy, New York Medical College, Valhalla, NY, USA*

KEVIN CONNOLLY • *Department of Molecular Biology, Cell Biology and Biochemistry, Center for Translational Neuroscience, Carney Institute for Brain Science and Brown Institute of Translational Science, Brown University, Providence, RI, USA*

PEPPER DAWES • *Program in Bioinformatics and Integrative Biology, University of Massachusetts Chan Medical School, Worcester, MA, USA; Department of Neurology, University of Massachusetts Chan Medical School, Worcester, MA, USA; NeuroNexus Institute, University of Massachusetts Chan Medical School, Worcester, MA, USA; Department of Molecular, Cell and Cancer Biology, University of Massachusetts Chan Medical School, Worcester, MA, USA*

KARMEN T. DIEP • *Neuroscience and Behavior Graduate Program, University of Massachusetts, Amherst, MA, USA*

JAY ENGLISH • *Department of Biochemistry and Molecular Biology, University of Massachusetts Amherst, Amherst, MA, USA; Molecular and Cellular Biology Graduate Program, University of Massachusetts Amherst, Amherst, MA, USA*

GUZIDE AYSE ERDEMIR • *Department of Molecular Biology, Cell Biology and Biochemistry, Center for Translational Neuroscience, Carney Institute for Brain Science and Brown Institute of Translational Science, Brown University, Providence, RI, USA*

LINGYI FAN • *Psychiatric Laboratory and Mental Health Center, The State Key Laboratory of Biotherapy, West China Hospital of Sichuan University, Chengdu, China*

KOJI L. FOREMAN • *Department of Chemical and Biological Engineering, Department of Neurological Surgery, University of Wisconsin–Madison, Madison, WI, USA*

RAFAEL GABRIEL III • *Child Health Institute of New Jersey, Robert Wood Johnson Medical School, New Brunswick, NJ, USA*

ETHAN HOWELL • *Department of Biochemistry and Molecular Biology, University of Massachusetts Amherst, Amherst, MA, USA*

YU-WEN ALVIN HUANG • *Department of Molecular Biology, Cell Biology and Biochemistry, Center for Translational Neuroscience, Carney Institute for Brain Science and Brown Institute of Translational Science, Brown University, Providence, RI, USA*

JUSTYNA A. JANAS • *Institute for Stem Cell Biology and Regenerative Medicine, Stanford University School of Medicine, Stanford, CA, USA; Department of Pathology, Stanford University School of Medicine, Stanford, CA, USA*

MIKAEL LEHOUX • *Department of Molecular Biology, Cell Biology and Biochemistry, Center for Translational Neuroscience, Carney Institute for Brain Science and Brown Institute of Translational Science, Brown University, Providence, RI, USA*

ELAINE T. LIM • *Program in Bioinformatics and Integrative Biology, University of Massachusetts Chan Medical School, Worcester, MA, USA; Department of Neurology, University of Massachusetts Chan Medical School, Worcester, MA, USA; NeuroNexus Institute, University of Massachusetts Chan Medical School, Worcester, MA, USA; Department of Molecular, Cell and Cancer Biology, University of Massachusetts Chan Medical School, Worcester, MA, USA*

JUDY S. LIU • *Department of Molecular Biology, Cell Biology and Biochemistry; and Carney Institute for Brain Science, Brown University, Laboratories for Molecular Medicine, Providence, RI, USA; Center for Translational Neuroscience, Carney Institute for Brain Science and Brown Institute for Translational Science, Brown University, Providence, RI, USA; Department of Neurology, Rhode Island Hospital, Providence, RI, USA*

SELENE LOMOIO • *Department of Neuroscience, Tufts University School of Medicine, Boston, MA, USA*

ZOE MATTINGLY • *Center for Regenerative Medicine, Massachusetts General Hospital, Boston, MA, USA*

DANNY MCSWEENEY • *Department of Biochemistry and Molecular Biology, University of Massachusetts Amherst, Amherst, MA, USA; Molecular and Cellular Biology Graduate Program, University of Massachusetts Amherst, Amherst, MA, USA*

LOUISE A. MESENTIER-LOURO • *Nash Family Department of Neuroscience, Ronald M. Loeb Center for Alzheimer's Disease, and Friedman Brain Institute, Icahn School of Medicine at Mount Sinai, New York, NY, USA; Black Family Stem Cell Institute, Icahn School of Medicine at Mount Sinai, New York, NY, USA*

ERIC M. MORROW • *Department of Molecular Biology, Cell Biology and Biochemistry, Brown University, Providence, RI, USA; Center for Translational Neuroscience, Carney Institute for Brain Science and Brown Institute for Translational Science, Brown University, Providence, RI, USA; Brown University, Laboratories for Molecular Medicine, Providence, RI, USA*

LIAM MURRAY • *Program in Bioinformatics and Integrative Biology, University of Massachusetts Chan Medical School, Worcester, MA, USA; Department of Neurology, University of Massachusetts Chan Medical School, Worcester, MA, USA; NeuroNexus Institute, University of Massachusetts Chan Medical School, Worcester, MA, USA; Department of Molecular, Cell and Cancer Biology, University of Massachusetts Chan Medical School, Worcester, MA, USA*

ASHLEY NEIL • *Conrad Prebys Center for Chemical Genomics, Sanford Burnham Prebys Medical Discovery Institute, La Jolla, CA, USA*

YI HAN NG • *Department of Molecular and Cellular Physiology, Stanford University School of Medicine, Stanford, CA, USA; Institute for Stem Cell Biology and Regenerative Medicine, Stanford University School of Medicine, Stanford, CA, USA*

PEIYAN NI • *Psychiatric Laboratory and Mental Health Center, The State Key Laboratory of Biotherapy, West China Hospital of Sichuan University, Chengdu, China*

RYAN O'ROURKE • *Department of Molecular Biology, Cell Biology and Biochemistry, Center for Translational Neuroscience, Carney Institute for Brain Science and Brown Institute of Translational Science, Brown University, Providence, RI, USA*

MEAGAN N. OLSON • *Program in Bioinformatics and Integrative Biology, University of Massachusetts Chan Medical School, Worcester, MA, USA; Department of Neurology, University of Massachusetts Chan Medical School, Worcester, MA, USA; NeuroNexus Institute, University of Massachusetts Chan Medical School, Worcester, MA, USA; Department of Molecular, Cell and Cancer Biology, University of Massachusetts Chan Medical School, Worcester, MA, USA*

QING OUYANG • *Department of Molecular Biology, Cell Biology and Biochemistry, Brown University, Providence, RI, USA; Center for Translational Neuroscience, Carney Institute for Brain Science and Brown Institute for Translational Science, Brown University, Providence, RI, USA; Brown University, Laboratories for Molecular Medicine, Providence, RI, USA*

CHANGHUI PAK • *Department of Biochemistry and Molecular Biology, University of Massachusetts Amherst, Amherst, MA, USA*

SEAN P. PALECEK • *Department of Chemical and Biological Engineering, Department of Neurological Surgery, University of Wisconsin–Madison, Madison, WI, USA*

Zhiping P. Pang • *Child Health Institute of New Jersey, Robert Wood Johnson Medical School, New Brunswick, NJ, USA; Department of Neuroscience and Cell Biology, Robert Wood Johnson Medical School, Piscataway, NJ, USA*

Narciso S. Pavon • *Neuroscience and Behavior Graduate Program, University of Massachusetts, Amherst, MA, USA*

Ana C. Pereira • *Nash Family Department of Neuroscience, Ronald M. Loeb Center for Alzheimer's Disease, and Friedman Brain Institute, Icahn School of Medicine at Mount Sinai, New York, NY, USA; Department of Neurology Icahn School of Medicine at Mount Sinai, New York, NY, USA*

Matthew F. Pescosolido • *Department of Molecular Biology, Cell Biology and Biochemistry; and Carney Institute for Brain Science, Brown University, Laboratories for Molecular Medicine, Providence, RI, USA; Center for Translational Neuroscience, Carney Institute for Brain Science and Brown Institute for Translational Science, Brown University, Providence, RI, USA*

Deborah Pré • *Conrad Prebys Center for Chemical Genomics, Sanford Burnham Prebys Medical Discovery Institute, La Jolla, CA, USA*

Fumiko Ribbe • *Department of Biochemistry and Molecular Biology, University of Massachusetts Amherst, Amherst, MA, USA*

Michael Schmidt • *Department of Molecular Biology, Cell Biology and Biochemistry, Brown University, Providence, RI, USA; Center for Translational Neuroscience, Carney Institute for Brain Science and Brown Institute for Translational Science, Brown University, Providence, RI, USA*

Rebecca Sebastian • *Department of Biochemistry and Molecular Biology, University of Massachusetts, Amherst, MA, USA; Neuroscience and Behavior Graduate Program, University of Massachusetts, Amherst, MA, USA*

Eric V. Shusta • *Department of Chemical and Biological Engineering, Department of Neurological Surgery, University of Wisconsin–Madison, Madison, WI, USA*

Yoonjae Song • *Neuroscience and Behavior Graduate Program, University of Massachusetts, Amherst, MA, USA*

Natalie Suhy • *Nash Family Department of Neuroscience, Ronald M. Loeb Center for Alzheimer's Disease, and Friedman Brain Institute, Icahn School of Medicine at Mount Sinai, New York, NY, USA; Black Family Stem Cell Institute, Icahn School of Medicine at Mount Sinai, New York, NY, USA*

Zijun Sun • *Department of Molecular and Cellular Physiology, Stanford University School of Medicine, Stanford, CA, USA; Howard Hughes Medical Institute, Stanford University School of Medicine, Stanford, CA, USA*

Giuseppina Tesco • *Department of Neuroscience, Tufts University School of Medicine, Boston, MA, USA*

Gabriella Vulakh • *Department of Molecular Biology, Cell Biology, and Biochemistry, Brown University, Providence, RI, USA*

Yu-Hui Wong • *Brain Research Center, National Yang Ming Chiao Tung University, Taipei, Taiwan*

Alexander T. Wooten • *Conrad Prebys Center for Chemical Genomics, Sanford Burnham Prebys Medical Discovery Institute, La Jolla, CA, USA*

Yuanwei Yan • *Waisman Center, University of Wisconsin-Madison, Madison, WI, USA*

Xin Yang • *Department of Molecular Biology, Cell Biology, and Biochemistry, Brown University, Providence, RI, USA*

Su-Chun Zhang • *Waisman Center, University of Wisconsin-Madison, Madison, WI, USA; Department of Neuroscience, School of Medicine and Public Health, University of Wisconsin, Madison, WI, USA; Department of Neurology, School of Medicine and Public Health, University of Wisconsin, Madison, WI, USA; Program in Neuroscience and Behavioral Disorders, Duke-NUS Medical School, Singapore, Singapore*
Haowen Zhou • *Conrad Prebys Center for Chemical Genomics, Sanford Burnham Prebys Medical Discovery Institute, La Jolla, CA, USA*
Amy Zinski • *Department of Cell Biology and Anatomy, New York Medical College, Valhalla, NY, USA*

Chapter 1

Generation of Cerebral Cortical Neurons from Human Pluripotent Stem Cells in 3D Culture

Yuanwei Yan and Su-Chun Zhang

Abstract

Human forebrain cortical neurons are essential for fundamental functions like memory and consciousness. Generation of cortical neurons from human pluripotent stem cells provides a great source for creating models specific to cortical neuron diseases and for developing therapeutics. This chapter describes a detailed and robust method for generating human mature cortical neurons from stem cells in 3D suspension culture.

Key words Pluripotent stem cell, Neural differentiation, 3D suspension culture

1 Introduction

Forebrain cortical neurons, or glutamatergic projection neurons, are the main cellular composition in the human cortex. They form neural networks with other cell subtypes for fundamental functions including memory, consciousness, language, and thought. The cortical projection neurons organize into six layers and integrate into cortical circuits during brain development [6, 10, 19, 20]. Studying cortical networks is essential to probing neurological diseases including, Alzheimer's, Huntington's disease, and autism spectrum disorders (ASD) [3, 11, 15]. This is, however, challenging due to the lack of a reliable model of living human cortical tissues. Human pluripotent stem cells (hPSCs) including human embryonic stem cells (hESCs) and human-induced pluripotent stem cells (hiPSCs) provide a promising tool to derive human forebrain cortical neurons and generate cortical tissues. The mammalian brain develops from the neural tube at the gastrula stage in a process called neural induction by the inhibition of bone morphogenetic protein (BMP) and/or activation of fibroblast growth factor (FGF) pathways [12, 17]. The neural progenitors in the neural tube are specified and give rise to specific neuronal subtypes along

Yu-Wen Alvin Huang and ChangHui Pak (eds.), *Stem Cell-Based Neural Model Systems for Brain Disorders*, Methods in Molecular Biology, vol. 2683, https://doi.org/10.1007/978-1-0716-3287-1_1,

the rostral-caudal axis by the regulation of Wnt and retinoic acid (RA) signaling, and along the dorsal-ventral axis mainly tuning by Wnts/BMPs from the dorsal and sonic hedgehog (SHH) from the ventral [18].

The cortical neuron differentiation from hPSCs was first developed by using the embryoid body (EB) method, which generated neural tube-like rosettes in 2001 [21]. It was noticed that the neural rosettes generated in the serum-free condition without the presence of morphogens are primarily of the dorsal forebrain identity [8, 13]. The EB method was a combined 2D monolayer and 3D suspension culture. In 2008, the Sasai lab modified the EB culture to a complete 3D culture system called the SFEBq method [2], which forms the basis of cortical organoids. Meanwhile, Studer's lab developed the dual SMAD method to induce neural conversion of >80% of hESCs using small molecules to inhibit the BMP and TGF-β signalings that are required for meso- and endo-derm development [1]. The dual SMAD method was initially developed for monolayer cultures, overcoming the technical challenges of producing healthy EBs, but it also enables neural induction in 3D cultures, including the generation of cortical organoids/spheroids [7, 14].

Given that the neuroepithelia generated in the presence or absence of dual SMAD are of forebrain origin, it is clear that the dorsally derived cortical neurons are readily generated by balancing the dorsal-ventral signaling, i.e., WNT and sonic hedgehog (SHH) signaling [9]. The human cortical progenitors tend to divide over a long period before exiting the cell cycle and becoming post-mitotic neurons [4]. By blocking MAP kinase as well as FGF and Notch signaling pathways with small molecules, the Studer group optimized the culture system to accelerate the generation of cerebral cortical neurons [16]. Currently, the available protocols yield primarily deep-layer cortical neurons.

We described the differentiation of cortical neurons from hPSCs in 2D cultures [5, 9]. Here, we describe a modified protocol for 3D suspension culture. The protocol provides a simple and robust method to generate human cortical neurons step-by-step. It allows the generation of mature cortical neurons in 2D cultures for cellular and functional analysis. It also enables generation of cortical organoids by extending the 3D suspension culture.

2 Materials

Prepare all solutions using deionized water unless indicated otherwise.

2.1 hPSC Maintenance

1. *hPSC line:* The cell line used in this study is H9 hESCs. The cell vials are cryopreserved in a liquid nitrogen tank for long-term storage.
2. *1% Matrigel-coated surface:* Matrigel LDEV-Free Reduced Growth Factor Basement Membrane Matrix is diluted at the ratio of 1:1 with cold Dulbecco's Modified Eagle's Medium (DMEM) at 4 °C. The solution is aliquoted and stored at −80 °C. Prior to coating, the diluted Matrigel is thawed at room temperature (RT). After thawing, Matrigel is mixed well with cold DMEM at the ratio of 1:50. The culture surface is coated with 1% Matrigel at 0.1 mL/cm^2 at 37 °C for at least 30 min before use (*see* **Note 1**).
3. *TeSR E8 medium for undifferentiated hPSC culture:* TeSR™-E8™ medium is used for undifferentiated hPSC culture. This medium is composed of TeSR™- E8™ basal medium and 25× TeSRTM-E8™ supplement. TeSR™- E8™ basal medium is stored at 2–8 °C and the supplement is stored at −20 °C. When used, the two components are mixed completely. The mixture can be stored at 2–8 °C for 1 week (*see* **Note 2**).
4. *Cell dissociation reagent:* Accutase is used to dissociate undifferentiated hPSCs grown on the Matrigel-coated surface and to dissociate cortical neural spheres in suspension culture. Accutase is stored at −20 °C. After thawing, it is stored at 2–8 °C for 2 weeks.
5. *ROCK inhibitor Y27632:* ROCK inhibitor Y27632 is used to protect hPSCs and cortical neurons from apoptosis after dissociation into single cells. For stock solution, 10 mM Y27632 is prepared by dissolving in dimethyl sulfoxide (DMSO) and is preserved at −80 °C for long-term usage. The final working concentration is 10 μM.
6. *PBS*: Gibco™ dulbecco's phosphate-buffered saline (DPBS) is used to wash samples. DPBS is stored at RT for long-term use.

2.2 Cerebral Cortical Neuron Differentiation from hPSCs

1. *Neural induction medium:* The neural induction medium contains Gibco™ DMEM-F12 plus 1% N2 supplement- and 1% Gibco™ MEM non-essential amino acids solution (NEEA). This medium is referred to as neural induction medium (NIM). After preparation, the medium is stored at 2–8 °C for 1 month.
2. *Dual SMAD inhibitors:* Neural induction is enhanced by the addition of two SMAD inhibitors, SB431542 and DMH1. Both inhibitors are dissolved in DMSO and kept at −80 °C for long-term use. For neural induction, 10 μM SB431542 and 2 μM DMH1 are added to neural induction medium (*see* **Note 3**).

3. *Growth factors:* Recombinant human basic fibroblast growth factor (FGF-2) and epidermal growth factor (EGF) are used for neural differentiation. Both FGF-2 and EGF are prepared in 0.1% bovine serum albumin (BSA) DI water. They are aliquoted and stored at −80 °C for long-term use. The working concentration of FGF-2 and EGF is 10 ng/mL.
4. *Neural differentiation medium:* The neural differentiation medium contains neurobasal™ medium plus 2% B27 minus vitamin A serum-free supplement, 1% Gibco™ MEM non-essential amino acids solution (NEEA), and 1% GlutMAX™ supplement. This medium is referred to as neural differentiation medium (NDM). After preparation, the medium is stored at 2–8 °C for 1 month.
5. *Neurotrophic factors:* Brain-derived neurotrophic factor (BDNF) and glial cell line-derived neurotrophic factor (GDNF) are used for neural differentiation. Both BDNF and GDNF are prepared in 0.1% bovine serum albumin (BSA) DI water. They are aliquoted and stored at −80 °C for long-term use. The working concentration of BDNF and GDNF is 10 ng/mL.
6. *Small molecules:* small molecules of ascorbic acid (AA) and cAMP are used. They are aliquoted and stored at −80 °C for long-term use. The working concentration of AA is 200 μM, and cAMP is 1 μM.

3 Methods

3.1 hPSC Maintenance

3.1.1 Thaw hPSCs from a Cell Bank

1. Coat a 6-well plate with 1% Matrigel at least 30 min before starting the experiment. Add 1 mL Matrigel for each well (*see* **Note 4**). Place the coated plate into a 37 °C, 5% CO_2 incubator.
2. Remove one H9 cryovial from the cell bank. Place the vial in a 37 °C water bath immediately. Gently swirl the vial to quickly thaw the cells until the ice has completely melted in the vial (*see* **Note 5**).
3. Take a 15 mL centrifuge tube and label it. Transfer the cell suspension from the vial into the centrifuge tube when the ice has completely melted.
4. Immediately add ten times the volume of TeSR™-E8™ medium into the tube (*see* **Note 6**) and gently mix. Centrifuge the tube at 300 g for 5 min.
5. Remove the supernatant from the tube and make sure a complete cell pellet is visible at the bottom of the tube.

6. Add ten times the volume of DPBS into the tube and mix well (*see* **Note 7**). Centrifuge the tube again at 300 g for 5 min.
7. Prepare the hPSC culture TeSR™-E8™ medium. Add 10 μM ROCK inhibitor Y27632 into E8 medium and mix well (*see* **Note 8**).
8. Add the above-prepared E8 medium to resuspend the cells. And distribute the cells evenly into the Matrigel-coated plate at 1.5×10^5 cells/cm^2. Add 3 mL of medium to each well. Place the plate in the 37 °C, 5% CO_2 incubator.

3.1.2 hPSC Culture and Passage

1. On day 1 after thawing, observe the cells under a phase contrast microscope to check the recovery of the cells.
2. Change the medium with fresh E8 medium without Y27632 (*see* **Note 9**). Add 3 mL of medium to each well. Place the plate back into the incubator.
3. Check the growth of cells and replace the culture media daily for 5–6 days. When the culture confluency reaches 80–90%, passage the cells.
4. For passaging, completely remove the medium and add 1 mL Accutase to each well. Incubate the plate at RT for 3–5 min (*see* **Note 10**).
5. Add 2 mL DPBS to each well and gently pipet the detached cell clumps for dissociation into single cells. Transfer the single-cell suspension to a 15 mL tube. Centrifuge the tube again at 300 g for 5 min.
6. Remove the supernatant from the tube and make sure all the cells settle down at the bottom of the tube. Add fresh DBPS to the wash again (*see* **Note 11**).
7. Take a small amount of sample and count the cells using a hemacytometer. Based on the cell count, calculate the number of new batches of H9 maintenance and cortical neuron differentiation.

3.2 Differentiation of Forebrain Cortical Neurons in 3D Suspension Culture (Fig. 1)

3.2.1 Neural Induction

1. After harvesting hPSCs, determine the initial cell number for neural differentiation with a seeding density of 5×10^5 cells/cm^2
2. Prepared NIM with 10 μM SB431542 and 2 μM DMH1. And add 10 μM ROCK inhibitor Y27632 into NIM and mix well (*see* **Note 12**).
3. Use the above prepared NIM to resuspend the cells. Seed the desired cells in a Corning™ Costar™ ultra-low attachment 6-well plate (Fisher Scientific, Cat#07-200-601). Add 3 mL NIM with cells in each well.

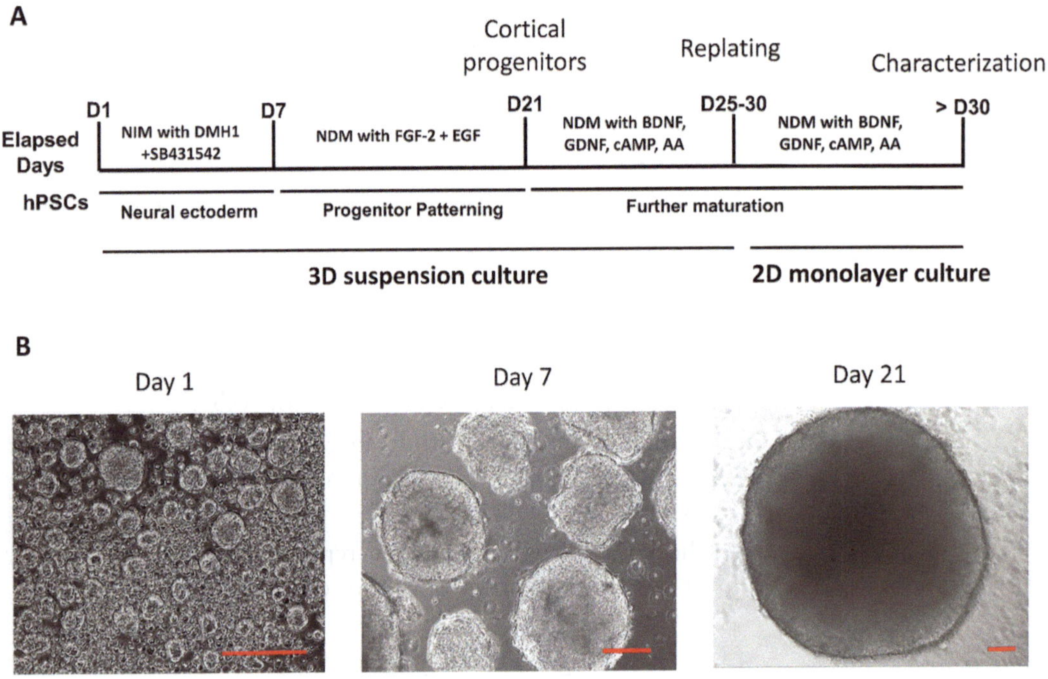

Fig. 1 Cortical neuron differentiation from hPSCs. (**a**) Schematic illustration of cortical neuron differentiation from hPSCs in 3D culture. (**b**) Morphology of neural spheres at different stages of differentiation (scale bar 200 μm)

4. Place the plate in the 37 °C, 5% CO_2 incubator. This is day 0 for cortical neuron differentiation from hPSCs.
5. On day 1, check the cell growth under the microscope (Fig. 1b). Change the medium to remove ROCK inhibitor Y27632 using fresh NIM with 10 μM SB431542 and 2 μM DMH1. To change the medium, transfer the spheres to a 50 mL centrifuge tube. Wait for a few minutes to let the spheres settle down at the bottom of the tube. Aspirate the supernatant and add the fresh NIM. Pipette up and down to break the aggregated spheres (*see* **Note 13**). Transfer the cells to an ultra-low attachment 6-well plate and place the plate in the incubator.
6. Feed the cells every 2–3 days with NIM with 10 μM SB431542 and 2 μM DMH1 until day 7. On day 7, the neural spheres become larger than the size of the spheres on day 1 (Fig. 1b).

3.2.2 Differentiation and Expansion of Cortical Progenitors

1. On day 7, withdraw the dual SMAD inhibitors and switch the medium to NDM with 10 ng/mL FGF-2 and EGF.
2. Feed the cells every 2–3 days with NDM with FGF-2 and EGF until day 21. To change the medium, transfer the spheres to a 50 mL centrifuge tube. Wait for a few minutes to let the spheres settle down at the bottom of the tube. Aspirate the supernatant

and add the fresh medium. Transfer the cells to an ultra-low attachment 6-well plate and place the plate in the incubator.

3. On day 21, withdraw the growth factors and switch NDM with 10 ng/mL BDNF, 10 ng/mL GDNF, 200 μM AA, and 1 μM cAMP.
4. Feed the cells every 2–3 days with NDM with BDNF, GDNF, AA, and cAMP until further use.

3.2.3 Differentiation and Maturation of Cortical Neurons

1. On days 25 to 30, the cortical spheres can be dissociated into single cells using Accutase. Before the experiment, coat a 6-well tissue culture plate with 1% Matrigel for at least 30 min in 37 °C incubator.
2. Collect cortical spheres and transfer the spheres to a 50 mL centrifuge tube. Wait for a few minutes until the spheres settle down. Remove the supernatant.
3. Wash the spheres with 5 mL DPBS and remove DPBS. Add 3 mL Accutase.
4. Incubate the sample in a 37 °C water bath for about 5 min, until the spheres become fuzzy.
5. After 5 min, add 5 mL DPBS. Pipette up and down to break spheres. Centrifuge the tube again at 300 g for 5 min.
6. Remove the supernatant after centrifuge. And wash again with 5 mL DPBS and centrifuge again.
7. Add fresh NDM with 10 ng/mL BDNF, 10 ng/mL GDNF, 200 μM AA, 1 μM cAMP, and 10 μM ROCK inhibitor Y27632 and resuspend the cells (*see* **Note 14**).
8. Transfer the cells into a Matrigel-coated 6-well plate at a cell density of 2.5–3 × 10^5 cells/cm^2. Add 3 mL medium to each well. Place the plate in the 37 °C, 5% CO_2 incubator.
9. The first day after replating, check the cells under the microscope. Remove 2.5 mL of medium from each well. And add 2.5 mL fresh NDM with 10 ng/mL BDNF, 10 ng/mL GDNF, 200 μM AA, and 1 μM cAMP without removing ROCK inhibitor Y27632 from each well (*see* **Note 15**).
10. Feed the cells every 2–3 days using NDM with BDNF, GDNF, AA, and cAMP.
11. The cells can be characterized by various assays, such as immunocytochemistry, confocal microscopy, electrophysiology, etc. The neural sphere on day 21 exhibited positive expression of FOXG1, a marker for forebrain cortical progenitors (Fig. 2a). On day 32, the cells showed expression of cortical transcriptional markers TBR1 and CTIP2 (Fig. 2b). And another cortical transcriptional markers, SATB2, was observed for long-term culture (Fig. 2c).

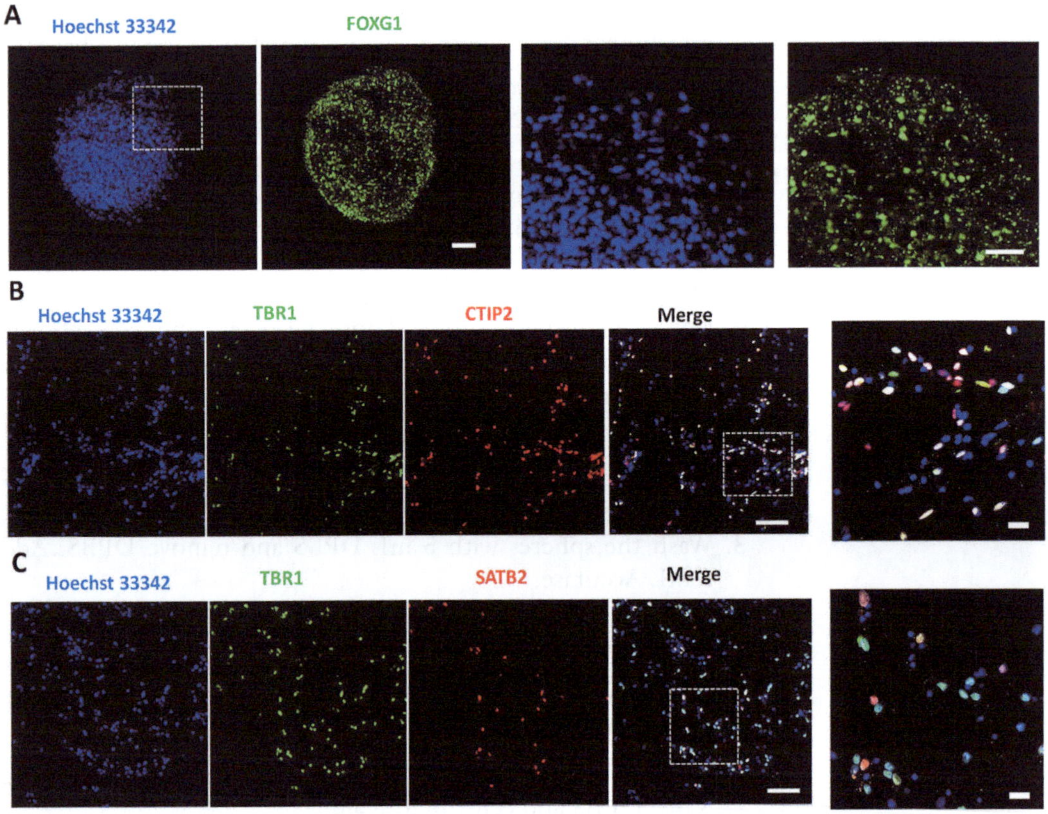

Fig. 2 Characterization of forebrain cortical neurons derived from hPSCs. (**a**) Immunostaining for FOXG1 of neural sphere on day 21. (**b**) Immunostaining for TBR1 and CTIP2 of neurons on day 32. (**c**) Immunostaining for TBR1 and SATB2 of neurons on day 56. Scale bar, 100 μm (**a**–**c**), 20 μm (zoomed-in images for **b**, **c**)

4 Notes

1. *Matrigel coating*: Matrigel coating is essential for hPSCs and cortical neuron growth on the surface of tissue culture plate. Matrigel is diluted with cold DMEM and mixed well. It can be stored at 2–8 °C for 1 month.
2. *E8 medium:* hPSCs are sensitive to the quality of culture medium and temperature of medium. E8 medium needs to be stored properly. The E8 25× supplement is very sensitive to temperature. The supplement is thawed either at RT or 4 °C overnight. When changing the medium each time, the complete E8 medium is warmed at RT, not at 37 °C. When E8 reaches RT, use it to feed hPSCs. The suboptimal E8 medium would result in cell detachment and death.
3. *Dual SMAD inhibitors:* the dual SMAD inhibitors DMH1 and SB431542 are very important for triggering neural induction from hPSCs. They need to be prepared and properly stored.

Another small molecule, LDN193189, could also be used to replace DMH1; it works with SB431542 to induce neural differentiation, and the working concentration is 100 nM.

4. *Matrigel Coating*: The culture plate is coated with 1% Matrigel at least 30 min before replating cells. Add at least 1 mL of Matrigel to each well for 6-well plates. Make sure the coating is evenly distributed.
5. *Thawing hPSCs*: The cryopreservation agent is toxic to hPSCs, especially at a high temperature. Try not to keep the cells in the agent for a long time. Completely thawing the cells in a water bath would be better for cell survival. Also, hPSCs are fragile and sensitive right after thawing. Transfer the cells to proper culture environment as soon as possible.
6. *Wash volume of E8 medium post-thaw*: Adding enough volume of fresh E8 medium could significantly reduce the toxicity of the cryopreservation agent. It is better to add at least ten times the volume of fresh E8 medium.
7. *Wash with DBPS post-thaw*: The second wash post-thaw could significantly reduce the residuals of cryopreservation agent and increase cell recovery. We use DBPS for the second wash, which is cheaper than using E8 medium.
8. *Addition of Y27632 post-thaw*: The addition of ROCK inhibitor Y27632 is essential for hPSC recovery post-thaw. Without adding the ROCK inhibitor, cells could not attach.
9. *Removal of Y27632*: Remove ROCK inhibitor from the medium one day after hPSC recovery. This would exclude the effects of ROCK inhibitor on cell growth.
10. *Accutase incubation*: There is no need to wash with DPBS before adding Accutase. Closely monitor as the cell colonies detach from the surface of the plate. Avoid over-incubation of Accutase. The incubation time may vary depending on different hPSCs cell lines.
11. *Wash with DBPS again*: The second wash with DBPS could significantly reduce the residuals of Accutase.
12. *Addition of Y27632 post-thaw*: The addition of ROCK inhibitor Y27632 is key for formation of neural spheres in 3D culture. Without adding the ROCK inhibitor, cells would not survive and form spheres.
13. *Breaking spheres*: Avoid pipetting heavily, which may damage the cells. Do the pipetting about five times to break large spheres.

14. *Addition of Y27632 for cortical neuron replating*: The addition of ROCK inhibitor Y27632 is not necessary at this step. But adding the ROCK inhibitor could enhance the cell's ability to attach and survive.
15. *Medium change after cortical neuron replating*: After replating, the neurons are very fragile and detach from the surface. Leave 0.5 mL medium in each well to avoid hydration. Add fresh medium very gently to avoid flushing away the cells from the surface.

Acknowledgments

This study was supported in part by NIH-NINDS (NS096282, NS076352, NS086604), NICHD (HD106197, HD090256), the National Medical Research Council of Singapore (MOH-000212, MOH-000207), Ministry of Education of Singapore (MOE2018-T2-2-103), Aligning Science Across Parkinson's (ASAP-000301), the Bleser Family Foundation, and the Busta Foundation.

References

1. Chambers SM, Fasano CA, Papapetrou EP et al (2009) Highly efficient neural conversion of human ES and iPS cells by dual inhibition of SMAD signaling. Nat Biotechnol 27:275–280
2. Eiraku M, Watanabe K, Matsuo-Takasaki M et al (2008) Self-organized formation of polarized cortical tissues from ESCs and its active manipulation by extrinsic signals. Cell Stem Cell 3:519–532
3. Hardiman O, Al-Chalabi A, Chio A et al (2017) Amyotrophic lateral sclerosis. Nat Rev Dis Primers 3:17071
4. Haupt S, Edenhofer F, Peitz M et al (2007) Stage-specific conditional mutagenesis in mouse embryonic stem cell-derived neural cells and postmitotic neurons by direct delivery of biologically active Cre recombinase. Stem Cells 25:181–188
5. Hu BY, Zhang SC (2010) Directed differentiation of neural-stem cells and subtype-specific neurons from hESCs. Methods Mol Biol 636:123–137
6. Huang ZJ, Paul A (2019) The diversity of GABAergic neurons and neural communication elements. Nat Rev Neurosci 20:563–572
7. Lancaster MA, Renner M, Martin CA et al (2013) Cerebral organoids model human brain development and microcephaly. Nature 501:373–379
8. Li XJ, Du ZW, Zarnowska ED et al (2005) Specification of motoneurons from human embryonic stem cells. Nat Biotechnol 23:215–221
9. Li XJ, Zhang X, Johnson MA et al (2009) Coordination of sonic hedgehog and Wnt signaling determines ventral and dorsal telencephalic neuron types from human embryonic stem cells. Development 136:4055–4063
10. Lim L, Mi D, Llorca A et al (2018) Development and functional diversification of cortical interneurons. Neuron 100:294–313
11. Marin O (2012) Interneuron dysfunction in psychiatric disorders. Nat Rev Neurosci 13:107–120
12. Munoz-Sanjuan I, Brivanlou AH (2002) Neural induction, the default model and embryonic stem cells. Nat Rev Neurosci 3:271–280
13. Pankratz MT, Li XJ, Lavaute TM et al (2007) Directed neural differentiation of human embryonic stem cells via an obligated primitive anterior stage. Stem Cells 25:1511–1520
14. Pasca AM, Sloan SA, Clarke LE et al (2015) Functional cortical neurons and astrocytes from human pluripotent stem cells in 3D culture. Nat Methods 12:671–678
15. Poduri A, Evrony GD, Cai X et al (2013) Somatic mutation, genomic variation, and neurological disease. Science 341:1237758

16. Qi Y, Zhang XJ, Renier N et al (2017) Combined small-molecule inhibition accelerates the derivation of functional cortical neurons from human pluripotent stem cells. Nat Biotechnol 35:154–163
17. Stern CD (2005) Neural induction: old problem, new findings, yet more questions. Development 132:2007–2021
18. Suzuki IK, Vanderhaeghen P (2015) Is this a brain which I see before me? Modeling human neural development with pluripotent stem cells. Development 142:3138–3150
19. Tremblay R, Lee S, Rudy B (2016) GABAergic interneurons in the neocortex: from cellular properties to circuits. Neuron 91:260–292
20. Wonders CP, Anderson SA (2006) The origin and specification of cortical interneurons. Nat Rev Neurosci 7:687–696
21. Zhang SC, Wernig M, Duncan ID et al (2001) In vitro differentiation of transplantable neural precursors from human embryonic stem cells. Nat Biotechnol 19:1129–1133

Chapter 2

Generation of Homogeneous Populations of Cortical Interneurons from Human Pluripotent Stem Cells

Peiyan Ni, Lingyi Fan, Amy Zinski, and Sangmi Chung

Abstract

Cortical interneurons (cINs), especially those that are derived from the medial ganglionic eminence (MGE) during early development, are associated with various neuropsychiatric disorders. Human pluripotent stem cell (hPSC)-derived cINs can provide unlimited cell sources for studying disease mechanisms and developing novel therapeutics. Here, we describe an optimized method to generate homogeneous cIN populations based on three-dimensional (3D) cIN sphere generation. This optimized differentiation system could sustain generated cINs relatively long term without compromising their survival or phenotypes.

Key words Cortical interneurons (cINs), Medial ganglionic eminence (MGE), Human pluripotent stem cells (hPSCs), cIN spheres

1 Introduction

Cortical interneurons (cINs), especially those that are derived from the medial ganglionic eminence (MGE) during early development, are associated with various brain disorders, such as schizophrenia [1], autism, and epilepsy [2]. MGE progenitor cells and MGE-type cINs can be generated from human pluripotent stem cells (hPSCs) [3], either by rapid induction using exogenous expressions of fate-inducing transcription factors [4] or by activating the developmentally relevant signaling pathways during differentiation [5, 6]. The latter approach is used in this protocol, which better simulates the natural developmental process of cINs. During early neurodevelopment, signaling molecules secreted by nearby organizers modulate relevant dorsoventral and rostrocaudal phenotype patterning. Inhibition of Wnt [7] and bone morphogenetic protein (BMP) [8] pathways, along with activation of sonic hedgehog (SHH) [9, 10] and FGF8 [11, 12] pathways, induces MGE phenotype in early neuroectodermal tissues by inducing key transcription factors such as Nkx2.1 [13–15]. Once they become post-mitotic in ventral

Yu-Wen Alvin Huang and ChangHui Pak (eds.), *Stem Cell-Based Neural Model Systems for Brain Disorders*, Methods in Molecular Biology, vol. 2683, https://doi.org/10.1007/978-1-0716-3287-1_2,

telencephalon, MGE-derived GABAergic cINs migrate tangentially to their target sites in the dorsal telencephalon, form local synaptic connections, and critically regulate cortical circuitry [16]. In this protocol, to induce MGE phenotypes, early neuroectodermal tissue was generated by double SMAD inhibition using SB431542 and LDN193189 (BMP inhibitors) [17]. This was followed by ventralization using SAG (SHH activator) and IWP2 (Wnt inhibitor), and further with rostralization with FGF8 signaling activation. This protocol resulted in differentiation progenies enriched with MGE progenitor cells with Nkx2.1 expression. This protocol requires neither the introduction of exogenous genes nor the support of feeder cells. The generated cINs could be maintained for a long period of time without affecting their survival and phenotype, could be passaged as needed, and could be cryopreserved efficiently [18]. This protocol will allow for the generation of unlimited quantities of homogeneous populations of MGE-type cINs that can be used for studying disease mechanisms and developing novel therapeutics.

2 Materials

2.1 Cells

Human embryonic stem cells (hESCs): H9 (WA09) cells.

2.2 hPSC Culture Media

1. Essential 8 media: Add 10 mL of Essential 8 supplement into 500 mL of Essential 8 Basal Media, and mix thoroughly.
2. Matrigel dilution (*see* **Note 1**): Thaw a bottle of Matrigel by keeping it overnight at 4 °C at least 1 day before the experiment. During the preparation, keep the diluted Matrigel on ice, since Matrigel solidifies at room temperature (RT). Dilute the Matrigel at 1:50 with DMEM. Coat a 3.5 cm dish with 1.5 mL of diluted Matrigel and shake well to cover the entire surface (Table 1). Aspirate the diluted Matrigel from the dish after incubation at 37 °C for at least 30 min (*see* **Note 2**).

Table 1
Coating media volume for plates with different surface areas

Plate size	Volume
24 well plate	0.5 mL/well
6 well plate	1.0 mL/well
3.5 well plate	1.5 mL/well
10 well plate	2.0 mL/well

2.3 Differentiation Media Preparation

1. *Basal media:*
   - PSC freezing solution: Add 1 mL of DMSO into 9 mL of Fetal Bovine Serum (FBS) and mix thoroughly.
   - Serum replacement media (SRM): Add 89 mL of knockout serum replacement (KSR) and 589 μL of 55 mM Beta-Mercaptoethanol into 500 mL of DMEM with GlutaMAX and mix thoroughly.
   - N2AA media [19]: Add 500 μL of 10 mg/mL ascorbic acid (Table 2) and 5 mL of N2 supplement into 500 mL of DMEM/F12 with GlutaMAX and mix thoroughly.
   - B27 media: Add 10 mL B27 supplement into 500 mL DMEM/F12 with GlutaMAX and mix thoroughly.
2. *Differentiation and maturation media:*
   - Neutralization media: Add 1 mL of FBS into 9 mL of DMEM/F12 with GlutaMAX and mix thoroughly.
   - Week 1 media: Add 5 μL of 1 mM LDN193189 (Table 2), 50 μL of 10 mM SB431542 (Table 2), 50 μL of 0.1 mM SAG (Table 2), and 50 μL of 5 mM IWP2 (Table 2) into 50 mL of SRM in a 50 mL conical tube and mix thoroughly.
   - Week 2 media: Add 5 μL of 1 mM LDN193189 and 50 μL of 0.1 mM SAG into 50 mL of SRM in a 50 mL conical tube and mix thoroughly.

Table 2
Preparation of chemicals and proteins in media

Chemical/ proteins	Function	Solvent	Weight of chemicals/ proteins	Volume of solvent	Stock concentration	Final concentration
Y-27632	ROCK inhibitor	DMSO	10 mg	3.1 mL	10 mM	10 μM
LDN193189	BMP inhibitor	DMSO	5 mg	12.3 mL	1 mM	100 nM
SB431452	BMP inhibitor	DMSO	50 mg	13 mL	10 mM	10 μM
IWP2	Wnt inhibitor	DMSO	50 mg	21.4 mL	5 mM	5 μM
SAG	SHH activator	DMSO	5 mg	94.9 mL	100 μM	100 nM
Ascorbic acid	Promotion of differentiation	PBS	5 g	500 mL	10 mg/mL	10 μg/mL
FGF8	Further rostralization to induce MGE phenotype	PBS	1 mg	10 mL	100 μg/mL	50 ng/mL
GDNF	Neurotrophic factor	PBS	10 μg	0.1 mL	100 μg/mL	5 ng/mL
BDNF	Neurotrophic factor	PBS	10 μg	0.1 mL	100 μg/mL	5 ng/mL

- Week 3 media: Add 50 μL of 0.1 mM SAG and 25 μL of 100 μg/mL FGF8 (Table 2) into 50 mL of N2AA media in a 50 mL conical tube, and mix thoroughly.
- Week 4 media: Add 25 μL of 100 μg/mL BDNF (Table 2), and 25 μL of 100 μg/mL GDNF (Table 2) into 500 mL of N2AA media and mix thoroughly.
- Maturation media: Add 25 μL of 100 μg/mL BDNF, and 25 μL of 100 μg/mL GDNF into 500 mL of B27 media and mix thoroughly.

3 Methods

3.1 hPSC Culture, Passaging, and Cryopreservation

1. Inspect the cells under an inverted microscope for normal hPSC colony morphology and confluency.
2. Take PBS and Essential 8 media from the refrigerator and equilibrate to RT. Vacuum off most of the media from the culture plate, but do not dry out the cells. Wash the cells with 2 mL of PBS.
3. Add fresh Essential 8 media to cultures by pipetting it to the side of the well. Pipetting media directly to the bottom will dislodge the attached cells from the dish.
4. Take PBS, Essential 8 media, trypsin-EDTA, neutralization media, and Y-27632 from the refrigerator, and equilibrate to RT.
5. Aspirate the media from the hPSC culture. Wash the dish with 2 mL of PBS (without Ca^{2+} and Mg^{2+}). Add PBS to the wall of the dish to avoid washing the colonies off the dish. Incubate the cells with 1 ml of 0.05% trypsin-EDTA at 37 °C for 5 min in the incubator.
6. Add 1 mL of neutralization media to neutralize the trypsin. Gently tap the bottom of the dish to lift the colonies off the dish, and break the collected colonies into single cells using a P1000 pipette (*see* **Note 3**).
7. Transfer the dissociated cells to a 15 mL tube and centrifuge at 1000 g for 3 min, then, carefully remove the supernatant. Resuspend the cells in fresh Essential 8 media with 10 μM Y-27632 (*see* **Note 4**).
8. Plate the trypsinized cell suspension onto new plates. For each 3.5 cm dish, add a total of 5 mL of Essential 8 media with 10 μM Y-27632. Immediately shake the dish back and forth and side to side, to spread cells evenly before placing the dish in the incubator (*see* **Note 5**).

9. Maintain the hPSCs in Essential 8 media. Because of the heat-labile nature of bFGF, make sure to change the media every day.
10. For cryopreservation, dissociate hPSCs from one 3.5 cm dish with 0.05% trypsin-EDTA, resuspend them in 3 mL of freezing solution, distribute them to cryovials (1 mL each) in a freezing container, and immediately store them in a −80 °C freezer. The next day, move them to a liquid N_2 tank.

3.2 Generation of cINs

1. In Week 1, take the Week 1 media, trypsin-EDTA, neutralization media, and Y-27632 from the refrigerator, and equilibrate to RT.
2. Vacuum out most of the media from the dish. Add 1 mL of trypsin-EDTA to a 3.5 cm dish with confluent hPSCs so that it covers the entire surface. Incubate the dish at 37 °C for 5 min.
3. Neutralize the trypsin by adding an equivalent volume of neutralization media. Gently tap the bottom of the dish to dislodge the cells. Transfer the dissociated cells to a 15 mL conical tube.
4. Centrifuge the tube at 1000 g for 3 min, then carefully remove the supernatant.
5. Resuspend the cell pellet in the Week 1 media with 10 μM Y-27632 and gently triturate using a P1000 pipette to obtain a homogeneous cell suspension. Transfer about five million cells in 6 mL of Week 1 media into a T25 flask. A schematic diagram of cINs differentiation procedures is presented in Fig. 1.
6. Change the media every other day or as needed (*see* **Note 6**). For media change, tilt the T25 flask for 1 min to allow the spheres to collect at the bottom of the flask. Once the spheres are collected at the bottom, carefully remove the media to avoid vacuuming up the spheres (*see* **Note** 7).
7. Add fresh media. If there are too many spheres and the media is turning yellow within 1 day, transfer half the spheres to a new flask with the appropriate media.
8. In Week 2, take the Week 2 media from the refrigerator and equilibrate to RT.
9. Change the media with fresh Week 2 media twice a week (*see* **Note 8**). Place the T25 flasks with the formed spheres on a shaker at 80 RPM.
10. In Week 3, take the Week 3 media from the refrigerator and equilibrate to RT.
11. Carefully inspect the spheres to ensure that the spheres are healthy and the media is not cloudy. Change the media with fresh Week 3 media twice a week (*see* **Note 8**).
12. Keep the flasks with spheres on the shaker at 80 RPM.

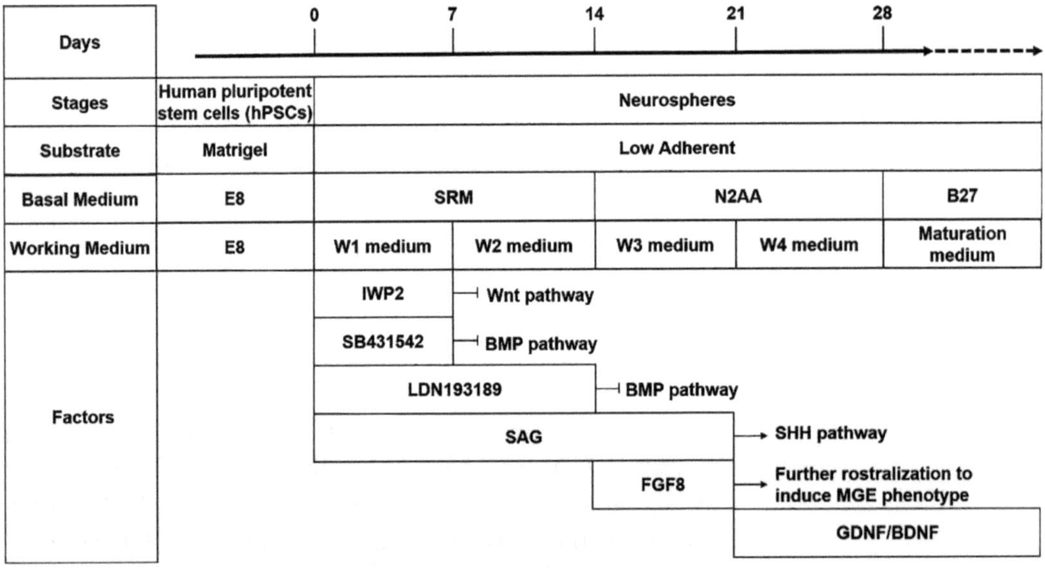

Fig. 1 Overview of the optimized cINs differentiation protocol

13. In Week 4, take the Week 4 media from the refrigerator and equilibrate to RT.
14. Carefully inspect the spheres to ensure that the spheres are healthy and the media is not cloudy. Change the media with fresh Week 4 media twice a week (*see* **Note 8**).
15. Keep the flasks with spheres on the shaker at 80 RPM.
16. Phenotypic identification can be carried out by immunofluorescent methods as required at this stage.
17. Take the Maturation media from the refrigerator and equilibrate to RT.
18. Carefully inspect the spheres to ensure that the spheres are healthy and the media is not cloudy. Change the media with fresh Maturation media twice a week for long-term maintenance (*see* **Note 8**).

4 Notes

1. Matrigel is used as a substrate to support the adherence and expansion of hPSC clones.
2. If used immediately after coating, make sure all the liquid is removed.
3. Be gentle with trituration to avoid cell damage and death.
4. Add Y-27632 (ROCK inhibitor, 10 μM) to prevent dissociation-induced death of hPSCs [20].

5. Spread cells with a back-and-forth and right-to-left motion only, not in a circular motion, to evenly distribute the cells and prevent the cells from aggregating densely in the center.
6. If some spheres adhere to the bottom of the flask, tap the flask gently to detach the spheres from the bottom.
7. Early MGE cultures may not have visible spheres. If the spheres are not easily visible, transfer the cells to a conical tube, centrifuge, and change the media, making sure not to suck up any cells.
8. The volume of the media needs to be adjusted according to the amount of spheres in the flasks. All ingredients within the effective concentration range are essential to support the healthy sphere cultures.

Acknowledgments

This work was supported by the National Natural Science Foundation of China (82071502 and 81871054; P.N.), MH107884 (S.C.), NS121541(S.C.), and NYSTEM C32607GG (S.C.).

References

1. Marin O (2012) Interneuron dysfunction in psychiatric disorders. Nat Rev Neurosci 13(2): 107–120
2. Zhu Q, Naegele JR, Chung S (2018) Cortical GABAergic interneuron/progenitor transplantation as a novel therapy for intractable epilepsy. Front Cell Neurosci 12:167
3. Liu Y, Weick JP, Liu H et al (2013) Medial ganglionic eminence-like cells derived from human embryonic stem cells correct learning and memory deficits. Nat Biotechnol 31(5): 440–447
4. Maroof AM, Keros S, Tyson JA et al (2013) Directed differentiation and functional maturation of cortical interneurons from human embryonic stem cells. Cell Stem Cell 12(5): 559–572
5. Kim TG, Yao R, Monnell T et al (2014) Efficient specification of interneurons from human pluripotent stem cells by dorsoventral and rostrocaudal modulation. Stem Cells 32(7): 1789–1804
6. Nicholas CR, Chen J, Tang Y et al (2013) Functional maturation of hPSC-derived forebrain interneurons requires an extended timeline and mimics human neural development. Cell Stem Cell 12(5):573–586
7. Li XJ, Zhang X, Johnson MA et al (2009) Coordination of sonic hedgehog and Wnt signaling determines ventral and dorsal telencephalic neuron types from human embryonic stem cells. Development 136(23):4055–4063
8. Gulacsi A, Anderson SA (2006) Shh maintains Nkx2.1 in the MGE by a Gli3-independent mechanism. Cereb Cortex 16(Suppl 1):i89–i95
9. Chiang C, Litingtung Y, Lee E et al (1996) Cyclopia and defective axial patterning in mice lacking Sonic hedgehog gene function. Nature 383(6599):407–413
10. Fuccillo M, Rallu M, McMahon AP et al (2004) Temporal requirement for hedgehog signaling in ventral telencephalic patterning. Development 131(20):5031–5040
11. Fukuchi-Shimogori T, Grove EA (2001) Neocortex patterning by the secreted signaling molecule FGF8. Science 294(5544): 1071–1074
12. Garel S, Huffman KJ, Rubenstein JL (2003) Molecular regionalization of the neocortex is disrupted in Fgf8 hypomorphic mutants. Development 130(9):1903–1914
13. Sussel L, Marin O, Kimura S et al (1999) Loss of Nkx2.1 homeobox gene function results in a

ventral to dorsal molecular respecification within the basal telencephalon: evidence for a transformation of the pallidum into the striatum. Development 126(15):3359–3370

14. Ma T, Wang C, Wang L et al (2013) Subcortical origins of human and monkey neocortical interneurons. Nat Neurosci 16(11): 1588–1597
15. Hansen DV, Lui JH, Flandin P et al (2013) Non-epithelial stem cells and cortical interneuron production in the human ganglionic eminences. Nat Neurosci 16(11):1576–1587
16. Nobrega-Pereira S, Kessaris N, Du T et al (2008) Postmitotic Nkx2-1 controls the migration of telencephalic interneurons by direct repression of guidance receptors. Neuron 59(5):733–745
17. Chambers SM, Fasano CA, Papapetrou EP et al (2009) Highly efficient neural conversion of human ES and iPS cells by dual inhibition of SMAD signaling. Nat Biotechnol 27(3): 275–280
18. Ni P, Noh H, Shao Z et al (2019) Large-scale generation and characterization of homogeneous populations of migratory cortical interneurons from human pluripotent stem cells. Mol Ther Methods Clin Dev 13:414–430
19. Chung S, Moon JI, Leung A et al (2011) ES cell-derived renewable and functional midbrain dopaminergic progenitors. Proc Natl Acad Sci U S A 108(23):9703–9708
20. Watanabe K, Ueno M, Kamiya D et al (2007) A ROCK inhibitor permits survival of dissociated human embryonic stem cells. Nat Biotechnol 25(6):681–686

Chapter 3

Generation and Co-culture of Cortical Glutamatergic and GABAergic-Induced Neuronal Cells

Jay English, Danny McSweeney, Fumiko Ribbe, Ethan Howell, and ChangHui Pak

Abstract

The study of neurological disorders requires experimentation on human neurons throughout their development. Primary neurons can be difficult to obtain, and animal models may not fully recapitulate phenotypes observed in human neurons. Human neuronal culturing schemes which contain a balanced mixture of excitatory and inhibitory neurons that resemble physiological ratios seen in vivo will be useful to probe the neurological basis of excitation-inhibition (E-I) balance. Here, we describe a method for directly inducing a homogenous population of cortical excitatory neurons and cortical interneurons from human pluripotent stem cells, as well as the generation of mixed cultures using these induced neurons. The obtained cells display robust neuronal synchronous network activity as well as complex morphologies that are amenable to studies probing the molecular and cellular basis of disease mutations or other aspects of neuronal and synaptic development.

Key words Induced neurons (iNs), Glutamatergic excitatory neurons, GABAergic interneurons, E-I cultures, Mixed iN cultures, Human pluripotent stem cells (hPSCs)

1 Introduction

Neurodevelopmental disorders, such as autism spectrum disorders (ASDs), are possibly linked to synaptopathy and a disruption of the balance of excitatory and inhibitory (E-I) signaling [1–3]. However, the cellular contributions to this disruption are not clearly elucidated. As such, determining the cell-type-specific contributions to the phenotypes seen in these disorders is a worthwhile endeavor. Although studies in well-adopted rodent models may provide insight into pathogenic mechanisms, the results may not necessarily translate directly to the human brain. Therefore, studies in disease-relevant human neurons are essential. The development of a system for the direct induction of excitatory and inhibitory cortical neurons from human pluripotent stem cells (hPSCs) has now made these studies achievable.

Yu-Wen Alvin Huang and ChangHui Pak (eds.), *Stem Cell-Based Neural Model Systems for Brain Disorders*, Methods in Molecular Biology, vol. 2683, https://doi.org/10.1007/978-1-0716-3287-1_3,

For this reason, the generation of neurons from hPSCs is an alluring approach to study the neuronal involvement in neurological disorders. In fact, multiple studies utilized genetically modified hPSC-derived neurons to model and understand the synaptic mechanisms underlying disease-related genes [4–8]. This effort has been very fruitful for the homogeneous population of cortical glutamatergic-induced neuronal cells (Ngn2 iNs) [9]. However, these cultures completely lack the proper inhibition and microcircuitry that result from the signaling between excitatory and inhibitory neurons (i.e., feedforward inhibition). Thus, improved methodologies that incorporate the mixed cultures using Ngn2 iNs and cortical inhibitory neurons (Ascl1/Dlx2 iNs; A/D iNs for short) [10] will provide an exciting opportunity to explore the mechanisms of these disorders in human neurons in a more direct and reproducible manner compared to chemical induction protocols. In fact, a recent disease modeling study using such an approach has revealed that there are neuronal subtype-specific differentially expressed genes in the neuroligin-3 R451C genetic background of ASD [8].

In this protocol, we provide a step-by-step method to induce, mix, and co-culture cortical excitatory and inhibitory iNs to maturity (Fig. 1). Using a controllable induction of relatively few transcription factors, Ngn2 and A/D iNs are generated separately, then combined in the desired ratio, and grown to maturity (Fig. 2). With this method, relevant cultures can be generated from a variety of patient-derived hPSCs, yielding insight into the pathogenic mechanisms of these disorders.

2 Materials

2.1 Reagents Necessary for hPSC Culture

Reagents needed for hPSC culturing	
Reagent/Solution	**Stock concentration**
mTeSR+ Basal Medium	N/A
5X mTeSR+ Supplement	N/A
Dulbecco's Phosphate-Buffered Saline; no calcium, no magnesium (DPBS)	N/A
ReLeSR	N/A
Y-27632 Dihydrochloride (Y Compound)	10 mM
Matrigel	8–12 mg/mL

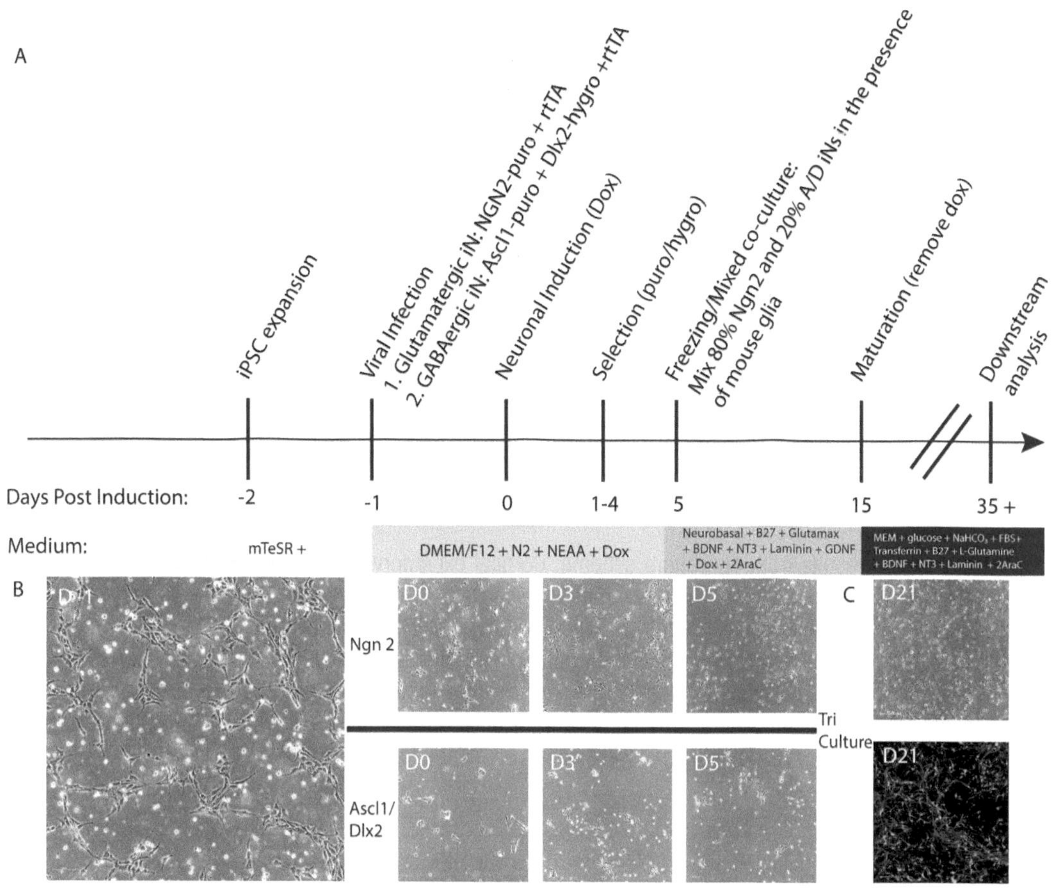

Fig. 1 Culture scheme. (**a**) Timeline of direct induction and media used for each part of differentiation. Human iPSCs were grown to approximately 90% confluency before being lifted and counted. One-to-three million cells were then infected with lentiviral vectors harboring the differentiation constructs in mTeSR+ supplemented with Y-27632. For cortical excitatory iNs, Ngn2, and rtTA lentiviruses were added, whereas hPSCs for cortical inhibitory iNs were infected with Ascl1, Dlx2, and rtTA constructs. For visualization, induced neurons were also infected with a TetO-EGFP construct. After 24 h of incubation, the media was changed with the indicated induction media. The cells were incubated for another 24 h, after which the media was changed daily with the appropriate selection media. On day 5, the cells were lifted, counted, and plated with WT mouse glia in plating media, and media changes were performed every 3 days thereafter. Doxycycline was removed from the media after day 14. (**b**) Representative images of the progress of the infection process (left panel), the induction process for Ngn2 (top row, middle panels), Ascl1/Dlx2 (bottom row, middle panels) on days 0, 1, 3, and 5 (scale bars 100 μM). (**c**) Representative image of an established tri-culture at day 21 post-infection in brightfield (top) and GFP (bottom) (scale bars 200 μM)

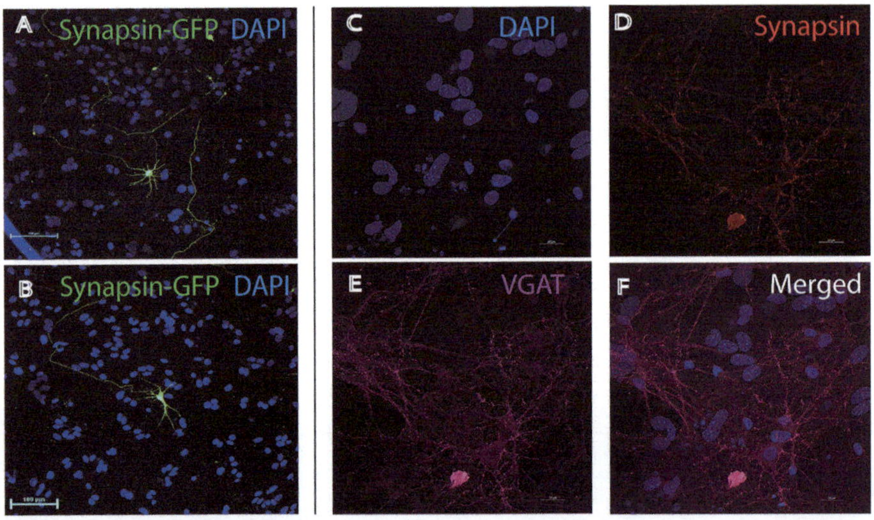

Fig. 2 Anticipated results from co-culture scheme. Cortical excitatory (Ngn2) and inhibitory (Ascl1/Dlx2) iNs were induced as described, and then co-cultured alongside mouse glia at 5 days post-induction (dpi). Co-cultures were sparsely transfected with Synapsin-driven GFP at 14 dpi, and at 35 dpi, images were taken at 20× magnification to visualize individual neuronal morphology. Panels (**a**, **b**) depict singular representative neurons with distinct neurite and axonal projections (scale bars 100 μM). (**c**–**f**) Co-cultures were imaged at 60× magnification in oil immersion at 35 dpi to visualize synaptic puncta. Both excitatory (Synapsin) and inhibitory (VGAT) synapses were observed (scale bars 20 μM)

2.2 Reagents Necessary for Lentivirus Production

Reagents/Consumables for HEK cell maintenance	
Reagent/Solution	**Stock concentration**
Human embryonic kidney (HEK) 293T cell line	N/A
Dulbecco's Modified Eagle Medium (DMEM) high glucose, pyruvate	N/A
10,000 U/mL Penicillin-Streptomycin	1% V/V
Fetal Bovine Serum (FBS)	10% V/V
Dulbecco's Phosphate-Buffered Saline; no calcium, no magnesium (DPBS)	N/A
0.05% Trypsin-EDTA	N/A
T75 cell culture flask	N/A

Reagents/consumables for transfection and lentivirus production	
Reagent/supplies	**Stock concentration**
pMDLg/pRRE (RRE) (Addgene Catalogue 12251)	N/A

(continued)

Reagents/consumables for transfection and lentivirus production	
Reagent/supplies	**Stock concentration**
pMD2.G vesicular stomatitis virus G protein expression vector (VSVG) (Addgene Catalogue 12259)	N/A
pRSV-REV (REV) (Addgene Catalogue 12253)	N/A
FUW-M2rtTA (rtTA) (Addgene Catalogue 20342)	N/A
pTet-O-Ngn2-puro (Ngn2) (Addgene Catalogue 52047)	N/A
TetO-Ascl1-puro (Ascl1) (Addgene Catalogue 97329)	N/A
TetO-DLX2-hygro (Dlx2) (Addgene Catalogue 97330)	N/A
HEPES buffered saline (HBS: 280 mM NaCl, 1.5 mM Na_2HPO_4, 50 mM HEPES) pH 7.1	2×
Calcium chloride ($CaCl_2$)	2.5 M
5 mL round-bottom polystyrene test tubes	N/A
0.22 μm sterile syringe filters	N/A
0.45 μm SFCA (Surfactant Free Cellulose Acetate) sterile syringe filters	N/A
10 mL syringes	N/A
Type SW32Ti rotor	N/A

2.3 Reagents Necessary for Mouse Glia Preparation

Reagents/tool for glial dissection and dissociation	
Reagent/solution	**Stock concentration**
Hank's balanced salts solution	N/A
Ethylenediaminetetraacetic acid (EDTA)	0.5 M
Ethanol	70% V/V
Papain	400 units/mL
Micro surgical scissors	N/A
Small surgical forceps (×2)	N/A
Stereoscopic scope	N/A
Dulbecco's Modified Eagle Medium (DMEM)	N/A
Penicillin-Streptomycin	10,000 units/mL
Fetal Bovine Serum (FBS)	N/A
0.22 μM sterile filter	N/A
Trypsin	0.25% W/V
Dimethyl sulfoxide (DMSO)	N/A

Brain dissociation mixture	
Reagent/solution	**Final concentration**
Hank's balanced salts solution	N/A
Ethylenediaminetetraacetic acid (EDTA)	500 μM
Papain	6.4 units/mL

Glia freezing media	
Reagent/solution	**Final concentration**
Dulbecco's Modified Eagle Medium (DMEM)	N/A
Penicillin-Streptomycin	1% V/V
Fetal Bovine Serum (FBS)	20% V/V
Dimethyl sulfoxide (DMSO)	10% V/V

2.4 Reagents Necessary for iN Induction

Reagents needed for neuronal induction and maturation	
Reagent/solution	**Stock concentration**
Dulbecco's Modified Eagle Medium/Nutrient Mixture F-12 (DMEM/F12)	N/A
Neurobasal Medium	N/A
Dulbecco's Phosphate-Buffered Saline; no calcium, no magnesium (DPBS)	N/A
Accutase	N/A
Glutamax supplement	100×
Non-Essential Amino Acids (NEAA)	100×
Puromycin	10 mg/mL
Hygromycin-B	50 mg/mL
Doxycycline	2 mg/mL
B27 supplement with vitamin A (B27)	50×
N2 supplement	100×
Brain-Derived Neurotrophic Factor (BDNF)	0.05 mg/mL
Mouse Laminin	1 mg/mL
Neurotrophin-3 (NT3)	0.05 mg/mL
Glial Cell-Derived Neurotrophic Factor (GDNF)	0.05 mg/mL
Cytosine β-D-Arabinofuranoside Hydrochloride (AraC)	4 mM
Y-27632 Dihydrochloride (Y Compound)	10 mM

2.5 Reagents Necessary for Ngn2, A/D Selection and Co-culture

Induction media	
Reagent/Solution	**Final concentration**
Dulbecco's Modified Eagle Medium/Nutrient Mixture F-12 (DMEM/F12)	N/A
Non-Essential Amino Acids (NEAA)	1×
Doxycycline	4 μg/mL
N2 supplement	1×

Ngn2 selection media	
Reagent/Solution	**Final concentration**
Dulbecco's Modified Eagle Medium/Nutrient Mixture F-12 (DMEM/F12)	N/A
Non-Essential Amino Acids (NEAA)	1×
Puromycin	10 μg/mL
Doxycycline	4 μg/mL
N2 Supplement	1×

A/D selection media	
Reagent/Solution	**Final concentration**
Dulbecco's Modified Eagle Medium/Nutrient Mixture F-12 (DMEM/F12)	N/A
Non-Essential Amino Acids (NEAA)	1×
Puromycin	10 μg/mL
Hygromycin-B	300 μg/mL
Doxycycline	4 μg/mL
N2 Supplement	1×

2.6 Reagents Necessary for Maturation of Mixed Cultures

Co-culture plating media	
Reagent/Solution	**Final concentration**
Neurobasal Medium	N/A
Glutamax supplement	1×
Doxycycline	4 μg/mL
B27 supplement with vitamin A (B27)	1×
Brain-Derived Neurotrophic Factor (BDNF)	10 ng/mL

(continued)

Co-culture plating media	
Reagent/Solution	**Final concentration**
Mouse Laminin	200 ng/mL
Neurotrophin-3 (NT3)	10 ng/mL
Glial Cell-Derived Neurotrophic Factor (GDNF)	10 ng/mL
Y-27632 Dihydrochloride (Y Compound)	1 mM

Maturation media	
Reagent/Solution	**Final concentration**
Minimal Essential Medium (MEM)	N/A
20% Glucose	0.5% v/v
8% $NaHCO_3$	0.05% v/v
Transferrin	100 μg/mL
200 mM L-Glutamine	500 μM
Doxycycline *(remove from media after D14)*	4 μg/mL
B27 Supplement with Vitamin A (B27)	1×
Brain-Derived Neurotrophic Factor (BDNF)	10 ng/mL
Mouse Laminin	200 ng/mL
Neurotrophin-3 (NT3)	10 ng/mL
Glial Cell-Derived Neurotrophic Factor (GDNF)	10 ng/mL
Cytosine β-D-Arabinofuranoside Hydrochloride (AraC)	2 μM
Fetal Bovine Serum (FBS)	5% v/v

3 Methods

3.1 hPSC Culture

Passaging of hPSCs:

1. Ensure cells have reached 70–100% confluence.
2. Coat a 6-well plate with Matrigel and incubate it for 30 min at 37 °C, 5% CO_2.
3. Warm mTeSR media in a water bath.
4. In a 15 mL conical, add 12 μL of Y compound to 12 mL of mTeSR media.

5. Vacuum out old media.
6. Add 2 mL of DPBS into a single well of cells.
7. Vacuum out the DPBS.
8. Add 1 mL of ReLeSR to the same well.
9. Immediately vacuum out the ReLeSR.
10. Incubate the cells for 5 min at 37 °C, 5% CO_2.
11. Add 1 mL of mTeSR + y compound into the well and transfer the media and cells solution into a 15 mL conical.
12. Add 11 mL of mTeSR + y compound into the 15 mL conical.
13. Remove plate coated with matrigel from the incubator.
14. Vacuum out the matrigel from the 6-well plate.
15. Distribute 2 mL of solution from the 15 mL conical into each of the wells.
16. Gently swirl the solution within the plate to spread the cells evenly.
17. Incubate overnight at 37 °C, 5% CO_2.

3.2 Lentivirus (LV) Production

Setup 1: Plasmid DNA

Grow bacterial stock in LB containing the appropriate selective antibiotic. Four plasmids (pMDLg/pRRE (RRE), vesicular stomatitis virus G protein expression vector (VSVG), pRSV-REV (REV), and FUW-M2rtTA (rtTA)) are required for packaging, along with whatever relevant plasmids required for induction should be prepared separately as described below. Use a midi scale plasmid purification kit to isolate plasmids. Measure the DNA concentration and purity and freeze at −20 °C.

Setup 2: Seeding HEK cells

Thaw HEK 293 T cells and seed into a T75 flask containing 12 mL HEK media. Maintaining an incubator at 37 °C, 5% CO_2.. Passage cells after a few days and allow the flask to reach confluency. Two confluent flasks for lentivirus production plus are needed per this protocol.

Day 1:

1. 24 h before transfection, passage confluent HEK flasks.
   (a) Remove media from flask.
   (b) Wash with DPBS.
   (c) Add 3 mL Trypsin to each flask. Incubate 37 °C, 5% CO_2 for 5 min.
   (d) Add 3 mL warm HEK media to each flask.

(e) Gather cells into a 15 mL conical.

(f) Pellet at 200–300 × *g*. Remove supernatant.

(g) Resuspend pellet in 3 mL per flask to be seeded (split 1:3).

(h) Distribute cells among new flasks.

Day 2:

1. Change media in all flasks to ensure a volume of 12 mL HEK media.
2. Thaw and prepare packaging plasmid master mix in a 1.5 mL tube.
   (a) Packaging plasmids: 8.1 μg, RRE, 6 μg VSVG and 3.9 μg REV per transfection (will make 6 transfections).
3. In a separate 1.5 mL tube, prepare transfection DNA mix. For each flask to be transfected, place:
   (a) 115 μL 2.5 M $CaCl_2$
   (b) 1/6 of the packaging plasmid master mix volume
   (c) 12 μg transfer plasmid DNA per transfection
   (d) Adjust volume to 900 μL with water
4. Add 900 μL 2× HBS into a 5 mL Falcon round-bottom test tube.
5. Add transfection DNA mix dropwise to each 2× HBS-containing tube while vortexing gently.
6. Incubate for 20 min at room temperature.
7. Add transfection mix to the appropriate flask dropwise. Swirl the flask to ensure even coverage before returning the flask to the incubator.
8. Incubate for 5 h at 37 °C with 5% CO_2, and then change the media.
9. Incubate for 48 h at 37 °C with 5% CO_2.

Day 3: If using a control plasmid that expresses some fluorescence, efficiency can be checked under a microscope.

Day 4:

1. Transfer media from each flask into a 15 mL conical tube.
2. Centrifuge for 5 min at 200–300 ×*g* at room temperature.
3. Filter supernatant through 0.45 μm cellulose acetate syringe filter into an ultracentrifuge tube.
4. Bring the volume of each ultracentrifuge tube up to 25 mL with DMEM.
5. Load ultracentrifuge tubes into the rotor and spin approximately 61,500 ×*g* for 2 h at 4 °C.

6. Carefully remove the supernatant by decanting. Discard the supernatant.
7. Invert ultracentrifuge tubes on a few layers of kimwipes to drain media. Remove any residual media from tubes using a sterile vacuum.
8. Add 100 μL cold MEM to each tube directly to the center of the bottom of the tube.
9. Seal the tubes with parafilm and incubate at 4 °C overnight.

Day 5

1. Pipette MEM up and down several times to mix.
2. Divide each tube into 10 μL aliquots over ice.
3. Store LV aliquots at −80 °C.

3.3 Mouse Glia Preparation

3.3.1 Dissection

1. Sterilize the dissection surface, scope, and all tools to be used for dissection.
2. Prepare Brain Dissociation Mixture as described and incubate it at 37 °C until the solution is homogeneous and no longer appears cloudy. 5 mL of this mixture is sufficient for the dissociation of two brains as prepared by this procedure.
3. Sterile filter the Brain Dissociation Mixture with a 0.22 μM filter into a clean container. Place back into incubation at 37 °C.
4. Prepare an appropriate number of tubes for brain collection by filling them with approximately 5 mL ice-cold HBSS. Keep on ice. One tube should hold a maximum of two brains as prepared in this procedure.
5. Place a sterile dish on the dissection surface and add enough ice-cold HBSS so that the bottom is covered.
6. Gather mouse pups aged P0–P3 and keep them in a bedding material until selected for dissection.
7. Anesthetize the first pup to be dissected with hypothermia and ethanol.
8. Quickly decapitate the pup with surgical scissors.
9. Position the head in a comfortable fashion, using light pressure with forceps through the orbit to hold the head in place.
10. Using a second pair of forceps, gently pull the skin from the cranium and remove any musculature from the cranium and base of the skull.
11. Insert the scissors carefully into the foramen magnum, cutting around the sides of the skull through the temporal bones on both sides and through the center of the sagittal line. Take care not to insert the scissors too deep, lest damage to the brain occur.

12. Using forceps, remove the upper half of the skull from both hemispheres.
13. Place the forceps under the brain, between the bottom of the brain and the inside of the lower cranium. Gently open the forceps to loosen the brain.
14. Extract the brain from the skull, placing it in a relatively clean region of the dish.
15. Remove all visible meninges from the brain.
16. Separate the cerebellum from the cerebrum. Discard the cerebellum.
17. Invert the cerebrum. Using the angled head of the forceps, scoop out the brain matter from the center of the brain, leaving only the cortex. Also, remove the olfactory bulbs if needed.
18. Place the remaining cerebral cortex matter in a tube of ice-cold HBSS. Store on ice until digestion.
19. Repeat the process for remaining pups.

3.3.2 Dissociation

1. Remove all but approximately 500 μL HBSS from the brain-containing tubes. Take care not to remove brain matter.
2. Add Brain Dissociation Mixture to each tube.
3. Incubate at 37 °C for 20 min, occasionally agitating the mixture.
4. Remove as much digestion mixture as possible.
5. Wash 3× with HEK media. In the last wash, leave approximately 2 mL HEK media in the tube.
6. Using a 1000 μL pipette, break up the remaining tissue clumps until the cells are homogeneously dissociated.
7. Plate cells onto an appropriate dish with HEK media. Incubate for approximately 24 h at 37 °C and 5% CO_2.
8. Perform a media change with fresh HEK media the next day.
9. Upon confluency, passage the glia:
   (a) Remove the media from the dish.
   (b) Wash with DPBS.
   (c) Add enough trypsin to cover the bottom of the dish.
   (d) Incubate at 37 °C for 5 min.
   (e) Add an equal volume of warm HEK media to quench trypsin.
   (f) Using a cell lifter, gather cells.
   (g) Pellet cells at 200–300 × *g*. Remove supernatant.

(h) Resuspend cells in an appropriate volume of warm HEK media and distribute into new dishes (1 confluent dish to three new dishes).

10. Grow to confluency and passage once more if necessary to increase yield or reduce neuronal contamination.
11. Upon final confluency, freeze the glia:
    (a) Trypsinize the glia as indicated in **steps 9a–g**.
    (b) Resuspend pellet in an appropriate volume of glia freezing media.
    (c) Count cells.
    (d) Place the desired number of cells into cryovials and freeze.

3.4 Ngn2 iN Differentiation

Setup: Grow hPSCs to confluency (e.g., cells grown and plated on a 6-well plate). Add 1–2 mL per well matrigel to a new 6-well plate prior to the infection process to be used in step A13. An aliquot of mTeSR + Y compound may be prepared beforehand to be used for the infection process.

3.4.1 Day 1: Infection

1. Remove media from confluent hPSCs.
2. Wash cells with 1 mL DPBS per well. Remove DPBS.
3. Add 1 mL Accutase per well.
4. Incubate cells at 37 °C for 5 min.
5. Add 1 mL mTeSR + Y compound per well.
6. Collect cells into a 15 mL centrifuge tube.
7. Pellet cells for 5 minutes at 200–300 × *g*.
8. Remove supernatant and resuspend in an appropriate volume of mTeSR + Y compound (2 mL is sufficient for one 6-well plate).
9. Count cells using either a hemocytometer or an automatic cell counter.
10. In a separate 15 mL centrifuge tube, for each 6-well plate of cells to be infected, place 4 mL of mTeSR + Y compound, one million cells*, 10 μL rtTA lentivirus, and 10 μL Ngn2 lentivirus.

    See* **Note 1
11. Bring infection tube volume to 12 mL total with mTeSR + Y compound.
12. Thoroughly mix the infection tube and distribute 2 mL per well of the infection mixture into a new 6-well plate (remove Matrigel before plating).

3.4.2 Day 0: Induction

1. Remove infection media from wells.
2. Add 2 mL per well induction media.

3.4.3 Day 1–4: Selection

1. Completely change media each day with Ngn2 Selection media.

   * *See* **Note 3**

3.4.4 Day 5: Collecting Cells

1. Remove media from all wells.
2. Wash cells with 1 mL per well DPBS.
3. Add 1 mL per well Accutase.
4. Incubate for 3 min at 37 °C.
5. Add 1 mL per well warm Plating Media.
6. Collect cells into a 15 mL centrifuge tube.
7. Pellet cells for 5 min at 200–300 × *g*.
8. For cryopreservation, remove supernatant and resuspend cells in 1 mL per plate of BamBanker. If cells are to be directly co-cultured upon collection, resuspend in 1 mL per plate of Plating Media.
9. Count Cells.
10. For cryopreservation, distribute the desired number of cells into a cryovial and freeze appropriately. Can be stored at −80 °C for a short amount of time but should be transferred to liquid nitrogen for long-term storage. For direct plating after lifting, proceed directly to step A-5 in Subheading 3.6: Maturation of Mixed Cultures.

3.5 A/D iN Differentiation

Setup Grow hPSCs to confluency (example is for cells grown and plated on a 6-well plate). Add 1–2 mL per well Matrigel to a new 6-well plate prior to the infection process to be used in **step 1**. A 50 mL aliquot of mTeSR + Y may be prepared beforehand to be used for the infection process.

3.5.1 Day −1: Infection

1. Remove media from cells
2. Wash cells with 1 mL DPBS per well. Remove DPBS.
3. Add 1 mL Accutase per well.
4. Incubate cells at 37 °C for 5 minutes.
5. Add 1 mL mTeSR + Y compound per well.
6. Collect cells into a 15 mL centrifuge tube.
7. Pellet cells for 5 min at 200–300 × *g*.

8. Remove supernatant and resuspend in an appropriate volume of mTeSR + Y compound (2 mL is sufficient for one confluent 6-well plate of hPSCs).
9. Count cells.
10. In a separate 15 mL centrifuge tube, for each 6-well plate of cells to be infected, place 4 mL of mTeSR + Y compound, three million cells, 20 μL rtTA lentivirus*, 20 μL Ascl1 lentivirus, and 20 μL Dlx2 lentivirus.

    See* **Note 2
11. Incubate cells at room temperature for 5 min.
12. Bring infection tube volume to 12 mL total with mTeSR + Y compound.
13. Thoroughly mix the infection tube and distribute 2 mL per well of the infection mixture into a new 6-well plate (remove Matrigel before plating).

3.5.2 Day 0: Induction

1. Remove infection media from wells.
2. Add 2 mL per well induction media.

3.5.3 Day 1–4: Selection

1. Completely change media each day with A/D Selection media.

3.5.4 Day 5: Collecting Cells

1. Remove media from all wells.
2. Wash cells with 1 mL per well DPBS.
3. Add 1 mL per well Accutase.
4. Incubate for 3 min at 37 °C.
5. Add 1 mL per well warm Plating Media.
6. Collect cells into a 15 mL centrifuge tube.
7. Pellet cells for 5 min at 200–300 × *g*.
8. For cryopreservation, remove supernatant and resuspend cells in 1 mL per plate of BamBanker. If cells are to be directly co-cultured upon collection, resuspend in 1 mL per plate of Plating Media.
9. Count Cells.
10. For cryopreservation, distribute the desired number of cells into a cryovial and freeze appropriately. Can be stored at − 80 °C for a short amount of time but should be transferred to liquid nitrogen for long term storage. For direct plating after lifting, proceed directly to step A-5 in Subheading 3.6: Maturation of Mixed Cultures.

3.6 Maturation of Mixed Cultures

Setup: Prepare the plate for the mixed culture. For this example, a 24-well plate with coverslips was treated with 0.5 mL Matrigel per well an hour before plating. The numbers provided are for one well of a 24-well plate. Pre-warm Plating Media.

3.6.1 Plating

1. Thaw the appropriate number of vials each of Ngn2 iNs, A/D iNs, and mouse glia (target number: 160,000 Ngn2 iNs, 40,000 A/D iNs, and 200,000 mouse glia).
   - (i) Remove vials from liquid nitrogen.
   - (ii) Place in a water bath at 37 °C until only a small ice core remains (approximately 45 s).
   - (iii) Introduce 1 mL Plating Media per vial and finish thawing if necessary by agitating until mixture is homogeneous.
   - (iv) Transfer thawed cells to a separate 15 mL centrifuge tube for each cell type.
2. Pellet cells at 200–300 × *g* for 5 min.
3. Remove supernatant. Resuspend cells in 1 mL Plating Media.
4. Count cells. Trypan blue was added to determine the percentage of live cells in each count. Calculate the volume necessary to achieve the target for each cell type.
5. In a new 15 mL centrifuge tube, place 160,000 live Ngn2 iNs, 40,000 live A/D iNs, and 200,000 live mouse glia per well to be plated.
6. Adjust volume with Plating Media so that there is 500 μL total media per well to be plated.
7. Remove matrigel from the prepared plate.
8. Distribute the 500 μL per well into the plate.

3.6.2 Establishing the Cultures

1. The day after plating, change all media in each well with fresh plating media. Every other day, remove half of the media per well and replace with an equal volume of fresh plating media.

3.6.3 Maturation

1. After day 14 post induction, replace the media with maturation media. Continue performing half media changes every 2–3 days.

4 Notes

1. If experiencing low yields, this may be due to line-by-line variation. Add up to 3 days of hPSC expansion in mTeSR+ media after LV infection but before induction. Alternatively, scale up the entire process.
2. The ratio of rtTA to other LVs may need to be increased for reliable expression of multiple transcription factors (such as in A/D iNs). Refining the ratio may be required before establishing cultures.

3. If you experience stem cell carryover, the LVs used may be ineffective. During production, monitor fluorescence in HEK cells to ensure proper LV production. Also, stem cell carryover may be due to insufficient selection time. Either increase the concentration of selection antibiotics or increase days of selection.

Acknowledgments

This work was supported by grants from the NIH. Pak lab is supported by NIMH R01 MH122519. J.E. is supported by NIGMS T32 GM135096.

References

1. Avazzadeh S, Ketya M, Reilly J et al (2019) Increased Ca2+ signaling in NRXN1α+/− neurons derived from ASD induced pluripotent stem cells. Mol Autism 10:52
2. Cast TP, Boesch D, Smyth K et al (2021) An autism-associated mutation impairs Neuroligin-4 glycosylation and enhances excitatory synaptic transmission in human neurons. J Neurosci 41:392–407
3. Chanda S, Aoto J, Lee S-J, Wernig M, Südhof TC (2016) Pathogenic mechanism of an autism-associated neuroligin mutation involves altered AMPA-receptor trafficking. Mol Psychiatry 21:169–177
4. Pak C et al (2015) Human neuropsychiatric disease modeling using conditional deletion reveals synaptic transmission defects caused by heterozygous mutations in NRXN1. Cell Stem Cell 17:316–328
5. Pak C, Danko T, Zhang Y et al (2021) Cross-platform validation of neurotransmitter release impairments in schizophrenia patient-derived *NRXN1* -mutant neurons. Proc Natl Acad Sci 118:e2025598118
6. Yi F, Danko T, Botelho SC et al (2016) Autism-associated SHANK3 haploinsufficiency causes I_h channelopathy in human neurons. Science 352:aaf2669
7. McSweeney D, Gabriel R, Jin K et al (2022) CASK loss of function differentially regulates neuronal maturation and synaptic function in human induced cortical excitatory neurons. iScience 25:105187
8. Wang L, Mirabella VR, Dai R et al (2022) Analyses of the autism-associated neuroligin-3 R451C mutation in human neurons reveal a gain-of-function synaptic mechanism. Mol Psychiatry. https://doi.org/10.1038/s41380-022-01834-x
9. Zhang Y, Pak C, Han Y et al (2013) Rapid single-step induction of functional neurons from human pluripotent stem cells. Neuron 78:785–798
10. Yang N, Chandra S, Marro S et al (2017) Generation of pure GABAergic neurons by transcription factor programming. Nat Methods 14:621–628

Chapter 4

Transcription Factor-Directed Dopaminergic Neuron Differentiation from Human Pluripotent Stem Cells

Yi Han Ng and Justyna A. Janas

Abstract

The ability to differentiate pluripotent stem cells and to generate specific cell types is a long-standing goal of regenerative medicine. This can be accomplished by recreating the developmental trajectories using sequential activation of the corresponding signaling pathways, or more recently—by direct programming of cell identities using lineage-specific transcription factors. Notably, to be functional in cell replacement therapies, generation of complex cell types, such as specialized neuronal sub-types of the brain, requires precise induction of molecular profiles and regional specification of the cells. However, the induction of the correct cellular identity and marker gene expression can be hampered by technical challenges, one of which is the robust co-expression of multiple transcription factors that is often required for correct cell identity specification. Here, we describe in detail a method for co-expression of seven transcription factors required for efficient induction of dopaminergic neurons with midbrain characteristics from human embryonic and induced pluripotent stem cells.

Key words Induced neuronal (iN) cells, Dopaminergic neurons, Transcription factor-based reprogramming, Pluripotent stem cells

1 Introduction

The ability of human pluripotent stem cells to proliferate and to differentiate into any cell type of the three germ layers has opened unlimited possibilities for cell replacement therapies [1–4]. Significant efforts in the field have been directed towards devising methods for differentiation and generation of specific cell types in a dish, with a goal to faithfully recapitulate the cell's molecular profile that is normally established during differentiation in vivo. Advances in generating neuronal subtypes, in particular, have revolutionized the field of cellular neurobiology. The ability to derive human neurons provides access to otherwise inaccessible cell types and thus, the opportunity to study disease-associated phenotypes or test the potential therapeutics in a relevant cellular context and pre-clinical setting [5]. Furthermore, capturing the potential of pluripotent

Yu-Wen Alvin Huang and ChangHui Pak (eds.), *Stem Cell-Based Neural Model Systems for Brain Disorders*, Methods in Molecular Biology, vol. 2683, https://doi.org/10.1007/978-1-0716-3287-1_4,

stem cells for derivation of clinical-grade neurons for cell-based replacement therapies has been a major focus in the case of dopaminergic neurons, given their proven potential to improve motor dysfunction associated with Parkinson's disease [5, 6].

One of the approaches for generating neuronal subtypes from pluripotent stem cells relies on the expression of lineage-specific transcription factors [7]. Accordingly, we developed methods for efficient and rapid conversion of ES cells or iPS cells into neurons, including glutamatergic, GABAergic, as well as midbrain dopaminergic neurons, by forced expression of a defined set of lineage-determining transcription factors [8–10]. Notably, we found that in the case of dopaminergic identity induction, the proper regional specification requires a more complex combination of transcription factors (Fig. 1). In this chapter, we describe the method for efficient delivery and expression of seven transcription factors required to generate functional dopaminergic neurons with midbrain characteristics from human pluripotent stem cells.

2 Materials

2.1 Transfection of PiggyBac Plasmids and Generation of the PiggyBac-6F Lines

1. PiggyBac plasmids: piggyBac-TetO-Ascl1-neomycin (Addgene, #176482), piggyBac-TetO-En1-t2a-Foxa2-hygro (Addgene, #176483), piggyBac-TetO-Lmx1a-t2a-Pitx3-puro (Addgene, #176484), piggybac-TetO-hNurr1-blast (Addgene, #176485).
2. Transposase expression plasmid.
3. Fugene 6.
4. Human ES or iPS cells: H1 (WiCell, WA01), H9 (WiCell, WA09), or other human iPS cell lines.

2.2 Lentivirus Production

1. Lentiviral plasmids: TetO-FUW-Myt1L (Addgene, #27152), TetO-FUW-Wnt1 (Addgene, #176486), TetO-FUW-EGFP (Addgene, #30130), FUW-M2rtTA (Addgene, #20342).
2. Lentivirus packaging plasmids: pMDLg/pRRE (Addgene, #12251), pRSV-Rev (Addgene, #12253), pMD2.G (Addgene, #12259).
3. HEK293T cells (ATCC, CRL-11268).
4. Polyornithine (100×): dissolve 100 mg of polyornithine in 67 mL of distilled water and store as 5 mL aliquots at −20 °C. To prepare a working solution (15 mg/L), dilute the 100× stock in distilled water and filter through 0.22 μm filter before use.

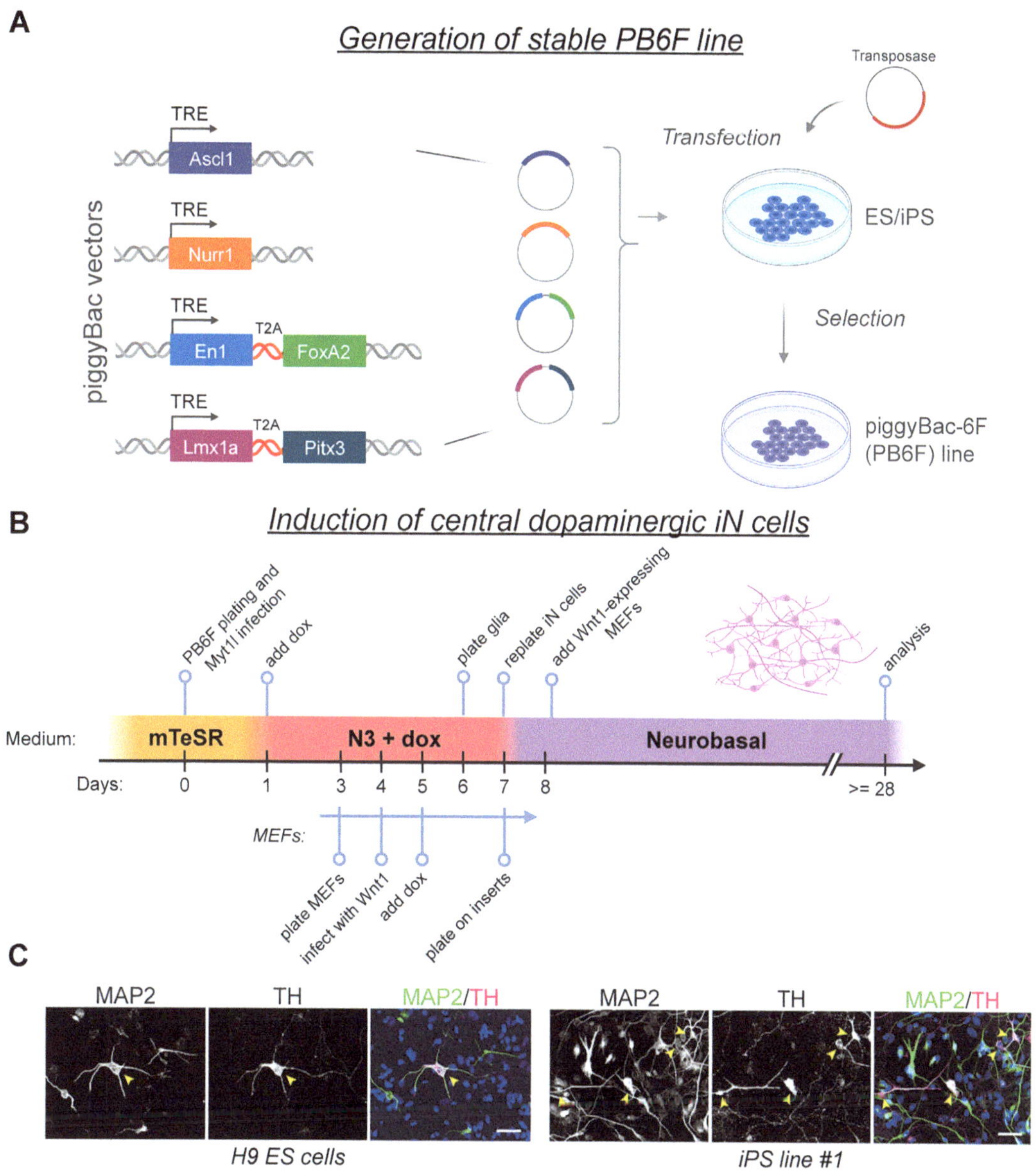

Fig. 1 Direct differentiation of human pluripotent stem cells (PSCs) into dopaminergic neurons. (**a**) Generation of stable piggyBac PSC line expressing six dopaminergic neuron lineage-specifying transcription factors (PB6F) under the control of doxycycline-inducible promoter. The human PSCs are subjected to four rounds of transfection with a transcription factors-expressing piggyBac vector followed by a corresponding antibiotic selection. (**b**) Timeline for dopaminergic iN cell generation. Stable PB6F PSCs are transduced with MYT1L-expressing lentivirus and treated with doxycycline to induce the expression of all seven transcription factors. iN cells are subsequently re-plated and co-cultured with primary mouse glia in the presence of Wnt1-expressing MEFs to deliver exogenous WNT1 and promote TH^{+} midbrain dopaminergic neuron induction and synaptic maturation. (**c**) Expression of MAP2 and TH—a dopaminergic neuronal marker, in dopaminergic iN cells that were derived from ES (left) or iPS cells (right) (scale bar 50 μm). (Created using BioRender.com)

5. 2 × BES-Buffered Saline (BBS): 50 mM N, N-bis(2-hydroxyethyl)-2-aminoethanesulfonic acid (BES), 280 mM NaCl, 1.5 mM Na_2HPO_4, pH 6.95. Sterilize using 0.22 μm filter before use. (*see* **Note 1**)
6. 2.5 M $CaCl_2$: Dissolve 73.5 g of $CaCl_2$ in 200 mL of distilled water. Sterilize with 0.22 μm filter, aliquot and store at −20 °C.

2.3 Cell Culture

1. MEF medium: DMEM supplemented with 1 mM Sodium Pyruvate, non-essential amino acids, 0.1 mM beta-mercaptoethanol, and 10% (vol/vol) Fetal Bovine Serum (FBS). (*see* **Note 2**).
2. Hank's Balanced Salt Solution (HBSS).
3. 0.25% Trypsin-EDTA.
4. mTeSR™ Plus medium.
5. Accutase.
6. Matrigel Basement Membrane Matrix.
7. Thiazovivin (5000×): add 3.2 mL of DMSO to 10 mg of Thiazovivin. Aliquot into 100 μL aliquots and store at −20 °C.
8. 0.4% Trypan Blue Solution.
9. Bambanker.
10. Polybrene (1000×): Dissolve 200 mg Hexadimethrine Bromide in 25 mL of distilled water to make 8 mg/mL stock solution. Sterilize by filtering through a 0.22 μm filter and store as 1 mL aliquots at −20 °C.
11. Doxycycline (1000×): To prepare 2 mg/mL stock solution, dissolve 40 mg of doxycycline in 20 mL of distilled water. Sterilize using 0.22 μm filter and store as 1 mL aliquots at −20 °C. Keep protected from light.
12. Puromycin (1000×): Dissolve 100 mg of puromycin dihydrochloride in 50 mL of distilled water to prepare a 2 mg/mL stock solution. Sterilize with 0.22 μm filter and store as 1 mL aliquots in −20 °C.
13. Neomycin (500×): Dissolve 5 g in 16.67 mL of distilled water to prepare a 300 mg/mL stock solution. Sterilize with 0.22 μm filter and store as 1 mL aliquots at −20 °C.
14. 50 mg/mL Hygromycin (250×).
15. Blasticidin S (500×): Dissolve 100 mg in 10 mL of distilled water to prepare a 10 mg/mL stock solution. Sterilize with 0.22 μm filter and store as 1 mL aliquots at −20 °C.
16. Insulin: Dissolve 250 mg of insulin in 40 mL of 10 mM NaOH and store 800 μL aliquots at −80 °C.

17. N3 medium: To 489 mL of DMEM/F12, add 5 mL of N2 supplement, 5 mL of 100× non-essential amino acids, and 1.6 mL of the prepared insulin stock.
18. Neurobasal medium: To 485 mL of Neurobasal™ medium, add 5 mL of 0.2 M GlutaMax™, and 10 mL of B27 supplement.
19. BDNF: Dissolve 50 μg of BDNF in 500 μL of 0.1% BSA/PBS and store as single-use 10 μL aliquots at −80 °C.
20. GDNF: Dissolve 50 μg of GDNF in 500 μL of 0.1%BSA/PBS and store as single-use 10 μL aliquots at −80 °C.
21. Ascorbic acid (1000×): Dissolve 1.19 g of ascorbic acid in 30 mL of distilled water to prepare a 200 mM stock. Aliquot and store at −80 °C.
22. Ara-C (2000×): Dissolve 25 mg of cytosine β-D-arabinofuranoside in 22.5 mL of distilled water to prepare a 4 mM stock solution. Store as 0.5 mL aliquots at −20 °C.
23. Neurobasal growth medium: To prepare 50 mL of neurobasal growth medium, combine 47.5 mL of neurobasal medium and 2.5 mL of FBS (final concentration 5% (vol/vol)). Add 50 μL of doxycycline (final concentration 2 μg/mL), 10 μL of BDNF (final concentration 20 ng/mL), 10 μL of GDNF (final concentration 20 ng/mL), 50 μL of ascorbic acid (final concentration 0.2 mM) and 25 μL of Ara-C (final concentration 2 μM). (*see* **Note 3**).
24. Papain dissociation solution: To 5 mL of HBSS add 80 μL of papain, 5 μL of 0.5 M EDTA, and 5 μL of 1 M $CaCl_2$. Incubate at 37 °C until solution turns clear. Filter using 0.22 μm filter before use.

2.4 Animals

1. Glia is isolated from *CD1* or *C57BL6* mice at the age of postnatal day 2–3 (P2–3).
2. MEF cells are prepared from embryonic day (E)12.5–14.5 old embryos obtained from 8 to 24-week-old *CD1* or *C57BL6* mice.

2.5 Equipment

1. Cell-freezing containers.
2. Ultracentrifuge (e.g., Sorvall WX ultracentrifuge with SureSpin 630 rotor and tube adapters).
3. Falcon® cell culture inserts, 24-well.
4. Cell strainers, 40 μm.
5. Dissecting microscope.
6. Glass Coverslips.

2.6 Immunofluorescence Analysis

1. Dulbecco's phosphate-buffered saline (D-PBS).
2. 4% paraformaldehyde (PFA) solution in D-PBS.
3. Permeabilization buffer: 0.2% Triton X-100 in D-PBS.
4. Blocking solution: 4% Bovine Serum Albumin (BSA), 1% serum in D-PBS. Filter through a 0.22 μM filter and store at 4 °C.
5. Primary antibodies: Tyrosine Hydroxylase (TH) (Pel-Freez, P60101), TUJ1 (Covance, MMS-435P or MRB-435P), MAP2 (Millipore, AB5392), DAT (Millipore, MAB369), VMAT2 (Millipore, AB1598P), Synapsin-1 (Synaptic Systems, 106 002).
6. DAPI (1000×): To 5 mg of DAPI add 5 mL of distilled water to prepare a 1 mg/mL stock solution.

3 Methods

3.1 Generation of PiggyBac-6F (PB6F) Lines

1. Day −1: Thaw Matrigel according to the manufacturer's instructions and dilute 1:200 in plain DMEM. Coat a 6-well plate and incubate at 37 °C for at least 1 h to overnight. (*see* **Note 4**).
2. Day 0: Plate H1 cells (or other human pluripotent stem cells) on Matrigel-coated 6-well plates. Remove cell media, add a sufficient amount of Accutase to cover the cell layer, and place the plate in 37 °C incubator for 3–5 min until cells begin to detach. Add equal volume of mTeSR™ Plus medium and dissociate the cells gently by pipetting several times using a P1000 tip. Collect the cells by centrifugation at 300 × *g* for 5 min. Resuspend the cell pellet in mTeSR™ Plus supplemented with thiazovivin (final concentration 2 μM). Count the number of viable cells using Trypan Blue and plate 5×10^5 cells/well.
3. Day 1: On the day of transfection, replace the medium in each well with fresh 1.5 mL of mTeSR™ Plus 30 min before transfection and return the plate to the incubator. Allow Fugene 6 reagent to reach room temperature. Prepare transfection mix using a 3:1 Fugene 6:DNA ratio as follows: for each well to be transfected combine in a 1.5 mL microfuge tube 90 μL of mTeSR™ Plus and 3 μg of total DNA mix, which contains 2 μg of the piggyBac plasmid and 1 μg of transposase plasmid. Lastly, add 9 μL of Fugene 6 and vortex for 5 s. Incubate at room temperature for 20 min and then add the transfection mix dropwise to the cells. Mix gently and return the cells to the incubator. (Strongly recommended: Include a positive control, i.e., a well transfected with a GFP-expressing plasmid. This

could serve as an indicator of the efficiency of transfection and as a control for antibiotic selection.)

4. Day 2: After 24 h replace the medium with mTeSR™ Plus supplemented with the appropriate antibiotic to select the resistant clones.
5. From day 3 onward: Replace medium with fresh mTeSR™ Plus supplemented with the appropriate antibiotics until selection is complete. Best indication of proper antibiotic selection is the absence of surviving cells in the non-antibiotic-resistant transfection control (i.e., the GFP-only transfected well). An approximate timing required for proper selection based on the antibiotic concentrations provided in this protocol is as follows: puromycin: 3 days, neomycin: 5 days, blasticidin: 5 days, hygromycin: 5–7 days.
6. When cells reach confluency, repeat **step 1** for the next piggyBac plasmid.
7. The piggyBac-6F (PB6F) stable line is generated after transfection of all four piggyBac plasmids (encoding 6 transcription factors). At this step, the cells should be frozen down to obtain a cell stock that is ready to be thawed and used when necessary. To cryopreserve the PB6F cells, dissociate the cells using Accutase and pellet by centrifugation at 300 × *g* for 5 min. Aspirate the supernatant and resuspend the cell pellet in Bambanker freezing medium. Aliquot into cryovials, transfer into freezing containers, and store at −80 °C overnight. Move the cryovials to the liquid nitrogen tank for long-term storage the following day.

3.2 Lentivirus Production

1. Coat 10 cm plates with polyornithine at 37 °C for at least 30 min. Wash the plates once with D-PBS before use.
2. 18–24 h before transfection, dissociate HEK293T cells using 0.25% Trypsin-EDTA. Pellet the cells by centrifugation at 300 × *g* for 5 min. Count viable cells using Trypan Blue and plate 3×10^6 cells in 10 mL of MEF medium per polyornithine-coated 10 cm plate. Cells should be 60–80% confluent at the time of transfection.
3. 1 h before transfection, replace the medium in each plate with 9 mL of pre-warmed MEF medium.
4. Prepare 1 mL of calcium phosphate DNA precipitate for transfection of one 10 cm plate of HEK293T as follows:
   (a) Combine in a sterile microfuge tube 5 μg of pMDLg/pRRE, 2.5 μg pMD2.G, 2.5 μg pRSV-Rev and 10 μg of the lentiviral plasmid. Add predetermined volume of 2.5 M $CaCl_2$ and adjust to 500 μL with sterile distilled water. (*see* **Note 1**).

(b) Add dropwise 500 μL of 2 × BBS to the calcium-DNA solution while vortexing. Let stand at room temperature for 15–20 min to allow the precipitate to form.

5. Distribute the 1 mL mixture dropwise over the cells in the culture dish. Mix gently by rocking the dish back and forth and return the plate to the incubator.
6. Replace the medium 18 h after transfection with 5 mL of fresh MEF medium.
7. Collect the medium containing viral particles 42–48 h after transfection.
8. Centrifuge viral supernatants at 1000 × *g* to remove cell debris and transfer the supernatant into ultracentrifuge tubes.
9. Concentrate the viral supernatant by centrifugation at 100,000 × *g* for 2 h at 4 °C.
10. Carefully aspirate the supernatant and resuspend the pellet in fresh DMEM to obtain 100× concentrated viral stock. Store at 4 °C for up to 2 weeks. (*see* **Note 5**).

3.3 Glia Isolation and Culture

1. Anesthetize P3 pups on ice for 3–5 min. Remove the heads with sharp surgical scissors and place into a 10 cm dish. To obtain a sufficient number of cells, 3 pups per 10 cm culture dish are recommended.
2. Slice back the skin and skull using a scalpel, open the skull with forceps, remove the brain with a narrow spatula, and place immediately into 10 cm dish containing cold HBSS.
3. Using a dissecting knife, remove the olfactory bulbs and separate the cortical hemispheres. Using fine forceps, carefully remove the meninges, hippocampus, and ventral telencephalon, and transfer the remaining tissue into a 15 mL falcon tube containing cold HBSS on ice.
4. After collecting cortical tissues from 3 pups, aspirate HBSS and add 5 mL of sterile-filtered papain dissociation solution. Incubate at 37 °C for 15 min inverting the tube every 5 min.
5. Carefully remove the papain solution using serological pipet. At this stage, the tube content will be viscous therefore, using vacuum to aspirate the supernatant should be avoided to prevent loss of the tissue material. Wash once with 15 mL of HBSS.
6. Add 1 mL of MEF medium. Gently triturate to dissociate the tissue into a single-cell suspension. Add 9 mL of MEF medium to the cell suspension and filter through a 40 μm cell strainer. Plate onto 10 cm tissue culture dish.

7. Replace the media with fresh 10 mL of MEF medium the following day. (Optional: Wash the plate gently once with MEF medium to remove the remaining cell debris).
8. When confluent, passage the glia in MEF medium in 10 cm dishes before using the cells for experiments to minimize neuronal contamination. Use the cultures within ~4 weeks.

3.4 MEF Isolation and Culture

1. Obtain and sacrifice a E12.5–E14.5 pregnant female mouse while adhering to the relevant institutional guidelines and protocols (*see* **Note 6**).
2. Remove uterine horns and place into 10 cm dish containing cold HBSS.
3. Cut open the uterine wall using fine scissors. Open amniotic sac and remove placental tissue with forceps to access the embryo.
4. Cut away the limbs of the embryo using fine scissors or forceps and transfer them to a clean 15 cm tissue culture dish. Collect limbs of 3 embryos (12 limbs) per 15 cm dish.
5. Add few drops of 0.25% Trypsin-EDTA to the limbs on the dish and mince them finely using a pair of curved scissors.
6. Add 1 mL of Trypsin-EDTA to the minced tissues and incubate at 37 °C for 15 min.
7. Add 20 mL of prewarmed MEF medium and pipet up and down several times to further dissociate the tissue. Return the plate to the incubator.
8. Replace the medium with fresh MEF medium the following day. Fibroblasts should be observed growing out from the tissue fragments. The culture should reach confluency in the next 2–3 days.
9. Passage confluent cultures for further expansion or cryopreserve. MEFs should not be used beyond 5 passages.

3.5 iN Cell Induction and Culture

1. Prior to iN induction, maintain the piggyBac-6F (PB6F) stable line (on H1 or other pluripotent stem cell background) in mTeSR™ Plus medium with antibiotic selection until 70–80% confluency.
2. Coat 6-well plates with Matrigel for a minimum of 1 h (overnight preferred).
3. Day 0: Dissociated the cells using Accutase as described in **step 2**, Subheading 3.1.
4. Count the viable cells using Trypan Blue and resuspend 5×10^5 cells per well in a total of 1 mL of medium. Add 1 μL of polybrene (final concentration 8 μg/mL), 3 μL of Myt1L, and 3 μL of rtTA-expressing lentivirus. Mix gently

but thoroughly. (Optional: transduce one well with TetO-EGFP and rtTA as control). (*see* **Note 7**).

5. Aspirate the Matrigel from the coated 6-well tissue culture dish and plate 1 mL of cell-lentivirus mix per well. Gently rock the plate back and forth to ensure even cell dispersion and return the plate to the incubator.
6. Day 1: Replace the medium with N3 medium supplemented with doxycycline (final concentration 2 μg/mL) the following day. Note that at this point cells should have attached to the bottom of the well.
7. Day 2–6: Replace the medium daily with fresh doxycycline-containing N3 medium.
8. Day 3: Dissociate MEFs using 0.25% Trypsin-EDTA and plate at a density of 2.5×10^5 cells per well onto polyornithine-coated 6-well plates.
9. Day 4: Replace MEF cell medium with 1 mL of fresh MEF medium containing 1 μL of polybrene, 2 μL of Wnt1, and 3 μL of rtTA lentivirus.
10. Day 5: Change MEF medium to fresh MEF medium supplemented with doxycycline (1.5 mL per well).
11. Day 6: Place sterile glass coverslips in the wells of a 24-well tissue culture plate and coat with Matrigel for at least 1 h. Dissociate glia (passage 1 or 2) with 0.25% Trypsin-EDTA. Pellet the cells by centrifugation at $300 \times g$ for 5 min, resuspend in MEF medium, and plate on the Matrigel-coated coverslips. One confluent 10 cm plate of glial culture is sufficient to plate on ~24 coverslips, which will be used for re-plating iN cells the following day.
12. Day 7: By this time, the PB6F iN cells from **step 7**, Subheading 3.5 should have adopted a distinctive neuronal morphology. Dissociate the cells using Accutase. Once the cells detach, add equal volume of medium and pellet the cells by centrifugation at $300 \times g$ for 5 min. Carefully aspirate the supernatant and taking care not to dislodge the cell pellet, resuspend the cells in 1–2 mL of neurobasal medium containing 5% serum and gently pipet up and down to obtain a single cell suspension. Caution: avoid harsh pipetting that would induce iN cell death. Count viable cells using Trypan Blue and plate 2×10^5 cell per well onto a glia plated in 24-well plates in **step 11**, Subheading 3.5.
13. Day 7: Place 0.5 mL of MEF medium per well of fresh 24-well tissue culture plate. Using sterile forceps, place Falcon® cell culture inserts into each well of the plate. Dissociate Wnt1-transduced MEF cells using 0.25% Trypsin-EDTA and centrifuge at $300 \times g$ for 5 min. Resuspend the cell pellet in MEF

medium, count viable cells using Trypan Blue, and plate 5×10^4 cells per insert.

14. Day 8: The PB6F iN cells from **step 12**, Subheading 3.5 should have attached to the glia cell layer, and Wnt1-transduced MEFs (**step 13**, Subheading 3.5) should have adhered to the membrane of the cell culture inserts. Replace the PB6F iN cell medium with fresh Neurobasal growth medium. Using sterile forceps, aseptically transfer the Wnt1-MEF containing inserts into the wells with the PB6F iN cells.
15. Day 8–28: Feed the PB6F iN cell cultures with fresh neurobasal growth medium every 3–4 days by replacing half of the existing medium with 0.3–0.5 mL of neurobasal growth medium. Pipet slowly along the side of the wall of the well to avoid dislodging the attached cells.

3.6 iN Cell Characterization by Immunofluorescence Analysis

After 4 weeks in culture, the PB6F iN cells acquire mature neuronal morphology and express dopaminergic neuron markers, including Tyrosine Hydroxylase (TH), DAT and VMAT2, in addition to the pan-neuronal markers, TUJ1 and MAP2, and synaptic markers, such as Synapsin-1.

1. Discard Wnt1-MEF culture insert and aspirate the medium.
2. Wash the plate with D-PBS. (*see* **Note 8**).
3. Add 4% PFA and incubate at room temperature for 10 min.
4. Aspirate the PFA and wash the coverslips with D-PBS.
5. Add permeabilization buffer and incubate for 5 min at room temperature.
6. Wash three times with D-PBS.
7. Add blocking solution and incubate at room temperature for at least 30 min to 1 h.
8. Add primary antibodies diluted in blocking solution and incubate at 4 °C overnight.
9. Remove the primary antibody and wash three times with D-PBS.
10. Add secondary antibodies diluted in blocking solution and incubate at room temperature for 30 min. (*see* **Note 9**).
11. Aspirate the secondary antibody solution, add D-PBS containing DAPI, and incubate at room temperature for 3 min.
12. Wash the plate three times with D-PBS. (*see* **Note 10**).
13. Using a pair of fine forceps remove the coverslips from the wells and mount onto microscope glass slides with mounting medium. Slides can be stored short term at 4 °C before imaging using fluorescence microscopy.

4 Notes

1. Accurate pH of 2 × BBS is critical for the transfection efficiency.
2. Fetal Bovine Serum varies between lots and manufacturers. It is advisable to test the FBS to identify the source and lot number that supports the neuronal cultures.
3. In case of insufficient density of the glia culture (observable gaps in the glial cell layer), Ara-C can be omitted during the initial rounds of media changes until the glial cell layer covers the entire area of the coverslip.
4. Thaw the Matrigel at 4 °C overnight. Matrigel is viscous and will gel at higher temperatures. Prepare single-use aliquots using pre-chilled tubes and tips, and freeze at −20 °C for future use.
5. Optimal volume of 2.5 M $CaCl_2$ to be used per transfection will change with each batch of the prepared 2 × BBS. It can be determined by testing the transfection efficiencies using $CaCl_2$ volumes in a range of 40–120 μL (in 10 μL increments). The volume of 2.5 M $CaCl_2$ that results in the highest transfection efficiency should be used for all subsequent transfections with this batch of 2 × BBS.
6. Aspirate the supernatant very carefully, as the precipitate is barely visible. 10 mL of viral supernatant produces 100 μL of concentrated virus.
7. Virus titers vary from batch to batch and the volume of virus to be used may require optimization.
8. iN cells detach easily from the surface of the coverslip. Do washes and add buffers by pipetting slowly against the wall of the well.
9. On occasion, aggregates that form in solutions of fluorescence-labeled secondary antibodies can be detected by fluorescence microscopy, contributing to the unspecific fluorescence signal. To remove any possible antibody precipitates, centrifuge the tube with the secondary antibody stock at 12,000 × *g* for 1 min using a benchtop centrifuge and use only the supernatant to prepare the dilutions used for staining.
10. Coverslips can be stored short-term at 4 °C in D-PBS supplemented with sodium azide (0.02% (w/v)) to prevent microbial growth.

References

1. Yamanaka S (2020) Pluripotent stem cell-based cell therapy—promise and challenges. Cell Stem Cell 27:523–531. https://doi.org/10.1016/j.stem.2020.09.014
2. Hanna JH, Saha K, Jaenisch R (2010) Pluripotency and cellular reprogramming: facts, hypotheses, unresolved issues. Cell 143:508–525. https://doi.org/10.1016/j.cell.2010.10.008
3. Blanpain C, Daley GQ, Hochedlinger K, Passegué E, Rossant J, Yamanaka S (2012) Stem cells assessed. Nat Rev Mol Cell Biol 13: 471–476. https://doi.org/10.1038/nrm3371
4. Okita K, Yamanaka S (2011) Induced pluripotent stem cells: opportunities and challenges. Philos Trans R Soc, B 366:2198–2207. https://doi.org/10.1098/rstb.2011.0016
5. Limone F, Klim JR, Mordes DA (2022) Pluripotent stem cell strategies for rebuilding the human brain. Front Aging Neurosci 14: 1017299
6. Kim TW, Koo SY, Studer L (2020) Pluripotent stem cell therapies for Parkinson's disease: present challenges and future opportunities. Front Cell Dev Biol 8:729
7. Wang H, Yang Y, Liu J, Qian L (2021) Direct cell reprogramming: approaches, mechanisms and progress. Nat Rev Mol Cell Biol 22:410–424. https://doi.org/10.1038/s41580-021-00335-z
8. Zhang Y, Pak C, Han Y, Ahlenius H, Zhang Z, Chanda S, Marro S, Patzke C, Acuna C, Covy J, Xu W, Yang N, Danko T, Chen L, Wernig M, Sudhof TC (2013) Rapid single-step induction of functional neurons from human pluripotent stem cells. Neuron 78: 785–798. https://doi.org/10.1016/j.neuron.2013.05.029
9. Ng YH, Chanda S, Janas JA, Yang N, Kokubu Y, Südhof TC, Wernig M (2021) Efficient generation of dopaminergic induced neuronal cells with midbrain characteristics. Stem Cell Rep 16:1763–1776. https://doi.org/10.1016/j.stemcr.2021.05.017
10. Yang N, Chanda S, Marro S, Ng YH, Janas JA, Haag D, Ang CE, Tang Y, Flores Q, Mall M, Wapinski O, Li M, Ahlenius H, Rubenstein JL, Chang HY, Buylla AA, Sudhof TC, Wernig M (2017) Generation of pure GABAergic neurons by transcription factor programming. Nat Methods 14:621–628. https://doi.org/10.1038/nmeth.4291

Chapter 5

Directed Differentiation of Human iPSCs into Microglia-Like Cells Using Defined Transcription Factors

Shih-Wei Chen and Yu-Hui Wong

Abstract

The generation of a homogeneous population of microglia from human induced pluripotent stem cells (hiPSCs) is crucial to modeling neurological disorders, as well as the carrying out of drug screening and toxicity testing. Here, we provide a stepwise protocol for the simple, robust, and efficient differentiation of hiPSCs into microglia-like cells (iMGs) by overexpression of SPI1 and CEBPA. This protocol details hiPSC culture, lentivirus production, lentivirus delivery, and, finally, the differentiation and validation of the iMG cells.

Key words Induced pluripotent stem cells, Microglia, SPI1, CEBPA, Transcription factor-based differentiation

1 Introduction

Microglia play many important roles in brain homeostasis and functioning. As the primary immune cells in the brain, microglia are a mesoderm-derived population from the blood island of the yolk sac that colonize the neuroepithelium before the formation of the blood-brain barrier [1].

Most existing methods for generation of human microglia from induced pluripotent stem cells (iPSCs) share several common features; these include specification of a mesodermal cell fate, development into primitive hematopoietic progenitors, and finally induction into mature microglia [2–5]. These protocols have shown the feasibility of this process but require several weeks or even months to obtain microglia. Therefore, our aim was to develop an efficient approach for generating human microglia via the forced expression of transcription factors (TFs). The concept of TF-based conversion has been widely utilized for the differentiation of various cell types in the brain [6–8]. After the screening of the critical TFs important to microglial development and

Yu-Wen Alvin Huang and ChangHui Pak (eds.), *Stem Cell-Based Neural Model Systems for Brain Disorders*, Methods in Molecular Biology, vol. 2683, https://doi.org/10.1007/978-1-0716-3287-1_5,

maintenance, we reported here an approach that can efficiently produce, within 10 days, microglia-like cells (iMGs) from hiPSCs by overexpression of both SPI1 and CEBPA [9]. This has created a useful tool for research into human microglia, both in the healthy brain and in the diseased brain.

Here, we describe the methods needed to generate human microglia-like cells in detail (Fig. 1a). The protocol is divided into four parts. The first part is lentiviral production, and this is followed by regular hiPSC culture and lentiviral delivery. The third part describes the detailed procedure to generate iMG cells from the lentiviral-infected hiPSC line. The final part is how we examine the iMG cells by immunocytochemistry and flow cytometry.

2 Materials

2.1 Lentivirus Generation

2.1.1 Cell Lines

1. HEK293FT cell line.

2.1.2 Media/Solutions

1. Dulbecco's Modified Eagle's Medium (DMEM).
2. Opti-MEM™ I reduced serum medium (Life Technologies, cat. no. 31985070).
3. Dulbecco's phosphate-buffered saline (DPBS), no calcium, no magnesium (Life Technologies, cat. no. 21600010).
4. Fetal bovine serum (FBS), heat-inactivated.
5. TrypLE™ Express Enzyme (Life Technologies, cat. no. 12605028).

2.1.3 Reagents

1. pTetO-CEBPA-T2A-SPI1-T2A-Puro.
2. pFUW-rtTA.
3. pCMV-Δ8.91 (gag, pol, and rev genes).
4. pVSV-G (VSV-G envelope).
5. Lipofectamine™ 3000 transfection reagent (Life Technologies, cat. no. L3000015).
6. Polybrene (Sigma-Aldrich, cat. no. 107689).

2.1.4 Consumables

1. A 0.45 μm syringe filter (low protein binding).
2. Amicon Ultra-15 centrifuge filter unit (100 kDa cutoff).
3. 100 mm cell culture dishes.

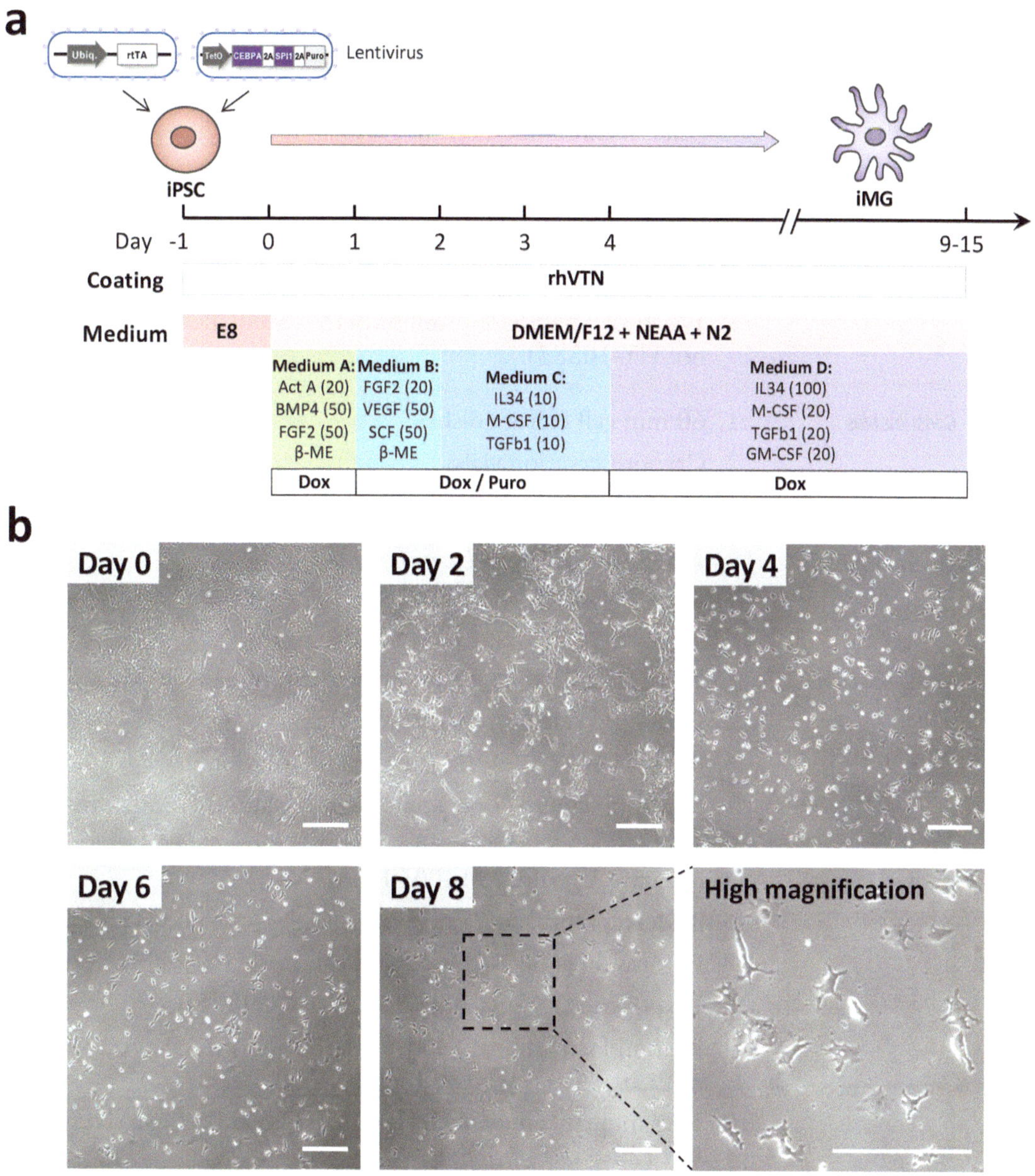

Fig. 1 Differentiation of human microglia-like cells from hiPSCs. (**a**) Flow diagram of the generation of induced microglia (iMG). Numbers in parentheses indicate the concentration of the human recombinant proteins in ng/mL. (**b**) Representative phase-contrast images of iMGs during differentiation. A typical iMG morphology can be observed from day 6 onward after differentiation (scale bar 200 μm)

2.2 Human iPSC Culture

2.2.1 Cell Lines

1. hiPSC line (Here we used NTUH-iPSC-02-02 (https://catalog.bcrc.firdi.org.tw/BcrcContent?bid=SC81117) as an example).

2.2.2 Media/Solutions

1. Essential 8™ medium (E8) (Life Technologies, cat. no. A1517001).
2. Vitronectin (VTN) recombinant human protein, truncated (Life Technologies, cat. no. A14700).
3. DPBS, no calcium, no magnesium.
4. 0.5 mM EDTA in DPBS (Life Technologies, cat. no. 15575020) (or Versene solution, Life Technologies, cat. no. 15040066).
5. Y-27632 (ROCK inhibitor) (MedChemExpress, cat. no. HY-10583).

2.2.3 Consumables

1. 60 mm cell culture dishes.
2. Cryopreservation vials.

2.3 Differentiation of iMG Cells from hiPSCs

2.3.1 Cell Lines

1. hiPSC line infected with lentivirus carrying a *doxycycline-inducible CEBPA-2A-SPI1 transgene.*

2.3.2 Media, Solutions, and Reagents

1. DMEM/F12 (Life Technologies, cat. no. 11330032).
2. DPBS, no calcium, no magnesium.
3. 0.5 mM EDTA/DPBS (or Versene solution).
4. N-2 supplement (Life Technologies, cat. no. 17502048).
5. MEM non-essential amino acids solution (NEAA, Life Technologies, cat. no. 11140050).
6. β-Mercaptoethanol (Life Technologies, cat. no. 31350010).
7. Puromycin (Life Technologies, cat. no. A1113803).
8. Doxycycline (Sigma-Aldrich, cat. no. D9891).
9. Recombinant human Activin A (50 μg/mL, PeproTech, cat. no. 120-14E) (*see* **Note 1**).
10. Recombinant human BMP4 (50 μg/mL, PeproTech, cat. no. 120-05ET).
11. Recombinant human FGF-basic (154 a.a.) (FGF2, 50 μg/mL, PeproTech, cat. no. 100-18B).
12. Recombinant human VEGF (50 μg/mL, PeproTech, cat. no. 100-20).
13. Recombinant human SCF (50 μg/mL, PeproTech, cat. no. 300-07).
14. Recombinant human IL-34 (100 μg/mL, PeproTech, cat. no. 200-34).
15. Recombinant human TGF-β1 (50 μg/mL, PeproTech, cat. no. 100-21).

16. Recombinant human M-CSF (50 μg/mL, PeproTech, cat. no. 300-25).
17. Recombinant human GM-CSF (50 μg/mL, PeproTech, cat. no. 300-03).

2.3.3 Compositions of the Differentiation Media

1. Medium A: DMEM/F12, N-2 supplement, NEAA, β-mercaptoethanol (50 μM), BMP4 (50 ng/mL), FGF2 (50 ng/mL), Activin A (20 ng/mL), doxycycline (2 μg/mL).
2. Medium B: DMEM/F12, N-2 supplement, NEAA, β-mercaptoethanol (50 μM), VEGF (50 ng/mL), SCF (50 ng/mL), FGF2 (20 ng/mL), doxycycline (2 μg/mL), puromycin (1 μg/mL).
3. Medium C: DMEM/F12, N-2 supplement, NEAA, IL-34 (10 ng/mL), M-CSF (10 ng/mL), TGF-β1 (10 ng/mL), doxycycline (2 μg/mL), puromycin (1 μg/mL).
4. Medium D: DMEM/F12, N-2 supplement, NEAA, IL-34 (100 ng/mL), M-CSF (20 ng/mL), TGF-β1 (20 ng/mL), GM-CSF (20 ng/mL), doxycycline (2 μg/mL).

2.4 Validation of the iMG Cells

2.4.1 Flow Cytometry

1. DPBS, no calcium, no magnesium.
2. TrypLE Express.
3. Home-made FACS buffer: 1% FBS in DPBS.
4. Human Fc block (0.5 mg/mL) (1:20 dilution, BD Biosciences, cat. no. 564220).
5. FITC conjugated anti-CD11b antibody (1:100 dilution, BioLegend, cat. no. 101206).
6. PE conjugated anti-TREM2 antibody (1:50 dilution, R&D Systems, cat. no. FAB17291P).
7. 5 mL round-bottom polystyrene test tubes with cell strainer snap cap.

2.4.2 Immunocytochemistry

1. Phosphate-buffered saline (PBS).
2. 4% paraformaldehyde (PFA).
3. 0.1% Triton X-100 in PBS.
4. Bovine serum albumin (BSA).
5. Normal goat serum (NGS).
6. Sodium azide (NaN_3).
7. Anti-IBA1 antibody, rabbit IgG (1:1000 dilution, Wako, cat. no. 019-19741).
8. Anti-TREM2 antibody, rat IgG (1:500 dilution, Merck Millipore, cat. no. MABN755).

9. Anti-PU.1 antibody, rabbit IgG (1:500 dilution, Abcam, cat. no. ab76543).
10. Anti-P2RY12 antibody, rabbit IgG (1:500 dilution, Sigma-Aldrich, cat. no. HPA014518).
11. Anti-TMEM119 antibody, rabbit IgG (1:500 dilution, Abcam, cat. no. ab185337).
12. Goat anti-rabbit IgG conjugated to Alexa Fluor 488 (1:250 dilution, Life Technologies, cat. no. A-11008).
13. Goat anti-rabbit IgG conjugated to Alexa Fluor 555 (1:250 dilution, Life Technologies, cat. no. A-21429).
14. Goat anti-rat IgG conjugated to Cy5 (1:250 dilution, Life Technologies, cat. no. A10525).
15. 4′,6-diamidino-2-phenylindole (DAPI).
16. Mounting medium: Fluoromount-G (Life Technologies, cat. no. 00-4958-02).
17. Coverslips.

3 Methods

3.1 Generation of Lentivirus for Gene Delivery

The following protocol describes the procedures used for producing infectious transgenic lentivirus carrying pFUW-rtTA and pTetO-CEBPA-T2A-SPI1-T2A-Puro (*see* **Note 2**). Production and usage of lentivirus should follow the biosafety level 2 guidelines. If the desired lentivirus is already prepared, skip this step and proceed to Subheading 3.2.

3.1.1 HEK293FT Seeding (Day − 1)

1. Maintain HEK293FT cells in DMEM with 10% heat-inactivated FBS in a 37 °C, 5% CO_2 incubator. Typically passage the cells every 2–3 days.
2. Passage and seed the cells at a density of 4–5 × 10^6 cells into a 100 mm cell culture dish to reach 80–90% confluency on the next day, which will allow efficient transfection.

3.1.2 Transfection (Day 0)

1. On the next day, for transfection in one 100 mm dish, prepare two different mixtures as follows:

   Mixture 1

   500 μL Opti-MEM

   24 μL Lipofectamine 3000 reagent

   Mixture 2

   500 μL Opti-MEM

   5 μg pCMV-Δ8.91 plasmid

   1 μg pVSV-G plasmid

6 μg pFUW-rtTA plasmid (or pTetO-CEBPA-2A-SPI1-2A-Puro plasmid)

24 μL P3000 reagent

2. Add mixture 2 into mixture 1, vortex to ensure there is a homogeneous mixture, and then incubate at room temperature (RT) for 15 minutes (min).
3. Add the DNA/Lipofectamine 3000 mixture drop-wise to the cells.
4. At 6 hours (h) after transfection, replace the old medium with 10 mL of 37 °C pre-warmed DMEM containing 10% FBS, and incubate the cells for a further 30 h.

3.1.3 First Medium Collection (Day 1.5)

1. Collect the viral supernatant at 30 h when changing the medium, and store it at 4 °C. Add 10 mL of 37 °C pre-warmed DMEM with 10% FBS carefully (*see* **Note 3**).

3.1.4 Second Medium Collection and Concentration (Day 3)

1. Collect the viral supernatant after a further 30 h, and pool it with the previously harvested medium.
2. Centrifuge at 1000 × *g* for 5 min to remove the cell debris.
3. Filter the supernatant with a 0.45 μm low-protein binding filter.
4. To concentrate the lentiviral particles, place 10 mL of filtered medium into an Amicon Ultra-15/100-kDa filter, and then centrifuge at 4000 × *g* for 20 min at 4 °C. Repeat this step again to obtain the final lentiviral medium at 20× concentration (~1 mL).
5. Use a 200 μL pipette tip to homogenize the concentrated medium and aliquot 200 μL into smaller storage tubes. This concentrate lentiviral medium can be used fresh or, alternatively, it can be stored for long terms at −80 °C.

3.2 Generation of hiPSC Lines Carrying the Doxycycline-Inducible SPI1 and CEBPA Transgenes (see Note 4)

The hiPSC lines should be obtained and used according to the appropriate legal and ethical guidelines. We use the standard protocol described by the Essential 8™ (E8) medium kit with vitronectin (VTN) for the hiPSC culture. The hiPSCs are regularly cultured, passaged, and maintained in 60 mm cell culture dishes in a humidified incubator at 37 °C and 5% CO_2. All cell culture procedures should be performed in a Class II biosafety cabinet under sterile conditions.

3.2.1 hiPSC Recovery

1. Prepare 60 mm dishes coated with 0.5 μg/cm^2 VTN according to manufacturer's instructions.
2. Prewarm the complete E8 medium to RT. Aliquot 4 mL of complete E8 medium into a 15 mL centrifuge tube. Aliquot

another 4 mL of complete E8 medium into another 15 mL centrifuge tube and add Y-27632 to a final concentration of 5 μM.

3. Remove the vial of cells from liquid nitrogen and immerse the cell vial in a 37 °C water bath until the last visible ice crystals have just disappeared.
4. Decontaminate the vial using 70% ethanol and place it into the hood.
5. Carefully and slowly transfer the thawed cells into the first 15 mL centrifuge tube containing 4 mL of complete E8 medium (without Y-27632).
6. Centrifuge the cells at 300 × *g* for 5 min.
7. Remove the supernatant and resuspend the cell pellet using the second 4 mL of complete E8 medium containing 5 μM Y-27632.
8. Remove the coating over the VTN and evenly disperse the 4 mL of cell suspension onto the dish. Avoid allowing the surface of the coating to dry.
9. Wait for 3 min until the cells attach to the dish and then carefully place the dish back into the incubator.
10. The next day, replace the Y-27632-containing medium with fresh complete E8 medium. Continue to replace the spent medium every day.

3.2.2 Regular Passaging

Once the cells reach 85% confluency, or 5 days since last passage, subculture is required to maintain healthy undifferentiated cultures.

1. Prepare 60 mm dishes coated with 0.5 μg/cm^2 VTN. Prewarm the complete E8 medium, DPBS, and EDTA/DPBS to room temperature.
2. Remove the old medium and rinse the cells using 3 mL DPBS once.
3. Add 1–2 mL of 0.5 mM EDTA/DPBS solution into each dish.
4. Incubate each dish at 37 °C for 3–5 min. When the cells start to separate and round up, the colonies appear to have holes in them when viewed under a microscope. They are now ready to be removed from the dish.
5. Aspirate the EDTA/DPBS solution from the plates.
6. Add 1 mL E8 medium, and this will detach the cells from the dish when gently pipetting without creating bubbles is carried out. Avoid excessive pipetting to keep small clumps of cells intact (*see* **Note 5**).

7. Transfer an ideal volume of the cells (usually about a 1:10 splitting rate) into a new dish. Evenly distribute the cells across the surface of the dish.
8. Wait for 3 min until the cells attach to the dish and then carefully place them back in the incubator.

3.2.3 Generation of hiPSC Lines Carrying the Doxycycline-Inducible SPI1 and CEBPA Transgenes (See **Note 6**)

1. One day before lentiviral infection, seed the hiPSCs into a 12-well plate coated with 0.5 μg/cm^2 VTN and then allow them to reach 60–70% confluency on the next day when infection will take place (*see* **Note 7**).
2. For infection with lentivirus carrying pFUW-rtTA, freshly prepare the infection medium as follows (*see* **Note 8**):

   200 μL of concentrated lentiviral medium of pFUW-rtTA

   800 μL of complete E8 medium

   8 μg/mL polybrene
3. Replace the spent medium of the hiPSCs with the infection medium. Incubate the cells overnight in a 37 °C, 5% CO_2 incubator.
4. After 16 h, replace the lentivirus-containing medium with complete E8 medium (*see* **Note 9**).
5. The next day, the infected hiPSCs can be further expanded by regular passaging and maintenance and then cryopreserved or prepared for another lentiviral infection step.
6. For infection of the lentivirus carrying pTetO-CEBPA-T2A-SPI1-T2A-Puro, freshly prepare the infection medium as follows:

   200 μL of concentrated lentiviral medium of pTetO-CEBPA-T2A-SPI1-T2A-Puro

   800 μL of complete E8 medium

   8 μg/mL polybrene
7. Replace the spent medium of the hiPSCs with the infection medium. Incubate cells overnight in the incubator.
8. After 16 h, replace the lentivirus-containing medium with complete E8 medium.
9. The infected hiPSCs can then be further expanded, cryopreserved, or differentiated into iMG cells (proceeding to Subheading 3.3). All experiments using infected hiPSCs are usually started after two passages, which is the time needed to allow cell adaptation.

3.3 Differentiation of Microglia-Like Cells from hiPSCs

The cryopreserved hiPSCs, which have been infected with both pFUW-rtTA and pTetO-CEBPA-T2A-SPI1-T2A-Puro, can be thawed as described in Subheading 3.2.1. Prior to starting the iMG differentiation (Fig. 1a), the thawed hiPSCs should be allowed to recover for 2–3 passages.

3.3.1 hiPSC Seeding (Day − 1)

1. Prepare 60 mm dishes with 0.5 μg/cm^2 VTN.
2. Prewarm the required volume of E8 medium at room temperature. Aliquot 4 mL E8 medium into a 15 mL centrifuge tube.
3. Aspirate the spent medium from the plate. Rinse the cells with 3 mL DPBS once.
4. Remove the DPBS. Add 2 mL EDTA/DPBS into the dish and incubate at 37 °C for 3 min.
5. Carefully remove the EDTA/DPBS.
6. Detach the cells by gently pipetting with 1 mL of complete E8 medium using a 1000 μL pipette tip; avoid pipetting up and down too much because this will break up the cell clumps into single cells. Transfer and evenly distribute the cells onto the plate so that these cells are able to reach 70–80% confluency on day 0 (*see* **Note 10**).
7. Wait for 3 min until the cells attach to the dishes, and then carefully place them back in the incubator.

3.3.2 hiPSC-to-iMG Induction (Day 0)

1. Prepare the differentiation medium A: DMEM/F12/NEAA/N2 containing activin A, BMP4, FGF2, β-mercaptoethanol, and doxycycline. Doxycycline is added to promote TetO downstream gene expression and is retained in the medium until the end of the experiment. The recombinant proteins, β-mercaptoethanol, and doxycycline need to be freshly added to the medium.
2. Replace the spent E8 medium with differentiation medium A.

3.3.3 hiPSC-to-iMG Induction and Selection (Day 1)

1. Prepare the differentiation medium B: DMEM/F12/NEAA/N2 containing FGF2, SCF, VEGF, β-mercaptoethanol, doxycycline, and puromycin. Puromycin is added to select cells expressing CEBPA and SPI1. The recombinant proteins, β-mercaptoethanol, doxycycline, and puromycin need to be freshly added to the medium.
2. On day 1, 24 h after starting induction, replace spent medium A with medium B.

3.3.4 hiPSC-to-iMG Differentiation (Days 2 and 3)

1. Prepare the differentiation medium C: DMEM/F12/NEAA/N2 containing IL-34, M-CSF, TGF-β1, and doxycycline. The recombinant proteins and doxycycline need to be freshly added to the medium.
2. On day 2, 24 h after puromycin selection, replace the spent medium B with medium C, or perform the replating steps (*see* **Note 11**).

3. When replating, the induced cells can be re-seeded onto the different cell cultureware, depending on different types of experiments that will be taking place (*see* **Note 12**).
4. To perform replating, prepare VTN-coated dishes.
5. Aspirate the spent medium. Rinse the cells with 3 mL DPBS once.
6. Aspirate DPBS. Add 2 mL EDTA/DPBS to the dish, and incubate at 37 °C for 3 min.
7. Carefully aspirate EDTA/DPBS.
8. Detach the cells by gently pipetting with 1 mL medium C using a 1000 μL pipette tip, this time pipetting up and down to break the cells into single cells (*see* **Note 13**).
9. Determine the cell number using a hemocytometer. Plate the cells at a density of 4–6 $\times$ 10^4 cells/cm^2 (*see* **Note 14**).
10. Wait for 3 min until the cells attach to the dishes and then carefully place them back in the incubator.

3.3.5 hiPSC-to-iMG Differentiation and Maturation (Day 4 to the End of Differentiation)

1. Prepare the differentiation medium D: DMEM/F12/NEAA/N2 containing IL-34, M-CSF, TGF-β1, GM-CSF, and doxycycline. The recombinant proteins and doxycycline need to be freshly added to the medium (*see* **Note 15**).
2. On day 4, replace the spent medium C with medium D.
3. Every other day, replace half of the medium with freshly prepared medium D until the end of differentiation (*see* **Note 16**).

3.4 Verification of the iMG Cells

A successful hiPSC-to-iMG differentiation means that the cells will exhibit changes in their morphology during differentiation (Fig. 1b). A typical iMG morphology is one with a small cell body and few branches and these can be observed after days 6. For the characterization and quality control of iMG cells, we usually analyze various microglial markers using either flow cytometry or immunocytochemistry.

3.4.1 Flow Cytometry

We characterize the iMG cell population at the end of differentiation using anti-CD11b and anti-TREM2 antibodies by flow cytometry. Typically, the iMG cells in one 35 mm or 60 mm dish are harvested on day 9 using TrypLE Express.

1. Remove the spent medium, and rinse the cells with 2 mL DPBS once.
2. Aspirate the DPBS. Add 1 mL TrypLE Express, ensuring the entire surface is covered by a thick layer of solution. Incubate the dish at 37 °C for 3 min.

3. Dissociate the cells using 1 mL basal medium (DMEM/F12 with N-2 and NEAA). Determine the cell number and adjust the density to 1×10^6 cells/mL.
4. Transfer 1 mL of the cell suspension into a new 1.5 mL microcentrifuge tube.
5. Centrifuge at $300 \times g$ for 5 min.
6. Remove the supernatant and resuspend the cells using 100 μL of blocking solution (*see* **Note 17**). Incubate the cells on ice for 20 min.
7. Add 1 mL of FACS buffer after blocking.
8. Distribute the cell suspension into three microcentrifuge tubes: one staining group and two control groups.
9. Centrifuge the cells at $300 \times g$ for 5 min. Discard the supernatant.
10. Resuspend the cell pellet of staining group in 100 μL of antibody solution, FITC-labeled anti-CD11b (1:100), and PE-labeled anti-TREM2 (1:50) antibodies in FACS buffer. Resuspend the cell pellet from two control groups in 100 μL of control solutions, with one group having an isotype control that contains a non-specific antibody with a fluorophore, and the other group being a negative control that doesn't have any antibodies or fluorophores.
11. Incubate for 20 min on ice. Protect from light.
12. Wash the cells by adding 1 mL FACS buffer. Centrifuge at $300 \times g$ for 5 min. Discard the supernatant.
13. Repeat **step 12** twice.
14. Resuspend the cell pellet using 500 μL FACS buffer, and then pass the cells through a 42 μm nylon mesh.
15. Analyze samples by flow cytometry analyzer (Fig. 2a).

3.4.2 Immunocytochemistry

For best quality imaging, the iMG cells should be seeded on a coverslip from the day of replating (generally on day 2–4 during differentiation).

1. Remove the spent medium. Fix the cells with 4% PFA in PBS at room temperature for 10–15 min.
2. Aspirate the fixative. Permeabilize the cells with 0.1% Triton X-100 in PBS at room temperature for 5 min.
3. Remove the penetration solution. Incubate the samples in the blocking solution (3% BSA and 3% NGS in PBS with 0.1% NaN_3) at room temperature for 1–2 h.
4. Aspirate the blocking solution and then incubate the samples with primary antibody solution at 4 °C overnight (*see* **Note 18**).

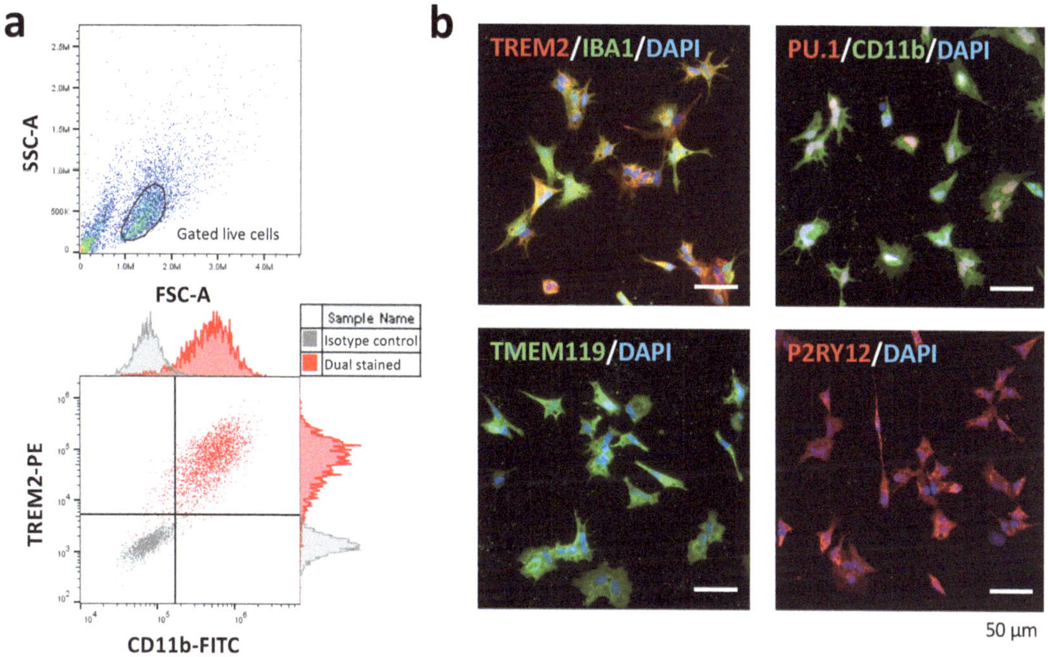

Fig. 2 Characterization of the iMG cells. (**a**) Flow cytometry analyses showing the expression of CD11b and TREM2 in the iMG cells on day 9 of differentiation. Red, CD11b and TREM2 antibody conjugated to FITC and PE, respectively; gray, isotype control. (**b**) Representative images of iMG cells immunostained for various microglia markers, CD11b, IBA1, P2RY12, PU.1, TMEM119, TREM2, and DAPI (scale bar 50 µm)

5. Wash the samples with PBS at room temperature for 5 min each. This should be done three times in total.
6. Incubate the samples with fluorescent dye conjugated secondary antibody and DAPI solution at room temperature for 1–2 h (*see* **Note 18**).
7. Wash each sample with PBS for 5 min. This should be done a total of three times (*see* **Note 19**).
8. Mount the cells by adding a drop of mounting medium onto the slide, carefully avoiding any air bubbles.
9. The slides can then be observed and images acquired using a fluorescence microscope (Fig. 2b).

4 Notes

1. All recombinant proteins are dissolved and aliquoted according to the manufacturer's instructions. Keep aliquots at −20 °C for further use for up to 6 months.
2. It should be possible to use any other delivery approach to express CEBPA and SPI1 from the perspective of

reprogramming. Here, we use the approach of lentiviral delivery to obtain hiPSC lines that can be induced to express CEBPA and SPI1.

3. Upon successful transfection, the HEK293FT cells show a more rounded morphology, have a higher contrast under microscope, and are easier to detach from the dish when changing medium. Therefore, refresh medium gently to avoid detachment.
4. This method has been validated using at least six different hiPSC lines regularly cultured in E8 or StemFlex medium on feeder-free VTN-coated dishes.
5. To avoid breaking the colonies too much when using a 1 mL pipette tip, we cut the head of the tip to make a larger hole at the tip. This reduces the mechanical forces present during pipetting.
6. To deliver two different lentiviral particles carrying rtTA and the CEBPA-2A-SPI1, we recommend infecting hiPSCs with rtTA lentivirus first and then infecting with CEBPA-2A-SPI1 lentivirus on the next passage.
7. It is recommended to use at least three different concentrations to seed the hiPSCs. This ensures that on the day of infection, there are cells at the optimal concentration for lentiviral delivery. It is best that these cells have about 60–70% confluency at the time of infection.
8. In order to minimize the effect of the lentiviral medium, which consists of DMEM and FBS, on the hiPSC lines, we used concentrated medium mixed with E8 (concentrated lentiviral medium: E8 = 1:4). Polybrene can increase the efficiency of infection.
9. At this point, these cells have a somewhat flat, fibroblast-like, morphology. Nevertheless, the infected cells still maintain their pluripotency if the cells almost reach full confluency. If the cell density is low (<50%), these cells will become differentiated more easily.
10. For optimal differentiation, the hiPSCs should be distributed as smaller, but not single, colonies at high confluency on the day that differentiation starts (Day 0).
11. We recommend redistributing the cells after puromycin selection, usually on day 2 or day 3. This is done by breaking the clones into single cells and this avoids a crowded culture. Aggregated clones during differentiation will result in poor results.
12. These cells will no longer increase their population significantly, thus, these cells are recommended to be seeded onto the new dishes or into multiple well plates for assays at this point. It should also be noted that if the infection efficiency is not high enough, which will mean that fewer cells survive, eventually this will lead to failure and no differentiated cells.

If this is the case, then increasing the starting number of the cells should help. For example, using 100 mm dishes to start the culture without lowering the confluency.

13. It is important to dissociate these cells into single cells and evenly disperse these cells onto the target plate. Leaving the cells in colonies will lead to undesired cell differentiation with a flat cell morphology.
14. To avoid too crowded or too isolated cell distributions, 6×10^4 cells/cm^2 is an optimal concentration, but this also depends on the survival rate after replating.
15. We stop the puromycin selection after 3 days of treatment. However, if undesired colonies appear after stopping the puromycin selection, it is possible to get rid of these cells by continuing puromycin selection.
16. These cells start to show microglial morphology after day 6 of induction (Fig. 1b). By convention, we assay the cells between day 9 and day 12 in most experiments.
17. Microglia cells are known to react with the Fc domain of antibodies. Thus blocking is necessary to avoid a false positive signal. Note that 100 uL of blocking buffer containing 2.5 μg Fc block in home-made FACS buffer is able to block at most 10^6 cells.
18. Prepare the primary and secondary antibody solution in 3% NGS/0.1% NaN_3/PBS buffer, and the recommended antibody dilution is described in Subheading 2.4.2.
19. After the second wash, it is recommended to observe the fluorescence in order to ensure immunostaining quality before mounting the samples.

Acknowledgments

This work was supported by the Brain Research Center of National Yang Ming Chiao Tung University from The Featured Areas Research Center Program within the framework of the Higher Education Sprout Project by the Ministry of Education (MOE) in Taiwan and by Yen Tjing Ling Medical Foundation (CI-111-4) to Y.-H. W.

References

1. Ginhoux F, Greter M, Leboeuf M et al (2010) Fate mapping analysis reveals that adult microglia derive from primitive macrophages. Science 330:841–845
2. Abud EM, Ramirez RN, Martinez ES et al (2017) iPSC-derived human microglia-like cells to study neurological diseases. Neuron 94:278–293.e279

3. Douvaras P, Sun B, Wang M et al (2017) Directed differentiation of human pluripotent stem cells to microglia. Stem Cell Rep 8:1516–1524
4. Haenseler W, Sansom SN, Buchrieser J et al (2017) A highly efficient human pluripotent stem cell microglia model displays a neuronal-co-culture-specific expression profile and inflammatory response. Stem Cell Rep 8:1727–1742
5. Muffat J, Li Y, Yuan B et al (2016) Efficient derivation of microglia-like cells from human pluripotent stem cells. Nat Med 22:1358–1367
6. Caiazzo M, Giannelli S, Valente P et al (2015) Direct conversion of fibroblasts into functional astrocytes by defined transcription factors. Stem Cell Rep 4:25–36
7. Yang N, Zuchero JB, Ahlenius H et al (2013) Generation of oligodendroglial cells by direct lineage conversion. Nat Biotechnol 31:434–439
8. Zhang Y, Pak C, Han Y et al (2013) Rapid single-step induction of functional neurons from human pluripotent stem cells. Neuron 78:785–798
9. Chen S-W, Hung Y-S, Fuh J-L et al (2021) Efficient conversion of human induced pluripotent stem cells into microglia by defined transcription factors. Stem Cell Rep 16:1363–1380

Chapter 6

The Generation and Functional Characterization of Human Microglia-Like Cells Derived from iPS and Embryonic Stem Cells

Mikael Lehoux, Kevin Connolly, Benedetta Assetta, and Yu-Wen Alvin Huang

Abstract

The following protocol describes the generation of microglia cells from human-induced pluripotent stem cells (hiPSCs) using commercially available kits by StemCell Technologies. This protocol consists of three major steps: (1) Differentiation of hematopoietic precursor cells, (2) Microglia differentiation, and (3) Microglia maturation. Assays are described to characterize hematopoietic precursor cells and mature microglia.

Key words Induced pluripotent stem (iPS) cells, Hematopoietic precursor cells, Microglia (iMG), Chemically defined differentiation

1 Introduction

Microglia are key players in brain homeostasis and maintenance. As the resident macrophage cell in the brain, these cells are implicated in the neurological immune response. Unlike neurons and other glia that are derived from ectoderm, microglia originate from yolk sac myeloid progenitors, which immigrate into the embryonic brain parenchyma and further differentiate and mature via chemokines, cytokines, and growth factors [1]. The protocol described here relies on commercially available kits (StemCell Technologies; adapted from the Blurton-Jones lab [2]), for the stepwise differentiation of microglia in 38–44 days.

To simulate the embryonic process, we first differentiate hematopoietic precursor cells (HPCs) that are comparable to yolk sac progenitors from human iPSCs using the STEMdiff™ Hematopoietic kit. First, STEMdiff™ Hematopoietic Medium A generates KDR+CD235A+ primitive hematopoietic hemangioblasts by activin-nodal signaling and inhibition of the Wnt–β-catenin

Yu-Wen Alvin Huang and ChangHui Pak (eds.), *Stem Cell-Based Neural Model Systems for Brain Disorders*, Methods in Molecular Biology, vol. 2683, https://doi.org/10.1007/978-1-0716-3287-1_6,

pathway [3]. Hematopoietic cytokines in STEMdiff™ Hematopoietic Medium B allow for the KDR+ CD235A+ primitive hematopoietic hemangioblast population to yield CD43+ CD235a+ CD41+ primitive HPCs. These cells are round and small and grow in semi-suspension. We assay these cells using anti-CD43 and anti-CD34 antibodies via flow cytometry.

HPCs are then cultured in STEMdiff™ Microglia Differentiation Media for 24 days. This media contains a cytokine cocktail of IL-34, Macrophage Colony Stimulating Factor (MCSF), and Transforming Growth Factor β (TGFβ), all required to induce expression of microglia markers and survival. After differentiation, microglia are allowed to mature for an additional 4–10 days in STEMdiff™ Microglia Maturation Media, composed of STEMdiff™ Microglia Differentiation Media with the addition of CD200 and CX3CL1 (STEMdiff™ Microglia Supplement 3), both neuronal ligands to microglia receptors. Mature microglia are assayed via immunohistochemistry for the validation of microglia markers and morphology.

2 Materials

2.1 Cultureware

1. 6-well plates.
2. 96-well plate.
3. 24-well plate (for immunohistochemistry).

2.2 Equipment

1. Lionheart FX Automated Microscope.
2. Hemocytometer or Bio-Rad TC20 Automated Cell Counter.
3. Rocker.

2.3 iPSC Lines

1. iPSC line (We used KOLF2.1 (JAX)).

2.4 Reagents

1. Gentle Dissociating Cell Reagent.
2. Corning® Matrigel® (hESC-qualified or Growth-Factor Reduced).
3. Poly-D-Lysine.
4. Trypan Blue.

2.5 Media

1. StemFlex™ Medium (for iPSC culture).
2. STEMdiff™ Hematopoietic Medium A:

   Mix 45 mL STEMdiff™ Hematopoietic Basal Medium with 225 uL STEMdiff™ Hematopoietic Supplement A.
3. STEMdiff™ Hematopoietic Medium B:

Mix 75 mL STEMdiff™ Hematopoietic Basal Medium with 375 uL STEMdiff™ Hematopoietic Supplement B.

4. STEMdiff™ Microglia Differentiation Medium:

   Mix 90 mL STEMdiff™ Microglia Basal Medium with 10 mL STEMdiff™ Microglia Supplement 1 and 400 uL STEMdiff™ Microglia Supplement 2.

5. STEMdiff™ Microglia Maturation Medium:

   Mix 90 mL STEMdiff™ Microglia Basal Medium with 10 mL STEMdiff™ Microglia Supplement 1, 400 uL STEMdiff™ Microglia Supplement 2, and 400 uL STEMdiff™ Microglia Supplement 3.

2.6 Flow Cytometry

2.6.1 Reagents

1. DPBS, no calcium, no magnesium.
2. FACS buffer.
3. Human Fc block (0.5 mg/mL).
4. FITC conjugated anti-CD34 antibody.
5. PE conjugated anti-CD43 antibody.

2.6.2 Consumables

1. 96-well V-bottom plate.
2. 5 mL round bottom polystyrene test tubes with cell strainer snap cap.

2.7 Immunohistochemistry

2.7.1 Reagents

1. Phosphate-buffered saline (PBS).
2. 4% paraformaldehyde (PFA).
3. 0.1% Triton X-100 in PBS.
4. Donkey serum.
5. Anti-IBA1 antibody.
6. Anti-TREM2 antibody.
7. Anti-CD11b antibody.
8. Anti-P2RY12 antibody.
9. Secondary antibody conjugated with fluorophore.

3 Methods

The following instructions are for a single well of a 6-well plate coated with Corning® Matrigel® according to the manufacturer's instructions.

3.1 Generation of Hematopoietic Precursor Cells (HPCs)

1. Passage human iPSCs using Gentle Dissociating Cell Reagent (*see* **Note 1**) and collect cell suspension on a 15 mL conical tube. Centrifuge at 300 × *g* for 5 min.

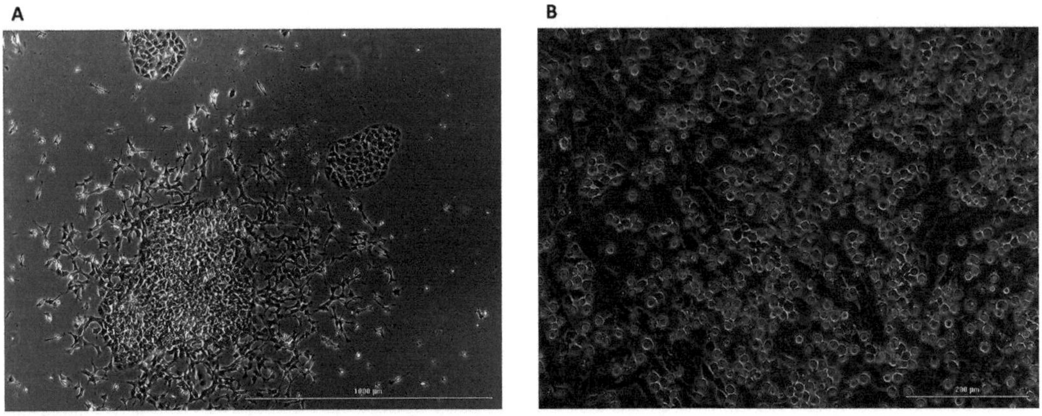

Fig. 1 Morphology of hematopoietic precursor cells. Bright field images of differentiating HPCs at day 2 and day 8. On day 2, you can observe colonies with dense centers from which HPCs will differentiate in semi-suspension. Cells in the perimeter will continue to expand and differentiate into other fates. On day 8, a clear population of small round cells is observed, these are HPCs

2. Gently resuspend the pellet in 5 mL of StemFlex™ Medium using a serological pipette.
3. Using a 96-well plate, count the number of aggregates in a 5 uL sample of the cell suspension (*see* **Note 2**).
4. Make the calculation to add 30–50 aggregates in 2 mL StemFlex™ Medium and incubate at 37 °C and 5% CO_2.
5. Confirm the appropriate number of colonies attached to the well. Proceed if there are 20–40 uniform colonies around 100 uM in diameter.
6. Aspirate StemFlex™ Medium and replace with 2 mL STEMdiff™ Hematopoietic Medium A (day 0).
7. Day 2: Add 1 mL of STEMdiff™ Hematopoietic Medium A (*see* **Note 3**).
8. Day 3: Aspirate STEMdiff™ Hematopoietic Medium A and replace with 2 mL STEMdiff™ Hematopoietic Medium B.
9. Days 5, 7, and 9: Add 1 mL of STEMdiff™ Hematopoietic Medium B (*see* **Note 4**).
10. HPCs in suspension can be harvested on day 10 (*see* **Note 5**) (Fig. 1).

3.2 Validation of Hematopoietic Precursor Cells (HPCs)

We characterize HPCs at days 10–12 using anti-CD43 and anti-CD34 antibodies by flow cytometry. All reagents are kept on ice, and cells are placed on ice after harvest.

1. Collect HPCs in suspension into a 15 mL conical tube, wash gently. Centrifuge at 300 × *g* for 5 min.
2. Without disturbing the pellet, and leaving 1 mL in the tube, transfer the conditioned media back into the well.

3. Count the cells, proceed if there are at least 10^5 cells per condition (5 total: unstained, isotype controls, anti-CD43, anti-CD34, anti-CD43 + anti-CD34).
4. Centrifuge at 300 × *g* for 5 min. Aspirate media and resuspend the pellet in 1 mL ice-cold FACS buffer.
5. Add 200 uL of the suspension to five separate wells of a v-bottom 96-well plate on ice. The following volumes correspond to individual conditions.
6. Centrifuge the plate at 300 × *g* for 5 min at 4 °C.
7. Aspirate, and resuspend in 100 uL Fc block. Incubate on ice for 15 min. Centrifuge the plate at 300 × *g* for 5 min at 4 °C.
8. Resuspend with the appropriate antibody solution and incubate on ice for 45 min.
9. Wash the cells twice more by centrifugation at 300 × *g* for 5 min at 4 °C.
10. Resuspend in 500 mL PBS, and place into FACS tubes. Keep samples in the dark and on ice until analysis (Fig. 2).

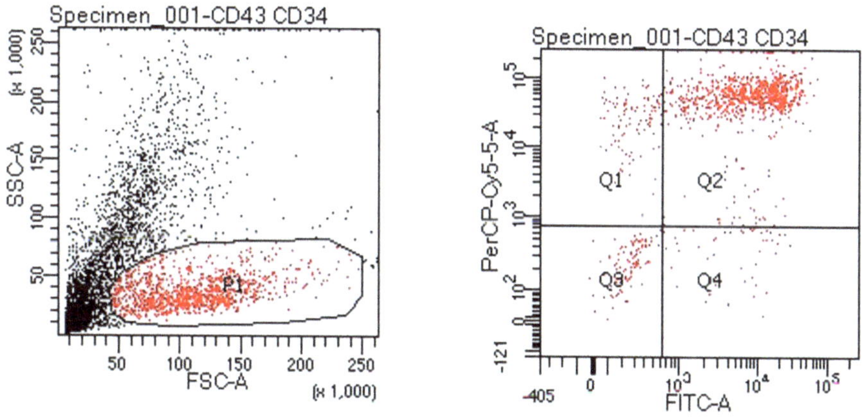

Tube: CD43 CD34

Population	#Events	%Parent	%Total
All Events	5,242	####	100.0
P1	997	19.0	19.0
Q1	104	10.4	2.0
Q2	744	74.6	14.2
Q3	100	10.0	1.9
Q4	49	4.9	0.9

Fig. 2 Flow cytometry data for D12 HPC. (Top Left) Dot Plot of forward scatter (FSC-A) against side scatter (SSC-A) allows for gating of the HPC population (red). (Top Right, Bottom) Double staining of the gated population for CD43 (PerCP-Cy5-5) and CD35 (FITC-A) with table

3.3 Microglia Differentiation

1. Centrifuge harvested HPCs at 300 × *g* for 5 min. Resuspend with STEMdiff™ Microglia Differentiation Medium and count cells using Trypan Blue and an automated cell-counter or hemocytometer.
2. Add 1×10^5–2×10^5 cells to one well of a Matrigel®-coated 6-well plate containing 2 mL STEMdiff™ Microglia Differentiation Medium (cell density 1.1×10^4–2.2×10^4 cells/cm^2). Incubate at 37 °C and 5% CO_2.
3. Add 1 mL STEMdiff™ Microglia Differentiation Medium to each well every other day (*see* **Note 6**).
4. On day 12, transfer the cell suspension to a 15 mL conical tube. Centrifuge at 300 × *g* for 5 min. Resuspend in 1 mL conditioned media and add to a new well containing 1 mL fresh STEMdiff™ Microglia Differentiation Medium (*see* **Note 7**).
5. Add 1 mL STEMdiff™ Microglia Differentiation Medium to each well every other day (Fig. 3).

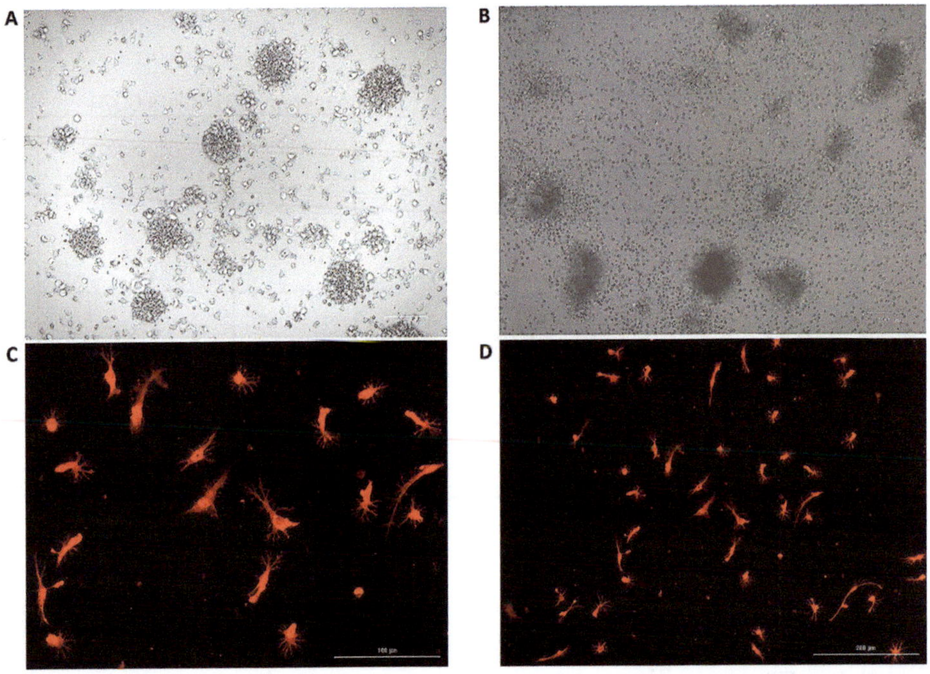

Fig. 3 Morphology of differentiating iMG. Bright field images of cells growing on Matrigel. (**a**) On day 3 of differentiation, round colonies are observed in suspension and attached, in addition to attached single cells with small projections. Significant changes to morphology in comparison to preceding HPCs. (**b**) On day 12 of differentiation, before replating, an expansion of yield can be observed. (**c**, **d**) Mature iMG plated on PDL at D32 were fixed on D33 and stained for Iba1. Clear projections characteristic of microglia morphology are observed (20× and 10×)

3.4 Microglia Maturation

1. On day 24, transfer the cell suspension to a 15 mL conical tube. Centrifuge at 300 × *g* for 5 min. Resuspend in 1 mL STEMdiff™ Microglia Maturation Medium, and count using Trypan Blue and an automated cell-counter or hemocytometer.
2. Add 1 × 10^6 cells to one well of a Matrigel®-coated 6-well plate to contain a total of 2 mL STEMdiff™ Microglia Maturation Medium. Incubate at 37 °C and 5% CO_2.
3. Add 1 mL of STEMdiff™ Microglia Maturation Medium every other day.
4. Microglia are mature starting at day 28 and can be harvested until day 34 (*see* **Note 8**).

3.5 Validation of Microglia-Like Cells

To assay mature microglia, we perform immunohistochemistry for microglia-specific markers. As a functional assay, we stimulate with LPS and IL-4 and analyze the expression of inflammatory genes via qPCR.

1. Coat 24-well plate with 250 uL of 10 mg/mL poly-D-lysine (PDL) for 3 h at room temperature or overnight at 4 °C.
2. Aspirate PDL and wash with water three times.
3. To harvest mature microglia, tilt the plate and collect cell suspension into a 15 mL conical tube. Wash with DPBS to ensure maximum yield.
4. Centrifuge at 300 × *g* for 5 min. Resuspend in 1 mL STEMdiff™ Microglia Maturation Medium, and count using Trypan Blue and an automated cell-counter or hemocytometer.
5. Add 6 × 10^4–9 × 10^4 cells per well of the PDL- coated 24-well plate to contain a total of 500 uL STEMdiff™ Microglia Maturation Medium per well. Incubate overnight at 37 °C and 5% CO_2.
6. The next day, ensure cells are attached and display microglia-like morphology (small cell body with branches) prior to fixing.
7. Aspirate media, and wash with DPBS.
8. Add 500 uL 4% PFA and incubate for 15–20 min at room temperature.
9. Wash 3 times with DPBS and either parafilm the plate and store at 4 °C for future staining or proceed.
10. Permeabilize cells with 0.1% Triton X-100 in PBS for 10 min at room temperature.
11. Block cells with 10% donkey serum in PBS for 30 min at room temperature in a rocker.
12. Remove blocking buffer, add primary antibody dilution, parafilm the plate, and incubate overnight on the rocker at 4 °C.
13. Wash three times with DPBS.

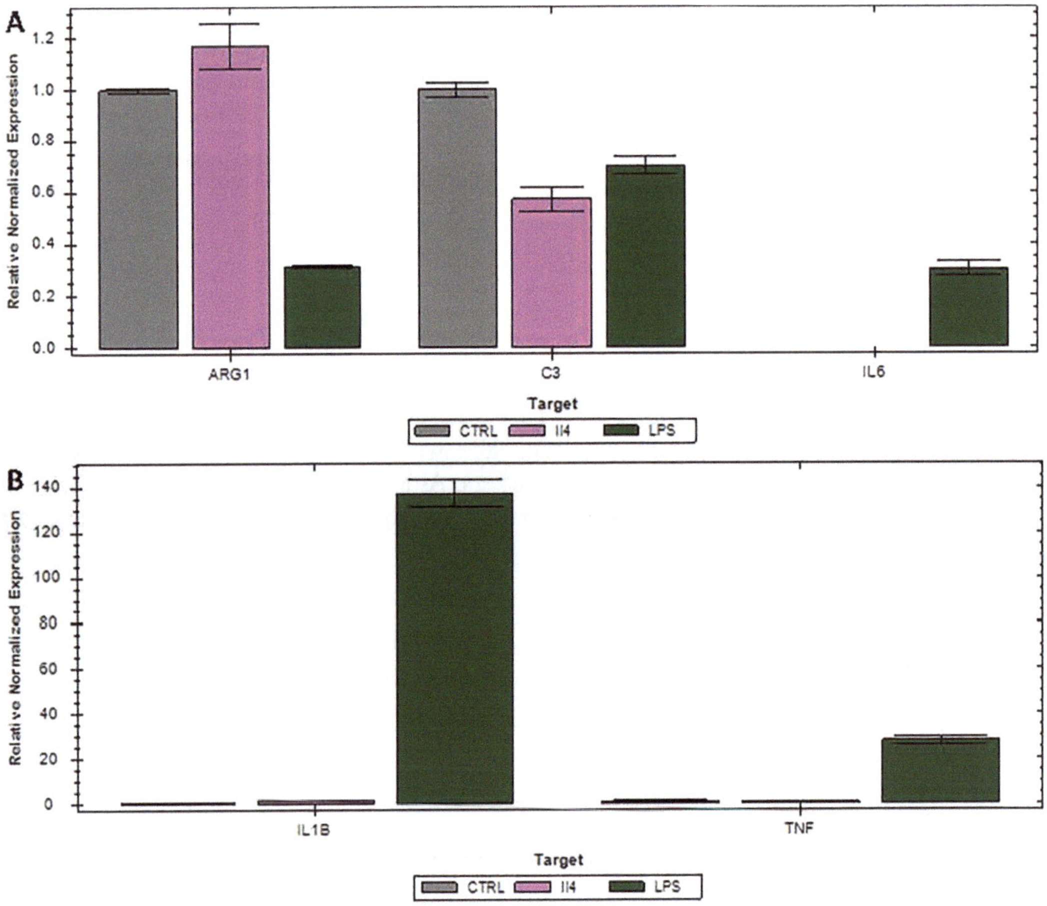

Fig. 4 Expression of inflammatory genes upon LPS and IL-4 stimulation of iMG. Mature iMG were treated with pro-inflammatory LPS for 4 h and anti-inflammatory IL-4 for 24 h. Relative normalized expression of (**a**) Arg-1, C3, and IL-6 and (**b**) IL-1β and TNFα is displayed

14. Prepare secondary antibody (1:500) dilution in DPBS and incubate for 2 h, covered from light.
15. Wash three times with DPBS and add DAPI (1:1000) in DPBS, incubate for 10 min, covered from light.
16. Wash three times with DPBS and proceed with imaging (Fig. 4).

4 Notes

1. Aspirate the spent media and add 1 mL of room temperature Gentle Dissociating Cell Reagent (GDCR) into one well of a six-well plate for 8 min. Aspirate off the GDCR and add 1 mL of StemFlex™ Medium and use a cell scraper to lift cells. Collect suspension in a 15 mL conical tube.

2. Draw a cross on the bottom of three separate wells of a 96-well plate with a fine-tip marker. Add 40 uL of StemFlex™ Medium to each well. Add 5 uL of cell suspension to each well. Count the number of aggregates ≥50 uM in size in each quadrant and calculate the average. Multiply this number by four to estimate viable aggregates in a 5 uL sample. Repeat for the remaining two wells, and find the average number of aggregates among the three wells. Divide by 5 uL to get the concentration of viable clumps in the cell suspension. During the time spent counting, the cell suspension may collect at the bottom of the tube. To redistribute, gently invert the tube.
3. At this stage (day 2), you should observe cell colonies that are dense in the middle with endothelial-like single cells surrounding at the edges.
4. On day 7, you should start to observe floating HPCs expanding over colonies. The HPC population will continue to expand for the remainder of this step.
5. To harvest HPCs, simply tilt the plate, collect the cell suspension, and wash gently to maximize yield (washing too vigorously might cause otherly differentiated cells from lifting and interfering with the subsequent microglia differentiation). For a greater yield, add half-medium of STEMdiff™ Hematopoietic Medium B and harvest on day 12.
6. During the first few days in STEMdiff™ Microglia Differentiation Medium, cells will recluster and display a larger cell-body, with small protrusions. Cell morphology does not vary significantly over the differentiation period.
7. After 12 days, the well should contain 7 mL of STEMdiff™ Microglia Differentiation Medium. This replating step eliminates conditioned/spent media. The new Matrigel-coated well should contain 1 mL of conditioned media and 1 mL new media per well of a 6-well plate.
8. In monoculture, microglia are viable for 10 days in STEMdiff™ Microglia Maturation Medium after day 24.

Acknowledgements

This work was supported by grants from National Institutes of Health (R21AG077697 and P20GM119943 [Subaward 7137296] to Y.A.H), Alzheimer's Association Strategic Fund (ABA-22-965518 to Y.A.H.), the Graduate Award in Brain Science from Brown's Carney Institute (to K.C), and the Postdoctoral Fellowship Program in Alzheimer's Disease Research from the BrightFocus Foundation (A2021002F to B.A.). Y.A.H. is a GFL Translational Professor from Center for Translational Neuroscience at the Brown Institute for Translational Sciences.

References

1. Wurm J, Konttinen H, Andressen C et al (2021) Microglia development and maturation and its implications for induction of microglia-like cells from human iPSCs. Int J Mol Sci 22:3088
2. Abud E, Ramirez R, Martinez E et al (2017) iPSC-derived human microglia-like cells to study neurological diseases. Neuron 94(2):278–293
3. Guttikonda S, Sikkema L, Tchieu J et al (2021) Fully defined human pluripotent stem cell-derived microglia and tri-culture system model C3 production in Alzheimer's disease. Nat Neurosci 24:343–354

Chapter 7

Modeling Cellular Crosstalk of Neuroinflammation Axis by Tri-cultures of iPSC-Derived Human Microglia, Astrocytes, and Neurons

Kevin Connolly, Mikael Lehoux, Benedetta Assetta, and Yu-Wen Alvin Huang

Abstract

Neuroinflammation is a common early pathological feature in many neurodegenerative disorders, including Alzheimer's disease (AD), which has been heavily implicated as a causative factor in disease pathology. However, the role neuroinflammation and inflammatory cells, including microglia and astrocytes, play in AD development and progression has not been fully defined. To try to better understand and study this neuroinflammatory role in AD pathogenesis, researchers use a variety of model systems, particularly in vivo animal models. Despite their usefulness, these models do come with a variety of limitations due to the inherent complexity of the brain and the human-specific nature of AD. Here, we describe a reductionist approach at modeling neuroinflammation by utilizing an in vitro tri-culture system of neurons, astrocytes, and microglia induced from human pluripotent stem cells. This tri-culture model is a powerful tool to dissect intercellular interactions that can facilitate future studies on neuroinflammation, particularly in the context of neurodegeneration and AD.

Key words Tri-culture, Induced pluripotent stem cells (iPSCs), Neurons, Astrocytes, Microglia, Neuroinflammation

1 Introduction

Microglia and astrocytes, the resident immune cells of the brain, are responsible for coordinating the inflammatory process in the central nervous system (CNS) known as neuroinflammation. In addition to neuroinflammation, these glial cells also play a role in neural development, synaptic pruning, plasticity, phagocytic activity, and trophic and neuroprotective support, among other functions [1–5]. Together, these processes, including neuroinflammation, are important for maintaining healthy brain function. However, the neuroinflammatory process can become imbalanced and go awry, resulting in continuous states of reactivity and toxicity in the glial

Yu-Wen Alvin Huang and ChangHui Pak (eds.), *Stem Cell-Based Neural Model Systems for Brain Disorders*, Methods in Molecular Biology, vol. 2683, https://doi.org/10.1007/978-1-0716-3287-1_7,

cells that have detrimental effects on neuroprotection and can damage neurons [4, 6–9]. This glial toxicity and decreased neuronal support, coupled with the inherent function of glial cells as well as the expression of risk factors, have heavily implicated neuroinflammation as a causative factor in neurodegenerative disorders, including Alzheimer's disease (AD) [1, 2]. Despite its implication in neurodegeneration, the mechanisms, signaling pathways, or the totality of factors triggered by neuroinflammation and how they may contribute to AD pathogenesis are still not well understood.

Animal models, namely transgenic mouse models, have been used in the field to study specific CNS cell types and potential mechanisms underlying neuroinflammation, particularly in the context of AD. While these models have been useful in disentangling potential factors influencing neuroinflammation and AD pathology, they are not without limitations. Limitations in these animal models extend to both in vivo work as well as in vitro work with primary cells isolated from the animals. Regarding in vitro primary cells, cells are commonly obtained from rodent brains via digestions and dissociations [3]. This dissection and processing, particularly in trying to isolate one specific cell type, can be quite difficult and meticulous, can be traumatic for the cells, and may not result in a pure culture as there may be lingering contamination by other cell types that were originally isolated in a mixed cell suspension [3, 10]. The inherent complexity of the brain, dynamic crosstalk between differing cell types, and their effect on each other make determining cell-specific effects not only difficult in primary cells in vitro, but especially in whole brains in vivo. In addition to difficulty in parsing out cell-specific effects in rodent models, studying neuroinflammation, especially in the context of AD, is limited in animal models due to the human-specific nature of the disease [2]. Evident species-specific differences between animal models and humans that contribute to limitations in using these animal models include: a lack of massive neuronal loss in some rodent models compared to what is observed in humans, structural differences in heavily implicated molecules between rodents and humans (including amyloid-β among others), differentially expressed receptors and downstream signaling and adaptor proteins in human cells versus animal models, and slightly different cell structure and function between certain human cell types and animal models [2–5, 10]. In order to circumvent the differences between human and rodent genetic backgrounds, some researchers utilize primary human cultures. While these cultures may address the species-specific differences, they are not without their own limitations. While human primary cultures allow for the study of human-specific factors that cannot be modeled in rodents, there is often limited access to human samples due to challenges in dissecting out and isolating specific cell types, the limited number of certain cell types in either adult or fetal samples, and the limited self-renewal or proliferative abilities of isolated cell types [5, 10–12].

To bypass the limitations of primary human cultures while still studying in a human context, researchers have turned to stem cell technology in order to induce specific human CNS cells from human stem cells. The use of either embryonic stem cells (ESCs) or induced pluripotent stem cells (iPSCs) are able to recapitulate the intrinsic human nature of AD and its associated genetics while still being proliferative [2]. This allows for a large, constant pool of cells that are able to be induced to the respective individual CNS cells. Not only is the induction of stem cells to CNS cells reproducible and scalable, but these induction protocols are also quite efficient as they can produce the desired cell type quickly with high purity [2, 13–15]. Additionally, these induced cells, whether it be neurons, astrocytes, microglia, or others, have been shown to express cell-specific markers, exhibit appropriate phenotypes, and display fundamental physiological functions that are able to recapitulate that which are observed in their in vivo counterparts [2, 5, 12, 14]. The utility of ES/iPSC derived CNS cells has been shown to be effective for drug screens of pharmaceutical compound libraries, in discovering biomarkers of disease (such as AD), and particularly in determining cell-specific contributions to healthy, inflamed, or diseased states [2, 12, 14]. Overall, the use of pure cultures of individual CNS cells induced from stem cell allows for a reductionist approach that enables the dissection of cell type-specific effects or events that can be difficult to study in the context of the whole brain due to its inherent complexity.

In addition to determining cell-specific effects by using pure cultures of induced cells, the differing induced cell types can be combined and cultured together as co- or tri-cultures in order to determine cellular cross-talk and interactions. While pure cultures are useful tools for studying individual cell types, these models alone do not address the important, intricate hetero-cellular interactions or the cross-talk interactions that may be implicated in health, neuroinflammation, or diseased states. Thus, a tri-culture approach allows for further probing into multiple cell types while still maintaining a reductionist approach that is again able to avoid issues posed by the inherent complexity inside the whole brain [14–16]. By combining induced neurons (iNs), induced astrocytes (iAs), and induced microglia (iMGs) together, one can study intercellular communications, interactions, underlying mechanisms, and their implication in the neuroinflammatory process in the context of AD. Identification of the function/dysfunction of these cell types together and the potential underlying mechanisms, whether it be in homeostatic or pathogenic conditions, in this tri-culture system can also enable development of targeted therapies that may help to combat neuroinflammation, AD, or other neurodegenerative disorders [15]. Here we detail a method to generate a tri-

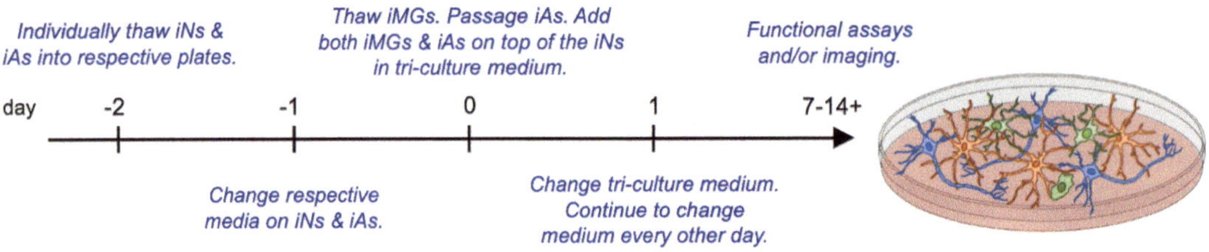

Fig. 1 Schematic overview of tri-culture preparation and maintenance. In brief, on day −2, iNs and iAs are thawed and plated individually. The respective media are changed the following day. On day 0, iMGs are thawed, iAs are passaged, and the cells are plated together with the iNs. The media is changed the next day and subsequently every other day

culture model of neurons, astrocytes, and microglia that will enable the study of these aforementioned intercellular interactions, particularly in the context of neuroinflammation and AD (Fig. 1).

2 Materials

1. 24-well tissue culture-treated cell culture plates.
2. 6-well tissue culture-treated cell culture plates.
3. Round cover slips (15 mm diameter, 0.19–0.22 mm thickness) (*see* **Note 1**).
4. Geltrex (*see* **Note 2**).
5. DMEM/F12 medium.
6. Induced Neurons [14, 17] (*see* **Note 3**).
7. Neuron medium: Neurobasal A medium, B27 supplement (50×).
8. Induced Astrocytes [5] (*see* **Note 3**).
9. Complete Astrocyte medium: Basal medium (500 mL), Fetal Bovine Serum (FBS; 10 mL), Astrocyte Growth Supplement (AGS; 5 mL), penicillin/streptomycin solution (P/S; 5 mL) (ScienCell; cat# 1801).
10. ROCK inhibitor Y-27632.
11. Accutase.
12. Induced Microglia [12] (*see* **Notes 3** and **4**).
13. Microglia Medium: DMEM/F12 medium, Non-Essential Amino Acids solution (NEAA; 100×), N2 supplement (100×), doxycycline (1 μg/mL), IL34 (100 ng/mL), M-CSF (20 ng/mL), TGFβ1 (20 ng/mL), GM-CSF (20 ng/mL).
14. Cell Counting Slides.
15. Trypan Blue Dye.
16. Ara-C.

3 Methods

3.1 Tri-culture Preparation and Maintenance

1. Thaw Geltrex on ice. Add 250 μL Geltrex to pre-chilled DMEM/F12 medium (25 mL). Add the Geltrex solution to the individual wells of a multi-well plate to coat (250 μL for a 24-well plate or 1 mL for a 6-well plate) (*see* **Notes 1** and **5**). Incubate plates at 37 °C for 1 h (*see* **Note 6**).
2. On day −2, thaw vials of iNs and iAs in a 37 °C water or bead bath until a small icicle is still remaining. Add 5 mL of neuron medium and astrocyte medium to individual 15 mL conical tubes. Transfer the iNs and iAs from the cryovials to the respective 15 mL conical. Centrifuge the tubes at 300 × *g* for 5 min. Aspirate off media leaving each pellet intact.
3. For iAs: Resuspend iAs in 2 mL of astrocyte medium with 2 μL of ROCK inhibitor. Aspirate the Geltrex solution off one well of a 6-well plate and transfer the iAs (*see* **Note** 7). Incubate the cells at 37 °C with 5% CO_2.
4. For iNs: Resuspend the iNs in 1 mL of neuron medium. Mix 10 μL of the neuron suspension with 10 μL of trypan blue dye. Add 10 μL of the mixture to a cell counting slide and count the number of neurons using an automated cell counter.
5. The iNs should be plated at 3×10^5 cells/well of a 24-well plate (*see* **Note 8**). Add enough neuron medium (and respective amount of ROCK inhibitor) to the iN suspension to fill the desired number of wells (500 μL medium/well of a 24-well plate). Aspirate the Geltrex solution off the necessary wells of the 24-well plate and transfer the iNs (*see* **Notes** 7 and **9**). Incubate the cells at 37 °C with 5% CO_2.
6. On day −1, aspirate the media off the iNs and iAs and replace with fresh respective media (500 μL/well of neuron medium; 2 mL/well astrocyte medium).
7. On day 0, aspirate the media off the iAs and add 1 mL of accutase to the well. Incubate at room temperature for 10–15 min.
8. While the iAs are incubating with the accutase, thaw vials of iMGs in a 37 °C water or bead bath until a small icicle is still remaining. Add 5 mL of microglia medium to a 15 mL conical tube. Transfer the iMGs from the cryovials to the 15 mL conical.
9. Transfer the iAs to a 15 mL conical tube containing 3 mL of astrocyte medium (*see* **Note 10**).
10. Centrifuge the iMG and iA tubes at 300 × *g* for 5 min. Aspirate off media leaving each pellet intact.
11. Resuspend the iMGs and iAs in 1 mL of respective medium. Individually mix 10 μL of each of the cell suspensions with 10 μL of trypan blue dye. Add 10 μL of the mixture to a cell

counting slide and count the number of iMGs and iAs using an automated cell counter.

12. While the cells are being counted, prepare the tri-culture medium: Mix the neuron medium, astrocyte medium, and microglia medium together in a 3:1:1 ratio.
13. The iMGs and iAs should each be plated at 5×10^4 cells/well (*see* **Note 8**). After counting, in a new conical tube, mix the appropriate amount of iMG and iA suspension together. Add enough tri-culture medium (and respective amount of ROCK inhibitor) to the iMG/iA suspension to fill the number of wells currently containing the plated iNs (500 μL medium/well of a 24-well plate).
14. Gently aspirate the media off of the iNs and add 500 μL of the iMG/iA suspension to each well. Incubate the cells at 37 °C with 5% CO_2 (*see* **Note 11**).
15. On day 1, gently aspirate the media off each well of the tri-cultures and replace with fresh tri-culture medium (500 μL/well). Continue to subsequently change the media every other day until needed for imaging or functional assay (Fig. 2) (*see* **Notes 11** and **12**).

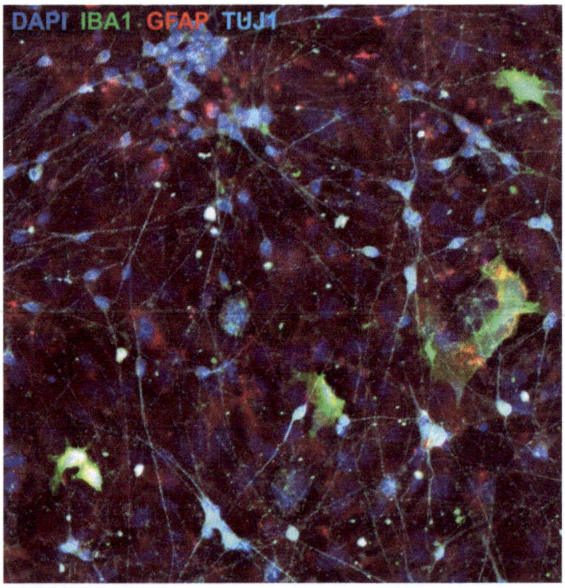

Fig. 2 Immunofluorescent staining of the human iPSC-derived tri-culture. Fixed tri-cultures were stained for Iba-1 (microglial marker; green), GFAP (astrocytic marker; red), and Tuj1 (neuronal marker; cyan). Donkey anti-rabbit 488, donkey anti-mouse 568, and goat anti-chicken 647 fluorophore-conjugated secondary antibodies were used, respectively. Tri-cultures on coverslips were mounted on a slide using fluoromount-G with DAPI (nuclei; blue). Tri-cultures were visualized and images were taken using a Zeiss LSM 880 Confocal Laser Scanning Microscope at 20× magnification

4 Notes

1. Cover slips are only needed for imaging purposes. If planning to do imaging on the tri-cultures, place cover slips at the bottom of the wells of the 24-well plate before adding the Geltrex solution.
2. Other extracellular matrix coating solutions, such as Matrigel, may be used instead of Geltrex. The protocol detailed above has been tested on Geltrex and Matrigel, both of which were able to maintain the three cell types.
3. The differing cell types can be induced from embryonic stem cells, such as H1 cells, or induced pluripotent stem cells. Cells induced from iPSCs with specific disease-relevant mutations may allow for better modeling in the context of AD. For example, here, iNs were derived from KOLF2.1 (control) iPSCs and APPV717I (mutant) iPSCs [18–21].
4. Microglia can be induced from ESCs or iPSCs via a variety of methods, including chemically defined methods or transcription factor-based methods [11, 12, 15]. Here, iMGs were derived using a transcription factor-based method [12].
5. Gently shake the plates to ensure the Geltrex solution is evenly distributed/coated.
6. Plates can be coated beforehand and stored at 4 °C after the 37 °C incubation until needed for use. Allow the plates to reach room temperature before use.
7. Swirl the plate around once cells are added to ensure even distribution of cells throughout the well.
8. The tri-culture will be plated roughly at a 6:1:1 ratio of iNs:iAs: iMGs.
9. The neurons will be thawed directly into the plate that will ultimately be used for the tri-culture.
10. After collecting the iAs in accutase, the well can be washed with astrocyte medium and added to the tube to collect any remaining iAs.
11. Be sure to be very gentle with taking off and adding media to the wells as being too harsh may cause the cells, particularly the iNs, to lift off the plate.
12. The iAs are still quite proliferative while the iMGs and iNs are not. If iAs proliferate too fast, reach confluency, or begin to overwhelm the tri-culture in less than 3 days, add Ara-C at a concentration of 2–5 μM.

Acknowledgments

This work was supported by grants from National Institutes of Health (R21AG077697 and P20GM119943 [Subaward 7137296] to Y.A.H), Alzheimer's Association Strategic Fund (ABA-22-965518 to Y.A.H.), the Graduate Award in Brain Science from Brown's Carney Institute (to K.C), and the Postdoctoral Fellowship Program in Alzheimer's Disease Research from the BrightFocus Foundation (A2021002F to B.A.). Y.A.H. is a GFL Translational Professor from Center for Translational Neuroscience at the Brown Institute for Translational Sciences.

References

1. Connolly K, Lehoux M, O'rourke R et al (2023) Potential role of chitinase-3-like protein 1 (CHI3L1/YKL-40) in neurodegeneration and Alzheimer's disease. Alzheimers Dement 19(1):9–24
2. Essayan-Perez S, Zhou B, Nabet AM et al (2019) Modeling Alzheimer's disease with human iPS cells: advancements, lessons, and applications. Neurobiol Dis 130:104503
3. Guttenplan KA, Liddelow SA (2019) Astrocytes and microglia: models and tools. J Exp Med 216:71–83
4. Paolicelli RC, Sierra A, Stevens B et al (2022) Microglia states and nomenclature: a field at its crossroads. Neuron 110:3458–3483
5. Tcw J, Wang M, Pimenova AA et al (2017) An efficient platform for astrocyte differentiation from human induced pluripotent stem cells. Stem Cell Rep 9:600–614
6. Clarke LE, Liddelow SA, Chakraborty C et al (2018) Normal aging induces A1-like astrocyte reactivity. Proc Natl Acad Sci U S A 115: E1896–E1905
7. Han X, Zhang T, Liu H et al (2020) Astrocyte senescence and Alzheimer's disease: a review. Front Aging Neurosci 12:148
8. Liddelow SA, Guttenplan KA, Clarke LE et al (2017) Neurotoxic reactive astrocytes are induced by activated microglia. Nature 541: 481–487
9. Niraula A, Sheridan JF, Godbout JP (2017) Microglia priming with aging and stress. Neuropsychopharmacology 42:318–333
10. Ray B, Chopra N, Long JM et al (2014) Human primary mixed brain cultures: preparation, differentiation, characterization and application to neuroscience research. Mol Brain 7:63
11. Abud EM, Ramirez RN, Martinez ES et al (2017) iPSC-derived human microglia-like cells to study neurological diseases. Neuron 94:278–293 e279
12. Chen SW, Hung YS, Fuh JL et al (2021) Efficient conversion of human induced pluripotent stem cells into microglia by defined transcription factors. Stem Cell Rep 16:1363–1380
13. Abu-Rumeileh S, Steinacker P, Polischi B et al (2019) CSF biomarkers of neuroinflammation in distinct forms and subtypes of neurodegenerative dementia. Alzheimers Res Ther 12:2
14. Assetta B, Tang C, Bian J et al (2020) Generation of human neurons and oligodendrocytes from pluripotent stem cells for modeling neuron-oligodendrocyte interactions. J Vis Exp
15. Guttikonda SR, Sikkema L, Tchieu J et al (2021) Fully defined human pluripotent stem cell-derived microglia and tri-culture system model C3 production in Alzheimer's disease. Nat Neurosci 24:343–354
16. Goshi N, Morgan RK, Lein PJ et al (2020) A primary neural cell culture model to study neuron, astrocyte, and microglia interactions in neuroinflammation. J Neuroinflammation 17: 155
17. Huang YA, Zhou B, Wernig M et al (2017) ApoE2, ApoE3, and ApoE4 differentially stimulate APP transcription and Abeta secretion. Cell 168:427–441 e421
18. Arber C, Toombs J, Lovejoy C et al (2020) Familial Alzheimer's disease patient-derived neurons reveal distinct mutation-specific effects on amyloid beta. Mol Psychiatry 25: 2919–2931
19. Frederiksen HR, Holst B, Ramakrishna S et al (2019) Generation of two iPSC lines with

either a heterozygous V717I or a heterozygous KM670/671NL mutation in the APP gene. Stem Cell Res 34:101368

20. Lloyd GM, Trejo-Lopez JA, Xia Y et al (2020) Prominent amyloid plaque pathology and cerebral amyloid angiopathy in APP V717I (London) carrier - phenotypic variability in autosomal dominant Alzheimer's disease. Acta Neuropathol Commun 8:31

21. Muratore CR, Rice HC, Srikanth P et al (2014) The familial Alzheimer's disease APPV717I mutation alters APP processing and tau expression in iPSC-derived neurons. Hum Mol Genet 23:3523–3536

Chapter 8

Generation of Oligodendrocytes from Human Pluripotent and Embryonic Stem Cells

Zoe Mattingly and Sundari Chetty

Abstract

Oligodendrocyte progenitor cells (OPCs) and mature oligodendrocytes (OLs) can be generated using human induced pluripotent stem cells (hiPSCs) and human embryonic stem cells (hESCs). By manipulating culture conditions, pluripotent cell types are serially guided through intermediary cell types, developing first into neural progenitor cells (NPCs) then OPCs before maturing as CNS-specific OLs. This procedure is conducted under adherent, feeder-free conditions to derive mature OLs in as few as 28 days.

Key words Induced Pluripotent Stem Cells (iPSCs), Human Embryonic Stem Cells (ESCs), Neural Progenitor Cells (NPCs), Neurons, Oligodendrocyte Progenitor Cells (OPCs), Oligodendrocytes (OLs), Differentiation

1 Introduction

Oligodendrocytes are the principal myelin-producing cells of the central nervous system (CNS), crucial for the maintenance of neuronal health and signal transduction. Axons are typically myelinated in a neuronal activity-dependent manner. Oligodendrocyte dysfunction, often resulting in demyelination, has been broadly implicated in neurodevelopmental and neurodegenerative diseases. Therefore, the generation of oligodendrocyte progenitor cells (OPCs) and mature oligodendrocytes (OLs) has utility in co-culture systems, therapeutic development, and regenerative medicine.

Here we outline a procedure to derive OPCs and mature OLs from human induced pluripotent stem cells (hiPSCs) and human embryonic stem cells (hESCs). For the original publication of this procedure, *see* Ref. [1]. Methods for deriving oligodendrocytes vary in culture style, choice of growth factors, procedure length, and purity of resulting cell population.

This procedure operates under adherent, monolayer culture conditions to reduce the potential for spontaneous differentiation

Yu-Wen Alvin Huang and ChangHui Pak (eds.), *Stem Cell-Based Neural Model Systems for Brain Disorders*, Methods in Molecular Biology, vol. 2683, https://doi.org/10.1007/978-1-0716-3287-1_8,

and avoid common limitations of three-dimensional culture. In concordance with normal oligodendrocyte ontogeny, hiPSCs, and hESCs serially transition first to neural progenitor cells (NPC) and then OPCs before maturing as OLs [2]. In 2 weeks, >95% of OPCs can stain O4+, and MBP+ positive OLs can be detected in as little as four.

Cell identity transitions are guided by supplementing cell culture media with small molecules that act as cues to trigger intracellular signaling patterns that aid cell-fate determination. To generate neural progenitor cells we use an accepted commercial kit based on the widely used dual SMADi procedure, STEMDiff SMADi Neural Induction Kit [3, 4]. Subsequent differentiation of NPCs into OPCs and OLs follows a simplified adaptation of a widely used protocol by Douvaras and Fossati [5].

At the end of each differentiation stage, cellular identity is characterized by specific transcription factor signatures. Immunohistochemistry (IHC) or qPCR can be used to detect presence of markers expressed by the target population and confirm transition from the cells' previous identity. The accelerated pace of oligodendroglia maturation in this procedure mimics in vivo development as evidenced by the sequential expression of endogenously represented stage-specific markers. Determination of CNS-specific lineage commitment relies on detection of markers such as PAX6, SOX1, and Nestin in NPCS and CSPG4, NKX2.2, and PDGFRα in OPCs [6].

2 Materials

All cell culture is maintained in a 37 °C, 5% CO_2 incubator. Pre-warm media before use and prepare fresh daily unless otherwise specified.

1. Basement membrane extracellular matrix for iPSCs/ESCs: Dilute frozen matrix stock in ice-cold, serum-free medium to the final concentration recommended by manufacturer (*see* **Notes 1** and **2**). Culture-ware can be coated the day of or pre-coated, sealed in parafilm, and stored at 4 °C for up to 1 week (*see* **Note 3**). Pre-coated culture-ware should always be prewarmed or brought to room temperature before use.
2. Basement membrane extracellular matrix for differentiated cell types: Growth factor reduced matrix (GFR-matrix) is critical for all culture-ware used in differentiation procedures. Dilute frozen stock in ice-cold, serum-free medium to the final concentration recommended by manufacturer (*see* **Note 4**). Pre-coated culture-ware should always be prewarmed or brought to room temperature before use.

3. Enzymatic cell dissociation reagent: Accutase or TrypLE (*see* **Note 5**).
4. Resuspension medium: Step-specific culture medium supplemented with Rho-kinase inhibitor Y-27632 (ROCK-i) at a final concentration of [10 μM] (*see* **Note 6**).
5. Sterile 1× D-PBS/PBS (calcium and magnesium free).
6. Stem cell maintenance medium: Complete mTeSR Plus (*see* **Note 7**).
7. STEMdiff SMADi Neural Induction Kit by STEMCELL Technologies (8581) is used to generate neural progenitor cells. Medium composition follows the kit manufacturer's protocol (*see* **Note 8**).
8. OPC differentiation medium: DMEM/F-12, 1% N2 supplement, 1% B27 supplement, bFGF at a final concentration of 20 ng/mL, SAG at a final concentration of 1 μM, PDGF-AA at a final concentration of 10 ng/mL.
9. OL maturation medium: Neurobasal-A medium, 2% B27 supplement, 1 μM cAMP, 200 ng/mL T3 triiodothyronine, 1 μM clemastine (*see* **Note 9**).
10. IHC blocking buffer: 1× PBS (calcium and magnesium free) with 5% Normal Donkey Serum and 0.3% Triton X-100. Can be prepared day before or morning of and stored at 4 °C until use. Omit use of triton if staining only cell surface markers. Composition of blocking buffer depends on cell type, sensitivity, and target antigen.
11. Primary and secondary antibody solutions: Dilute antibody stock in IHC blocking buffer to the working concentrations as recommended by manufacturer.
12. Dapi solution: Dilute Dapi [1:1000] in 1× PBS (calcium and magnesium free).
13. NPC freezing buffer: Medium from STEMdiff SMADi Neural Induction kit supplemented with 10% DMSO and [10 μM] ROCK-i.
14. Co-culture medium: Neurobasal A medium, 2% B27 supplement, 100 ng/mL T3 triiodothyronine.

3 Methods

Undifferentiated primary cells are maintained in serum-free stem-cell specific maintenance medium on culture-ware coated in iPSC/ESC qualified basement membrane matrix (*see* **Note 1**). Starting culture should be of high quality and validated pluripotency (*see*

Note 10). Identify and remove regions of spontaneous differentiation before use of SMADi Neural Induction Kit (*see* **Note 11**).

3.1 Prepare for iPSC/ESC Differentiation into Neural Progenitor Cells (NPC) (Day 1)

(Day 1) Passage when confluency reaches 70–80%. Passaging stem cells as single cells and seeding onto culture-ware coated in growth factor reduced (GFR) matrix initiates the transition of culture from pluripotency-favoring conditions to a neural conducive environment (*see* **Note 3**).

1. Bring pre-coated plate, sufficient volumes of stem cell maintenance medium, 1× PBS, and enzymatic dissociation reagent to room temperature. Prepare resuspension solution with stem-cell maintenance medium (*see* **Note 6**).
2. Aspirate excess coating solution on freshly coated plates and add 1 mL/well of resuspension solution to prevent wells from drying before cells are seeded.
3. Aspirate medium from hiPSC/ESC culture. Wash each well once with 1 mL room temperature 1× PBS dispensed on the side of each well. Gently tilt the plate in a circular motion. Aspirate the 1× PBS.
4. Add 1 mL/well (6-well plate) of enzymatic cell dissociation reagent (*see* **Note 5**).

   Incubate at 37 °C for 6–8 min (*see* **Notes 11** and **12**).
5. Remove plate from incubator and neutralize the dissociation enzyme by adding an equivalent volume of stem cell maintenance medium to each well (ex: add 1 mL medium for 1 mL dissociation enzyme).
6. Gently collect cells and add to one 15 mL conical tube.
7. Dislodge any remaining adherent cells by gently rinsing the well once with maintenance medium. Add the rinse to the same conical tube containing cells.
8. Centrifuge at 300 × *g* for 5 min. Visually ensure the cells have collected as a pellet at the bottom of the tube before proceeding.
9. Taking care to not disturb the cell pellet, aspirate the supernatant. Resuspend the pellet in 1 mL 1× PBS by gently triturating 1–2 times (*see* **Note 13**).
10. Centrifuge at 300 × *g* for 5 min. Again, confirm the cells have collected as a pellet at the bottom of the tube before proceeding.
11. Taking care to not disturb the pellet, aspirate the supernatant. Resuspend the cell pellet in 1 mL resuspension medium. Gently triturate until no large clumps remain (*see* **Note 14**).

12. Conduct a cell count with Trypan Blue and a hemocytometer or with an automated cell counter. Calculate the volume of cells required to seed 0.5–1 × 10^6 cells/well (per 6-well plate) (*see* **Note 15**). Add appropriate volume of resuspension medium to the 15 mL conical tube to obtain this concentration. Triturate gently to homogenize solution.
13. Add 1 mL/well cell suspension solution to the GFR-matrix precoated 6-well plate.
14. Set the plate in a 37 °C, 5% CO_2 incubator. Quickly and carefully move the plate back and forth, side to side, to distribute cells evenly across the growth surface area (*see* **Note 16**).
15. Take care to replace culture media with fresh maintenance medium without ROCK-i within 24 h. Continue to feed cells daily. Proceed to Subheading 3.2 when iPSCs/ESCs are 85–90% confluent.

3.2 Neural Progenitor Cell Generation (Days 0–7)

Refer to flow diagram in Fig. 1a.

NPC differentiation follows the monolayer protocol from STEMDiff SMADi Neural Induction Kit and requires use of proprietary kit reagents. Prepare neural induction medium per manufacturer protocol (*see* **Note 8**).

1. (Day 0) Option to incubate cells in 1–2% DMSO solution for 24 h prior to adding neural induction medium with SMAD inhibitors (*see* **Note 17**).
2. (Day 1–6) Pre-warm neural induction medium to 37 °C in a water bath. Aspirate spent media from cell culture plate and replace it with 2 mL/well (6-well plate) of neural induction medium containing SMAD inhibitors from the commercial kit. Replace medium in each well daily for days 1–6 (*see* **Notes 18** and **19**).
3. (Day 6) Designate a few wells for IHC or qPCR to characterize differentiation efficiency (*see* **Note 20**). If greater than 90% of the cell population is positive for neural progenitor cell markers, proceed to passage culture on day 7 (*see* **Notes 21** and **22**). Alternatively, NPCs at this stage can be frozen for continuation of procedure later (*see* **Note 23**).
4. (Day 7) Prepare to passage neural progenitor cells for OPC differentiation. Coat a 24-well plate with GFR-matrix. Detach NPCs from desired number of wells using an enzymatic dissociation reagent (*see* **Note 5**). Passage procedure is identical to that performed on day 1, Subheading 3.1, with the exception being the use of neural induction medium to resuspend the centrifuged cell pellet just prior to plating (**step 11**). Seed cells at a density of 1–2 × 10^5 cells/well (24-well plate). Bring final volume to 500 μL/well with neural induction medium supplemented by SMAD inhibitors from the commercial kit.

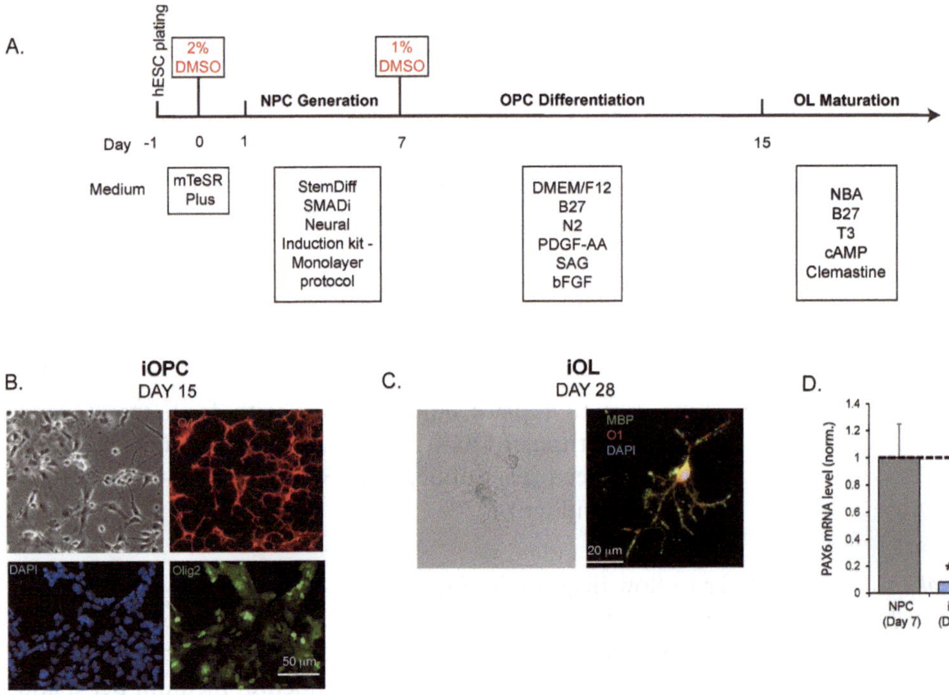

Fig. 1 (**a**) Flow diagram of entire procedural timeline. (**b**) Bright field and immunofluorescence images representative of OPCs at day 15. Olig2, the pan-oligodendroglia marker, is green; O4, the OPC marker is red; and DAPI blue. (**c**) Bright field and immunofluorescence images representative of OLs at day 28. MBP is tagged green, O1 in red, and DAPI blue. (**d**) Exemplifies the trend expected of type specific markers for determination of population homogeneity and robustness of differentiation efficiency. Shown here, NPC marker PAX6 is progressively downregulated as cells transition to OPCs and subsequently OLs around days 14 and 28 respectively

3.3 Oligodendrocyte Progenitor Cell Generation (Days 8–14)

Refer to flow diagram in Fig. 1a.

1. The option remains to incubate cells in 1–2% DMSO solution for 24 h just prior to initiating OPC generation (*see* **Note 17**).
2. (Day 8–13) Replace culture media with fresh OPC differentiation medium. Feed cells every other day or daily if necessary (*see* **Note 24**). Prepare OPC differentiation medium fresh daily.
3. (Day 14) Designate a few wells to characterize OPC identity via IHC or qPCR (*see* **Note 25**). Figure 1b provides a visual reference for OPC morphology and proportional expression of Olig2 and O4 markers at day 15. Day 14 OPCs can be used to set up a co-culture system with neurons, outlined in the next step. Otherwise, proceed to Subheading 3.5, Oligodendrocyte Maturation. Passage and seed cells in OPC differentiation medium at a density of 1–2 × 10^5 cells/well (24-well plate) (*see* **Note 26**).

3.4 OPC-Neural Co-culture

1. Pre-coat a 24-well plate with GFR matrix. Passage oligodendrocyte progenitor cells (OPCs) from day 14, Subheading 3.3, and seed at a density of 1×10^5 cells/well in OPC differentiation medium onto the prepared 24-well plate. Let OPCs settle for 24 h before adding Neurons (*see* **Note 27**).
2. Prepare co-culture medium. Passage neuronal culture and resuspend the pellet in co-culture medium for plating with OPCs (*see* **Note 28**). Seed neurons at the density of 2×10^5 cells/well (24-well plate) onto the same culture-ware as the day 14 OPCs. Incubate at 37 °C, 5% CO_2. Replace medium with fresh co-culture medium daily.
3. Pre- and post-synaptic markers can be detected via IHC after co-culture day 16 (*see* **Note 20**).
4. By co-culture day 21, abundant synaptic puncta can be observed and neuronal activity recorded. Perform IHC to measure mature oligodendrocyte markers such as MBP and PLP1 (*see* **Note 20**).
5. By co-culture day 28, oligodendrocyte myelination of neuronal axons can be observed. IHC can be used to stain both neuron and OL-specific markers (*see* **Note 20**).

3.5 Oligodendrocyte Maturation (Days 15–28)

1. (Day 15) Prepare fresh OL maturation medium daily. Replace culture media with 500 μL/well (24-well plate) prewarmed OL maturation medium. Repeat daily for days 15–28.
2. (Days 15–28) Passage at any point in which confluency is 90%, splitting culture at a 1:3 ratio (i.e., cells collected from one well are distributed into three new wells). Continue to maintain culture for up to two passages, replacing media with fresh OL maturation medium daily (*see* **Note 29**). Within this two-week span periodically perform IHC (*see* **Note 20**) or qPCR to monitor the progressive expression of OL-specific markers, such as MBP and PLP1, of whose expression can be expected around day 21 (*see* **Note 26**). Figure 2 depicts a time-course expression profile of common OPC and OL markers for reference.
3. (Day 28) The characteristically complex morphology of mature oligodendrocytes should be easily observable at this stage (*see* Fig. 1c).

4 Notes

1. Basement membrane extracts are extracellular matrices that support cell adhesion, proliferation, and differentiation. Stem cell qualified matrices are specifically formulated to aid

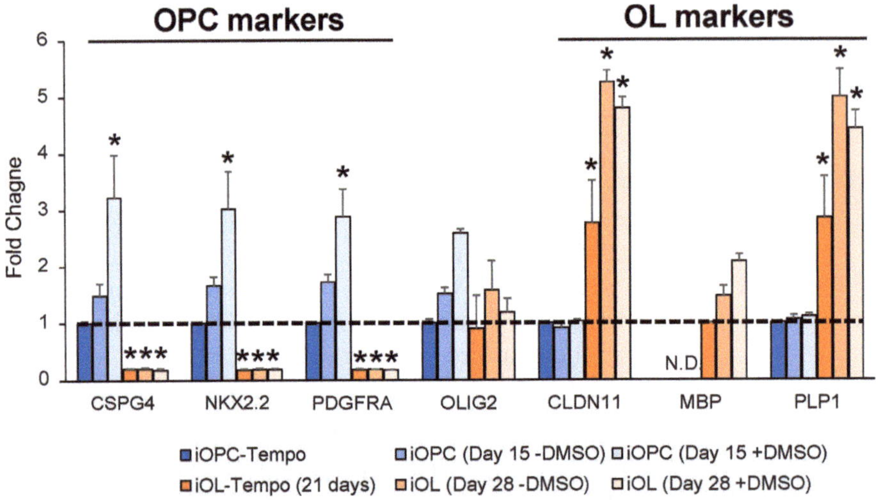

Fig. 2 Exemplified is a time course expression profile for common OPC and OL markers, showing progression through early, middle, and late maturation stages. Cells were sampled on baseline day 15 when the maturation procedure begins, and again on day 28. Fold change of marker expression expected with pre-treatment of 1–2% DMSO on day 0 and day 7 is also demonstrated for reference. iOPC-Tempo and iOL-Tempo refer to commercial oligodendrocytes (Tempo BioScience, SKU102) that were matured and maintained according to the manufacturer's instructions for direct comparison with the OPCs and OLs derived from our procedure

maintenance of pluripotency. Optimization of this protocol expanded stem cell culture on hESC Matrigel diluted in DMEM/F12 medium.

2. When coating culture-ware, a thin gel method is considered optimal for primary cell types. Stock protein concentration will vary between suppliers and batches, adjust dilution volumes accordingly.
3. When using a gel matrix, keep stock and medium on ice and pre-chill pipet tips to prevent premature gelling. Take care to ensure even distribution across culture-ware's growth surface area when coating. When using pre-coated plates, warm them in a 37 °C incubator or to room temperature under a tissue-culture hood before seeding cells. Aspirate excess coating solution immediately before plating to prevent wells from drying out.
4. For all differentiation steps, coat culture-ware in growth factor reduced (GFR) basement membrane extract. This procedure was optimized with GFR-Matrigel diluted in DMEM/F12 media.
5. Pluripotent stem cells grow in colonies. Enzymatic dissociation reagents are employed to dissolved cell aggregates for plating as single cells to encourage differentiation. Alternative enzymatic cell dissociation reagents may vary slightly in passaging protocol. If a different enzymatic reagent is preferred, refer to product-specific procedure.

6. The rho-kinase inhibitor (ROCK-i) Y-27632 is added to the culture medium to enhance survival when passaging single cells [7]. Take care to replace any media containing rho-kinase inhibitors within 24 h. Use of Y-27632 can be substituted with Thiazovivin.
7. Thaw mTeSR Plus 5× Supplement at 4 °C and add entire contents to mTeSR Plus Basal Medium. Complete mTeSR Plus can be stored for 2 weeks at 4 °C or aliquoted and stored at −20 °C for 6 months. Although this procedure was optimized with mTeSR Plus, an alternative stem cell-qualified media may be used.
8. Neural progenitor cell generation requires STEMdiff SMADi Neural Induction Kit [4]. Methods 3.2 requires adherence to this kit's monolayer induction protocol and uses plates prepared with GFR-Matrigel. Refer to manufacturer's technical manual for preparation and use of neural induction medium. If culture was recently thawed, passage at least twice to allow sufficient recovery from cryopreservation before proceeding with guided differentiation.

   STEMdiff SMADi protocol was developed based on the highly cited dual SMAD inhibition publication by Lorenz Studer's group, *see* Ref. [3]. The resulting cell population can be expected to be enriched for central nervous system (CNS)-type NPCs, confirmed by assaying expression levels of SOX1, Nestin, and PAX6.
9. Clemastine is a muscarinic and antihistaminic compound originally identified in drug screenings for treatment of multiple sclerosis [8]. Clemastine has been evidenced to shorten the oligodendrocyte maturation timeframe in both iPSCs and in vivo models [9, 10]. OL maturation medium composition is adapted from a protocol by Douvaras and Fossati [5].
10. Transition from colonies to single cells is necessary for inducing differentiation but can stress pluripotent cell types. It is important to start with a high-quality culture, characterized by: low passage number, typically below 50; morphological characteristics of pluripotency such as tightly packed colonies with round, defined edges, and large nuclei and nucleoli [11].

    Pluripotency can be assayed via immunohistochemical staining of well-characterized pluripotency markers such as intracellular NANOG, OCT3/4, or cell surface markers like SSEA3 or TRA-1-60.
11. Use a microscope to identify regions of spontaneously differentiated cells for removal prior to passaging. Outline areas of differentiation by marking the bottom of the plate with a felt tip marker. Aspirate or scrape indicated areas with a pipette tip.

12. Length of incubation can vary depending on cell line or use of an alternative cell dissociation reagent. Take care to optimize the incubation period by monitoring under a microscope. Cells visibly lift from the culture-ware, evidenced by greater light penetration beneath the colonies.
13. The purpose of this step is to wash the cells, removing residual detachment solution and debris which would ultimately impact cell health. If after centrifugation the supernatant remains cloudy, centrifuge for an additional 1–2 min.
14. Dissolution of clumps can be gauged by eye or a small volume can be pipetted onto a slide for confirmation under a microscope.
15. It is critical to perform a cell count for precision plating. Seeding density recommendations are made so that culture reaches target confluency the following day, however, density is provided as a range. It may be necessary to modify plating density and passage timing to optimize procedure for different cell lines. A cell density titration series may be of benefit before proceeding.
16. Once set in the incubator, agitate the plate with a back-and-forth, then side-to-side motion. This is critical to distribute cells evenly. Swirling the plate or dish has the opposite effect, bringing cells to the center or along the edges of their growth area. It is worth attending to the distribution of cells in each well. Dense clumps cause uneven consumption of nutritional resources. Conversely, sparse distribution can increase the risk of spontaneous differentiation.
17. Subjecting cells to a DMSO pre-treatment for 24-h before guided differentiation has been repeatedly evidenced to significantly improve differentiation efficiency across many cell lines, enhancing their propensity to generate various specialized cell types.

    Figure 2 depicts the benefits of DMSO pretreatment in context of cell-type specific gene expression for deriving robust populations of OPC and OLs. For more information on the molecular effects of DMSO pre-treatment and evidence of its efficacy, *see* Ref. [12–14].

    To pre-treat, prewarm the culture media as normal and add DMSO so that it comprises 1–2% of the final solution (Ex. 100 μL DMSO in 10 mL culture media = 1% DMSO solution. 200 μL DMSO in 10 mL media = 2% DMSO solution). Incubate cells in DMSO solution for 24-hours. Replace with appropriate volume of culture media *without* DMSO the following day.
18. To compensate for the high density of cells it may be necessary to adjust the neural induction medium feed volume from 2 mL/well to 3–4 mL/well (6-well plate) daily. Make this

adjustment if there is an atypical quantity of dead cells in suspension the day after feeding or if the media color changes color drastically.

19. If rapid proliferation is observed and high confluency is impacting cell health and viability, passage with an enzymatic dissociation reagent and replate culture at a density of 0.5×10^6 per well (6-well plate). Use neural induction medium for resuspension during passaging, otherwise, the passaging procedure is stepwise identical to day 1, Subheading 3.1.
20. Immunohistochemistry can be performed to validate cell type specific markers. If a compatible microscope is available, cells can be fixed, stained and imaged in a multi-well plate. Start by washing wells once with 1× PBS then incubate in fresh 4% PFA for 20 min at room temperature. Aspirate the PFA and wash the wells three times with 1× PBS. Incubate with IHC blocking buffer for 1 h at room temperature or overnight at 4 °C. Aspirate blocking buffer and incubate well with primary antibody solution for 1 h at room temperature or overnight at 4 °C. Wash three times with 1× PBS and add secondary antibody solution. Protect the plate from light and incubate secondary antibody for 1 h at room temperature. Aspirate antibody solution from wells, rinse again three times with 1× PBS, and add diluted Dapi solution for a 5-min incubation. Aspirate Dapi and wash one final time with 1× PBS. Add fresh PBS to stained wells and image immediately or wrap in parafilm and store at 4 °C.
21. Analysis of differentiation efficiency is recommended at each cell-type transition before proceeding. Immunohistochemistry can be used to detect presence of common identity-specific markers and confirm absence of residual pluripotency markers. Only a high purity population is ready for continued differentiation. The acceptable differentiation efficiency threshold is (90%) or more cells testing positive as the target cell type.
22. NPC markers include Nestin, PAX6, and SOX1. NKX2.1 is a CNS-specific forebrain marker [15].
23. Day 7 NPCs can be frozen and stored at −80 °C. Collect NPCs for cryopreservation when confluency is 80–90%. Detach NPCs with an enzymatic dissociation reagent and collect in a conical tube to obtain a pellet. Aspirate the supernatant and resuspend cells in NPC freezing buffer. Thawed NPCs retain multipotency and can generate mature neurons, astrocytes, and OPCs. NPCs can be thawed onto a GFR-matrix coated plate and incubated in Neural Induction Medium from STEMdiff SMADi Neural Induction Kit. Supplement NPC medium with ROCK-i for the first 24-h post thaw, then replace medium without ROCK-i.

24. If high confluency and rate of cell proliferation is appearing to affect the integrity or viability of the culture before day 15, passage and seed 1–2 × 10^5 cells/well (24-well plate).
25. The maturation of OPCs is demarcated by temporal variations in marker expression. OPC-specific markers should outweigh NPC markers by day 14 (Fig. 1d, e).

    Lineage-committed pre-OPCs express NKX2.2. Early OPCs express Olig1/2 and CSPG4/Ng2. PDGFRα is a classical marker of late developmental timepoints, as is O4.

    Mature oligodendrocytes (OLs) are primarily identified by upregulation of myelin basic protein (MBP) marker of myelinating oligodendrocytes. Additional markers for testing could include the abundant CNS-specific oligodendrocyte transmembrane proteins CLDN11 and PLP1 [2, 6, 15, 16]. Upregulation of oligodendrocyte specific markers is expected to be most significant between days 21–28. For a more extensive list of developmental signatures *see* Ref. [15].
26. Careful attention to the seeding density of OPCs prior to OL induction is critical. High seeding density can lead to an overgrowth of remaining NPC populations, with potential to interfere with OPC responsivity to differentiation stimuli. Low seeding density enables OPCs to extend and develop their characteristic arborizations (branching) (Fig. 1c).
27. Co-culturing pluripotent stem cell-derived neurons with OPCs can support activity and functional development of both cell types [17]. Co-culture procedure was optimized with neurons generated with a direct neuronal induction procedure [18]. Neurons derived from alternative procedures may require additional optimization.
28. Composition of co-culture medium used here has been tested with induced neurons generated by the Ngn2 single step forced transcription method [18]. Co-culture medium to support neurons from alternative differentiation methods may require additional optimization.
29. Rapid proliferation is indicative of OPC immaturity. Cell division is expected to slow significantly as the oligodendrocytes mature. If, during the OL maturation period, the culture has been passaged twice at a 1:3 split ratio and a notably high proliferation rate continues, briefly treat with Ara-C for 1–3 days until division slows. Add [2–5 μM] of Ara-C to OL maturation medium.

References

1. Assetta B, Tang C, Bian J, O'Rourke R, Connolly K, Brickler T, Chetty S, Huang YA (2020) Generation of human neurons and oligodendrocytes from pluripotent stem cells for modeling neuron-oligodendrocyte interactions. J Vis Exp: JoVE 165. https://doi.org/10.3791/61778
2. Bradl M, Lassmann H (2010) Oligodendrocytes: biology and pathology. Acta Neuropathol 119(1):37–53. https://doi.org/10.1007/s00401-009-0601-5
3. Chambers SM, Fasano CA, Papapetrou EP, Tomishima M, Sadelain M, Studer L (2009) Highly efficient neural conversion of human ES and iPS cells by dual inhibition of SMAD signaling. Nat Biotechnol 27(3):275–280. https://doi.org/10.1038/nbt.1529
4. STEMCELL Technologies. Monolayer culture protocol. In: STEMdiff SMADi neural induction kit technical manual. Available via https://www.stemcell.com/products/stemdiff-smadi-neural-induction-kit.html#section-protocols-and-documentation
5. Douvaras P, Fossati V (2015) Generation and isolation of oligodendrocyte progenitor cells from human pluripotent stem cells. Nat Protoc 10(8):1143–1154. https://doi.org/10.1038/nprot.2015.075
6. Wilson HC, Scolding NJ, Raine CS (2006) Co-expression of PDGF alpha receptor and NG2 by oligodendrocyte precursors in human CNS and multiple sclerosis lesions. J Neuroimmunol 1-2:162–173
7. Watanabe K, Ueno M, Kamiya D, Nishiyama A, Matsumura M, Wataya T, Takahashi JB, Nishikawa S, Nishikawa S, Muguruma K, Sasai Y (2007) A ROCK inhibitor permits survival of dissociated human embryonic stem cells. Nat Biotechnol 25(6): 681–686. https://doi.org/10.1038/nbt1310e0208110. https://doi.org/10.1371/journal.pone.0208110
8. Mei F, Fancy S, Shen YA, Niu J, Zhao C, Presley B, Miao E, Lee S, Mayoral SR, Redmond SA, Etxeberria A, Xiao L, Franklin R, Green A, Hauser SL, Chan JR (2014) Micropillar arrays as a high-throughput screening platform for therapeutics in multiple sclerosis. Nat Med 20(8):954–960. https://doi.org/10.1038/nm.3618Pan
9. Pan S, Mayoral SR, Choi HS, Chan JR, Kheirbek MA (2020) Preservation of a remote fear memory requires new myelin formation. Nat Neurosci 23(4):487–499. https://doi.org/10.1038/s41593-019-0582-1
10. Chanoumidou K, Mozafari S, Evercooren A, Kuhlmann T (2019) Stem cell derived oligodendrocytes to study myelin diseases. Glia 68. https://doi.org/10.1002/glia.23733
11. Chen KG, Mallon BS, McKay RD, Robey PG (2014) Human pluripotent stem cell culture: considerations for maintenance, expansion, and therapeutics. Cell Stem Cell 14(1):13–26. https://doi.org/10.1016/j.stem.2013.12.005
12. Li J, Narayanan C, Bian J, Sambo D, Brickler T et al (2018) A transient DMSO treatment increases the differentiation potential of human pluripotent stem cells through the Rb family. PLoS One 13(12):e0208110
13. Chetty S et al (2013) A simple tool to improve pluripotent stem cell differentiation. Nat Methods 10(6):553–556. [PubMed:23584186]
14. Sambo D, Li J, Brickler T, Chetty S (2019) Transient treatment of human pluripotent stem cells with DMSO to promote differentiation. J Vis Exp (149). https://doi.org/10.3791/59833
15. Alsanie WF, Niclis JC, Petratos S (2013) Human embryonic stem cell-derived oligodendrocytes: protocols and perspectives. Stem Cells Dev 22(18):2459–2476. https://doi.org/10.1089/scd.2012.0520
16. Douvaras P, Wang J, Zimmer M, Hanchuk S, O'Bara MA, Sadiq S, Sim FJ, Goldman J, Fossati V (2014) Efficient generation of myelinating oligodendrocytes from primary progressive multiple sclerosis patients by induced pluripotent stem cells. Stem Cell Rep *3*(2):250–259. https://doi.org/10.1016/j.stemcr.201
17. Thornton MA, Hughes EG (2020) Neuron-oligodendroglia interactions: activity-dependent regulation of cellular signaling. Neurosci Lett 727:134916. https://doi.org/10.1016/j.neulet.2020.134916
18. Zhang Y, Pak C, Han Y, Ahlenius H, Zhang Z, Chanda S, Marro S, Patzke C, Acuna C, Covy J, Xu W, Yang N, Danko T, Chen L, Wernig M, Südhof TC (2013) Rapid single-step induction of functional neurons from human pluripotent stem cells. Neuron 78(5): 785–798. https://doi.org/10.1016/j.neuron.2013.05.029

Chapter 9

Characterizing the Neuron-Glial Interactions by the Co-cultures of Human iPSC-Derived Oligodendroglia and Neurons

Gabriella Vulakh and Xin Yang

Abstract

Induced pluripotent stem cell (iPSC) techniques have had considerable breakthroughs in modeling human neurological diseases. Multiple protocols inducing neurons, astrocytes, microglia, oligodendrocytes, and endothelial cells have been well-established thus far. However, these protocols have limitations, including the long time period to get cells of interest or the challenge of culturing more than one cell type simultaneously. Protocols for handling multiple cell types within a shorter time period are still being established. Here we describe a simple and reliable co-culture system to study interactions between neurons and oligodendrocyte precursor cells (OPC) in health and in disease.

Key words Induced pluripotent stem (iPS) cells, Oligodendrocyte precursor cells, Neurons, Co-culture, Neuron-glial interactions

1 Introduction

Oligodendrocytes, differentiated from oligodendrocyte precursor cells (OPCs), are crucial for many important aspects of the brain including myelination, axonal integrity, and cognitive functions [1–3]. Communication between neurons and oligodendrocytes is being extensively studied [3–7] in neurodegenerative diseases such as Alzheimer's Disease where defects of myelin have reproducibly been reported. However, there is also evidence that the novel functions of OPCs in brain homeostasis and dynamics extend beyond serving as precursors for oligodendrocytes [8]. Synaptic [9–11] and non-synaptic [12, 13] connections between neurons and OPCs have been reported, but the underpinnings of these connections and to what extent they contribute to neurodevelopment and neurodegeneration remain largely elusive. A large caveat is the lack of a reliable study system to simplify the complexity of in vivo conditions. This chapter aims to summarize a detailed and

Yu-Wen Alvin Huang and ChangHui Pak (eds.), *Stem Cell-Based Neural Model Systems for Brain Disorders*, Methods in Molecular Biology, vol. 2683, https://doi.org/10.1007/978-1-0716-3287-1_9,

simple protocol for inducing neurons and OPC in vitro and establishing a co-culture system to study neuron-OPC interactions.

Regular hiPSC culture, lentiviral production and lentiviral delivery, neuron (iNs) induction [14], and iOPC induction [15] have been thoroughly discussed. Briefly, H1 stem cells were differentiated into neurons (Fig. 1a) by forced expression of Ngn2 with lentivirus and puromycin selection [14]. We induced iOPC (Fig. 1b) using a two-steps protocol [15]. The first step was inducing stem cells to neural progenitor cells (NPCs) with commercially available SMADi differentiating medium [15]. Successfully induced NPC expressed markers such as PAX6, Nestin, and Sox1, but not stem cell markers such as OCT4 [15]. The second step was inducing NPC into iOPC by chemically activated transcription factors that are critical for OPC generation in vivo. The iOPC markers such as O4, Olig1/2, CSPG4/Ng2, NKX2.2, PDGFRα can be detected by immunofluorescence staining (Fig. 2a, b) and quantitative-PCR (q-PCR, Fig. 2d) but not NPC markers PAX6 [15] (Fig. 2c). Here, we will focus on a protocol for neuron and OPC co-culture.

In this protocol, we have developed two different co-culture systems. One combines the iNs and iOPCs immediately (Figs. 1c and 3a) after successful induction [15], and the other cultures iNs for approximately 10 days before iOPC addition (Fig. 1d). Both systems are effective, however, arabinoside (AraC) can be added in the first protocol to suppress iOPC from over-proliferation. We can also detect synapse formation in iNs by staining pre- and postsynaptic markers (Synapsin1 and PSD95 staining in Fig. 3b) and also observe iOL processes surrounding iN axons [15]. The iNs, iNPCs, and iOPCs can be frozen in commercial freezing media (e.g., CS10) and stored in liquid nitrogen for up to 6 months. While the current protocol utilized the OPC and neuron differentiation from H1 ES cells, it can also be recaptured with other stem cells such as KOLF2.1J (data not shown). This protocol ultimately allows us to study neuron-OPC communication without contamination from other glial cell types.

2 Materials

2.1 ES Cell Medium

1. mTeSR™ Plus medium (STEMCELL Technologies #100-0276).

2.2 iN Culture Medium

1. Neurobasal A medium (Thermo Fisher Scientific # 10888022).
2. 1% N2 supplement (Thermo Fisher Scientific #17502048).
3. 2% B27 supplement (Thermo Fisher Scientific #17504044).

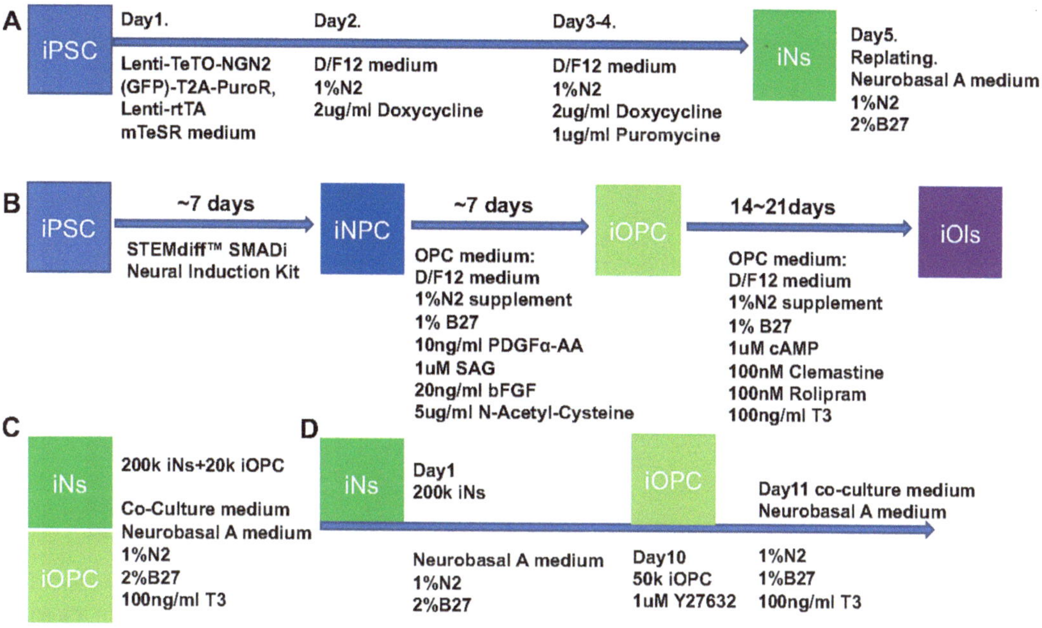

Fig. 1 Diagram of iNs and iOPC induction, and co-culture system: the simplified induction process for (**a**) neurons (**b**) NPC and OPC, and (**c** and **d**) co-culture system with key components in each medium is listed

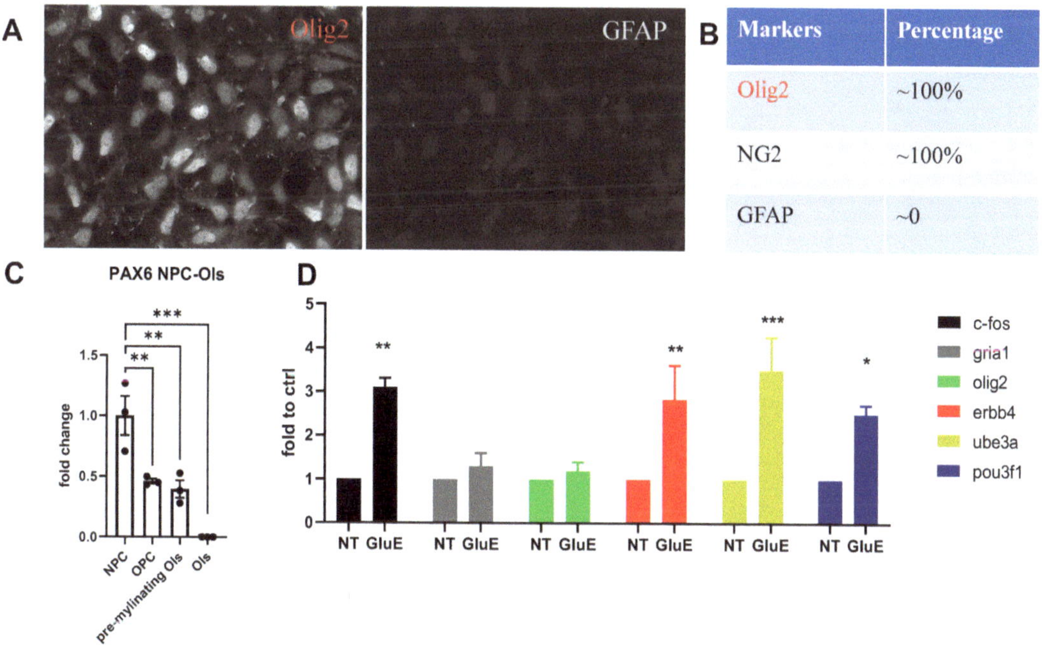

Fig. 2 Validation of induced oligodendrocyte precursor cells (iOPC). (**a**) The typical immunofluorescence staining of Olig2 and GFAP at day 7 post iOPC induction; nuclei were stained with DAPI (Blue). (**b**) The percentage of OPC markers in the cells was quantified. (**c**) The NPC marker PAX6 was undetected in mature oligodendrocytes by q-PCR, 3 wells per each condition. (**d**) Glutamate upregulates immediate early gene expressions in iOPC by q-PCR, 3 wells per condition and two-way ANOVA test. $^{*}P < 0.05$, $^{**}P < 0.01$, $^{***}P < 0.001$

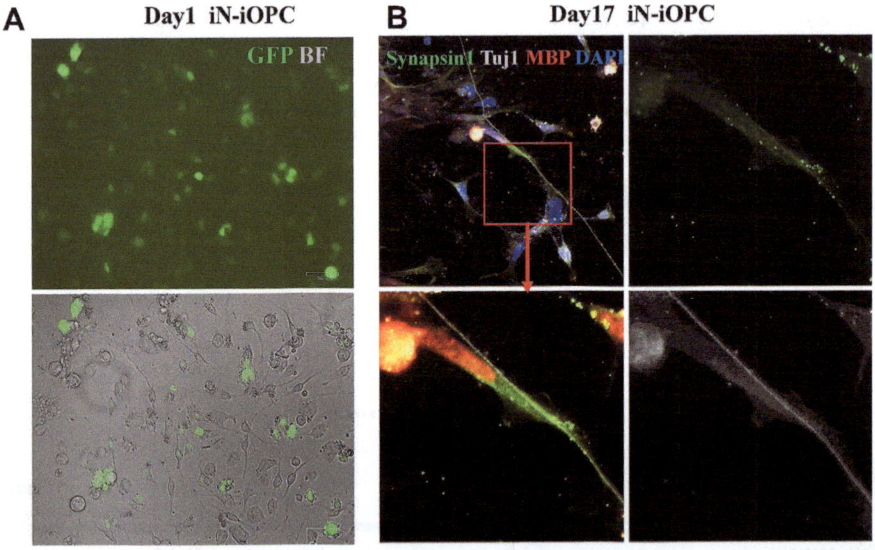

Fig. 3 Induced neurons (iN) and oligodendrocyte precursor cells (iOPC) co-culture. (**a**) Typical images of immediate co-culture at day 1, with GFP labeling iNs and brightfield showing both iNs and iOPC in the same well. (**b**) Immunostaining of delayed co-culture at day 17. Synpsin1/2 (Green), MBP (red), and Tuj1 (Gray) co-localize to the synaptic connections between neurons and early-myelinating Ols. Nuclei were stained with DAPI (Blue)

2.3 NPC Induction Medium

1. STEMdiff™ SMADi Neural Induction Kit (STEMCELL Technologies # 08581).

2.4 OPC Induction Medium

1. DMEM/F12 medium (Thermo Fisher Scientific #11320033).
2. 10 ng/mL PDGFα-AA (R&D Systems #221-AA-025).
3. 1 uM SAG (Tocris # 4366).
4. 5 ug/mL N-Acetyl-Cysteine (Millipore Sigma # A9165-5G).
5. 20 ng/mL bFGF (R&D Systems #233-FB-025/CF).
6. 1% N2 supplement (Thermo Fisher Scientific #17502048).
7. 1% B27 supplement (Thermo Fisher Scientific #17504044).

2.5 Ols Induction Medium

1. DMEM/F12 medium (Thermo Fisher Scientific #11320033).
2. 1% N2 supplement (Thermo Fisher Scientific #17502048).
3. 1% B27 supplement (Thermo Fisher Scientific #17504044).
4. 1 uM cAMP (MedChem Express # HY-B1511).
5. 100 nM Clemastine (MedChem Express # HY-B0298A).
6. 100 uM Rolipram (MedChem Express # HY-16900).
7. 100 ng/mL T3 (MedChem Express # HY-A0070A).

2.6 Co-culture Medium

1. Neurobasal Plus medium (Thermo Fisher Scientific # A3582901).
2. 1% N2 supplement (Thermo Fisher Scientific #17502048).
3. 2% B27 supplement (Thermo Fisher Scientific #17504044).
4. 100 ng/mL T3 (MedChem Express # HY-A0070A).

2.7 Other Chemicals, Kits, and Reagents

1. L-Glutamic acid (GluE) (Millipore Sigma # G1251-100G).
2. Y-27632 (Dihydrochloride) (STEMCELL Technologies # 72302).
3. Corning® Matrigel® hESC-Qualified Matrix (Sigma-Aldrich # CLS354277).
4. ACCUTASE™ (STEMCELL Technologies #07920).
5. Poly-D-Lysine (PDL) (Thermo Fisher Scientific #A3890401).
6. QiagenTM RNeasyTM RNA Isolation Mini Kit, 50 rxns (1762/ NC9677589).

2.8 qPCR Primers and Probes

1. All probes for human genes were purchased from iDT company with FAM fluorescent.

2.9 Antibodies

1. Chicken anti GFAP (Aves labs #ab53554, dilution 1:5000).
2. Rabbit anti Olig2 (Millipore Sigma #AB9610, dilution 1: 1000).
3. Rabbit anti NG2 (Millipore Sigma #AB5320, dilution 1:500).
4. Rabbit anti NG2 (Millipore Sigma #AB5320, dilution 1:500).
5. Mouse anti Anti-PSD-95 (NeuroMab clone K28/43, AB_2292909, dilution 1:1000).
6. Guinea pig anti Synapsin 1/2 (Synaptic System # 106004, dilution 1:1000).
7. Rat anti MBP (Novus #NB600-717, dilution 1:250).
8. Chicken Tuj1 (GeneTex GTX85469, dilution 1:1000).

2.10 Apparatus

1. CFX Opus 96 Real-Time PCR System or similar system.
2. Bio Rad ZOE images system or similar system.
3. BioTek Gen5 images system or similar system.

3 Methods

3.1 iN Induction

We use the same protocol as Zhang et al., 2013 to generate iNs from cultured H1 human ES cells [14] except we selected cells by 1 ug/mL Puromycin (up to 5 ug/mL final dose) for 2 days. The diagram can be found in Fig. 1a.

3.2 NPC Induction

We use a widely accepted commercial SMADi kit and follow the monolayer protocol provided by the manufacturer (*see* Materials). Induced neural progenitor cells (NPCs) from cultured H1 human ES cells were generated followed Assetta et al. [15]. The diagram is shown in Fig. 1b.

3.3 iOPC Generation (~7 days) and Validation (~7–14 days)

1. Passaged NPCs using Accutase (*see* Materials) or thawed from liquid nitrogen were plated at a seeding density of 1–2 × 10^5 cells per well in a 24-well plate in warm (37 °C) neural induction medium plus SMAD inhibitors from the commercial kit (*see* Materials). The next day, medium was refreshed with 1% DMSO containing OPC differentiation medium (*see* Materials) for 24 h before refreshing OPC differentiation medium without DMSO. Medium was then changed daily for at least 7 days. If cells reached confluence during this period, they were reseeded after Accutase detachment.
2. 7 days post induction, iOPCs could be further matured to Oligodendrocytes (iOls), frozen with CS10 freezing medium or used in functional validation or co-culture.

3.4 iOls Maturation

1. iOPCs were seeded at a density of 1–2 × 10^5 cells per well in a 24-well plate in warm (37 °C) iOL maturation medium (containing 1% N2 supplement, 1% B27, 1 uM cAMP, 100 nM Clemastine, 100 nM Rolipram, 100 ng/mL T3 triiodothyronine) (*see* Materials). For ~14 days, the efficiency of oligodendroglial maturation could be assessed by the expression of OL markers such as PLP1, MBP by qPCR, IHC staining or immunoblotting [15]. By this time, almost no PAX6 should be detected (Fig. 2c).
2. By 7 days in iOPC differentiation medium, the OPC specific markers (e.g., O4, Oligl/2, CSPG4/NG2, NKX2.2, PDGFRa; Fig. 2a, b) but not NPC markers (Pax6 or Nestin; Fig. 2c) could be detected by IHC staining or qPCR. We typically detected the Olig2 and NG2 immunoreactivity in more than 95% of the cells without GFAP labeling (Fig. 2a, b). In particular, it has been shown that neurons inter-connect and regulate OPC in vivo [3–11, 13]. OPC have also showed upregulation of early immediate genes upon stimulation by neurotransmitters [16]. We applied 300 uM Glutamate into iOPCs culture for 2 h before RNA was extracted (QIAGEN RNA extraction kits, *see* Materials) and detected approximately 3 folds increase of c-Fos expression [16] (Fig. 2d), 2.5 folds in increase of POU3F1 [17] (Fig. 2d), together with ERBB4 and Ube3a (Fig. 2d). Olig2 has no change (Fig. 2d).

3.5 iN-iOPC Co-culturing

3.5.1 Immediate Co-culture

This protocol is generally follows a published protocol from the Huang lab [15] with subtle modification. We chose a seeding density of 2×10^5 iNs and 2×10^4 iOPC (ratio 10:1) due to the long-term culture period and iOPC proliferation. Both Matrigel and PDL-coated plates work well. To start with, we used iN with GFP expression. A typical co-culture image at the second day of co-culture is shown in Fig. 3a (neurons show strong GFP expression, GFP negative cells are iOPCs). If the iOPC proliferate too fast, we can add 1–5 uM AraC into the culture.

3.5.2 Delayed Co-culture

The use of AraC and uncontrolled proliferation of iOPC prompted us to modify our co-culture. Furthermore, to better understand neuron-OPC synaptic interaction, we chose to add iOPC to the iN culture at day 10. We began with a seeding density of 200 K for iN and 50 K for iOPC. After 7 days in co-culture, we were able to detect typical synaptic protein (Synapsin1 here) expression from iN (Tuj1 labeling) and MBP expression from differentiating iOPCs (Fig. 3b).

3.6 Summary

Neuron-OPC interactions have been established in vivo for more than two decades, but how the synapses form and contribute to normal brain functions remains largely elusive. We recognized an urgent need to establish a reliable and simple paradigm to answer such biological questions. Here we describe a method to induce iOPCs in less than 2 weeks. Importantly, the iOPCs express Glutamergic receptors (GluA1, Fig. 2d), respond to Glutamate stimulation in pure culture (Fig. 2d), and form synaptic connections with neurons in our co-cultures (Fig. 3b). This protocol could be further utilized to understand how genetic changes in either neurons or iOPCs contribute to neurological diseases. 300 uM Glutamate stimulation, as reported [16], upregulated c-Fos expression to three folds by q-PCR (Fig. 2b). Surprisingly, we detected more immediate early genes upregulated, which themselves showed critical roles in brain functions. For example, ERBB4, implicated in familial amyotrophic lateral sclerosis (ALS) [18], showed ~three folds increase upon Glutamate stimulation. Similarly, Ube3a, which is the causal gene for Angelman syndrome [19] and activity regulated [17, 20], showed ~three folds increase. AMPA receptors such as GluA1 were also expressed in our iOPCs, although no significant expression changes upon Glutamate treatment.

The co-culture system described here is beneficial in understanding the underpinnings of neuron-OPC connections during development and dissecting the neuropathology behind neurological diseases.

4 Notes

1. Both Matrigel and PDL work well for coating plates, and theoretically, more coating materials should work as well with/without optimization.
2. Matrigel coating may encounter floating issues in long-term experiments.
3. All recombinant proteins are dissolved and aliquoted according to the manufacturer's instructions. Keep aliquots at −20 °C or lower for further use for up to 6 months.
4. It is preferred for all medium containing recombinant proteins to be freshly prepared, but they can be kept at 4 °C for up to 1 month.
5. There may be cell density concerns for each co-culture system. We used neuron to OPC at 10:1 and 4:1 for immediate and delayed co-culture respectively.
6. Theoretically, functional studies such as pair-recording, live imaging and calcium imaging could be applicable to our co-culture system.

Acknowledgments

We would like to thank Dr. Yu-Wen Alvin Huang at Brown University for the helpful discussions. This work was supported by the Postdoctoral Fellowship Program from the Foundation for Angelman Syndrome Therapeutics (FAST; grant number PD2023-001 to X.Y.) and the Brown SPRINT|UTRA Summer Award (to G.V.).

References

1. Lee Y, Morrison BM, Li Y et al (2012) Oligodendroglia metabolically support axons and contribute to neurodegeneration. Nature 487:443–448
2. Funfschilling U, Supplie LM, Mahad D et al (2012) Glycolytic oligodendrocytes maintain myelin and long-term axonal integrity. Nature 485:517–521
3. Simons M, Nave KA (2015) Oligodendrocytes: myelination and axonal support. Cold Spring Harb Perspect Biol 8:a020479
4. Marques S, Zeisel A, Codeluppi S et al (2016) Oligodendrocyte heterogeneity in the mouse juvenile and adult central nervous system. Science 352:1326–1329
5. Nave KA, Werner HB (2014) Myelination of the nervous system: mechanisms and functions. Annu Rev Cell Dev Biol 30:503–533
6. Pease-Raissi SE, Chan JR (2021) Building a (w)rapport between neurons and oligodendroglia: reciprocal interactions underlying adaptive myelination. Neuron 109:1258–1273
7. Tognatta R, Miller RH (2016) Contribution of the oligodendrocyte lineage to CNS repair and neurodegenerative pathologies. Neuropharmacology 110:539–547
8. Akay LA, Effenberger AH, Tsai LH (2021) Cell of all trades: oligodendrocyte precursor cells in synaptic, vascular, and immune function. Genes Dev 35:180–198

9. Lin SC, Bergles DE (2004) Synaptic signaling between GABAergic interneurons and oligodendrocyte precursor cells in the hippocampus. Nat Neurosci 7:24–32
10. Bergles DE, Roberts JD, Somogyi P et al (2000) Glutamatergic synapses on oligodendrocyte precursor cells in the hippocampus. Nature 405:187–191
11. Ge WP, Yang XJ, Zhang Z et al (2006) Long-term potentiation of neuron-glia synapses mediated by Ca2+–permeable AMPA receptors. Science 312:1533–1537
12. Bai X, Kirchhoff F, Scheller A (2021) Oligodendroglial GABAergic signaling: more than inhibition! Neurosci Bull 37:1039–1050
13. Wake H, Ortiz FC, Woo DH et al (2015) Nonsynaptic junctions on myelinating glia promote preferential myelination of electrically active axons. Nat Commun 6:7844
14. Zhang Y, Pak C, Han Y et al (2013) Rapid single-step induction of functional neurons from human pluripotent stem cells. Neuron 78:785–798
15. Assetta B, Tang C, Bian J et al (2020) Generation of human neurons and oligodendrocytes from pluripotent stem cells for modeling neuron-oligodendrocyte interactions. J Vis Exp
16. Pende M, Holtzclaw LA, Curtis JL et al (1994) Glutamate regulates intracellular calcium and gene expression in oligodendrocyte progenitors through the activation of DL-alpha-amino-3-hydroxy-5-methyl-4-isoxazolepropionic acid receptors. Proc Natl Acad Sci U S A 91:3215–3219
17. Hrvatin S, Hochbaum DR, Nagy MA et al (2018) Single-cell analysis of experience-dependent transcriptomic states in the mouse visual cortex. Nat Neurosci 21:120–129
18. Takahashi Y, Fukuda Y, Yoshimura J et al (2013) ERBB4 mutations that disrupt the neuregulin-ErbB4 pathway cause amyotrophic lateral sclerosis type 19. Am J Hum Genet 93:900–905
19. Yang X (2020) Towards an understanding of Angelman syndrome in mice studies. J Neurosci Res 98:1162–1173
20. Greer PL, Hanayama R, Bloodgood BL et al (2010) The Angelman Syndrome protein Ube3A regulates synapse development by ubiquitinating arc. Cell 140:704–716

Chapter 10

Defined Differentiation of Human Pluripotent Stem Cells to Brain Microvascular Endothelial-Like Cells for Modeling the Blood-Brain Barrier

Koji L. Foreman, Eric V. Shusta, and Sean P. Palecek

Abstract

The blood-brain barrier (BBB) comprises brain microvascular endothelial cells (BMECs) that form a high-resistance cellular interface that separates the blood compartment from the brain parenchyma. An intact BBB is pivotal to maintaining brain homeostasis but also impedes the entry of neurotherapeutics. There are limited options for human-specific BBB permeability testing, however. Human pluripotent stem cell models offer a powerful tool for dissecting components of this barrier in vitro, including understanding mechanisms of BBB function, and developing strategies to improve the permeability of molecular and cellular therapeutics targeting the brain. Here, we provide a detailed, step-by-step protocol for differentiation of human pluripotent stem cells (hPSCs) to cells exhibiting key characteristics of BMECs, including paracellular and transcellular transport resistance and transporter function that enable modeling the human BBB.

Key words Human pluripotent stem cells, Blood-brain barrier, In vitro model

1 Introduction

Diseases of the central nervous system (CNS) can be devastating for patients, both in terms of prognosis and quality of life [1]. However, novel CNS therapeutics are much less likely to make it to the clinic than those drugs developed for other indications due to a host of factors that include the presence of the blood-brain barrier (BBB) [1–3]. Delivering therapeutic doses across the BBB is thought to hamper over 98% of all new small molecules, and most current antibody therapies [3]. For instance, BMECs express polarized efflux transporters that semi-selectively pump lipophilic compounds back into the bloodstream, thereby restricting transcellular transport across the BBB [4, 5]. BMECs also express transmembrane proteins that mediate tight junctions between the cells which resist paracellular transport [4, 6]. A better understanding of this

Yu-Wen Alvin Huang and ChangHui Pak (eds.), *Stem Cell-Based Neural Model Systems for Brain Disorders*, Methods in Molecular Biology, vol. 2683, https://doi.org/10.1007/978-1-0716-3287-1_10,

BMEC transport interface is vital to designing approaches to deliver neurotherapeutics across the BBB.

Researchers have developed several cell-based systems to model the BBB. The most direct method is to study molecular penetration and BBB behavior in a live animal [7–10]. Animal models capture the physiological aspects of the BBB, but there are known differences in BBB molecular and functional phenotypes, including differential expression of efflux transporters [11, 12], between humans and experimental animals. Animal models are low throughput, limiting their applications. By contrast, in vitro models consisting of primary BMECs can offer a more controlled and scalable platform for evaluating compound transport through the BBB, with human cells offering human relevance. However, these models suffer from BMEC dedifferentiation and poor paracellular permeability after culture ex vivo [7, 10]. Immortalized human BMECs provide a highly scalable model for BBB research with some barrier and efflux protein expression, but do not typically form substantial barriers amenable to drug permeability screening [10].

The discovery of human pluripotent stem cells (hPSCs) [13] has opened the possibility of developing a human BBB model for drug screening and permeability studies. Several BBB models containing hPSC-derived BMEC-like cells that demonstrate high paracellular and functional transcellular resistance have been reported [14–22]. Like all models, these cells do not capture all aspects of the BBB, for example, with regards to endothelial gene and protein expression [23–26]. However, these hPSC-derived BMEC-like cells have been shown to have continuous junctional staining for various tight junction related proteins, and expression and activity of several BBB efflux transporters such as P glycoprotein (P-gp) and Breast Cancer Resistance Protein (BCRP). Hence, these cells models have been used to predict permeability of various small molecules and blood-borne components [19, 27, 28].

In addition to predicting molecular permeability, in vitro BBB models can be used to study the effects of disease on BBB structure and function. Induced PSCs (iPSCs) reprogrammed from differentiated cell types in adults can be used to study the impact of genetic modifications in some diseases. For example, researchers have explored the effects of genetic defects known to cause neurodegenerative diseases such as Allan Herndon Dudley Syndrome, Alzheimer's disease, and Huntington's diseases on BBB properties using patient-derived iPSC sources differentiated to BMEC-like cells [19, 29, 30].

Below, we detail a defined protocol for differentiation of hPSC-derived BMEC-like cells previously reported [31] (Fig. 1a). The differentiation process applies CHIR99021, a GSK-3 inhibitor used to activate Wnt signaling, as well as supplements and signaling factors like B-27, retinoic acid, and FGF2 to further induce the observed BMEC-like phenotypes. Importantly, like any hPSC differentiation, some

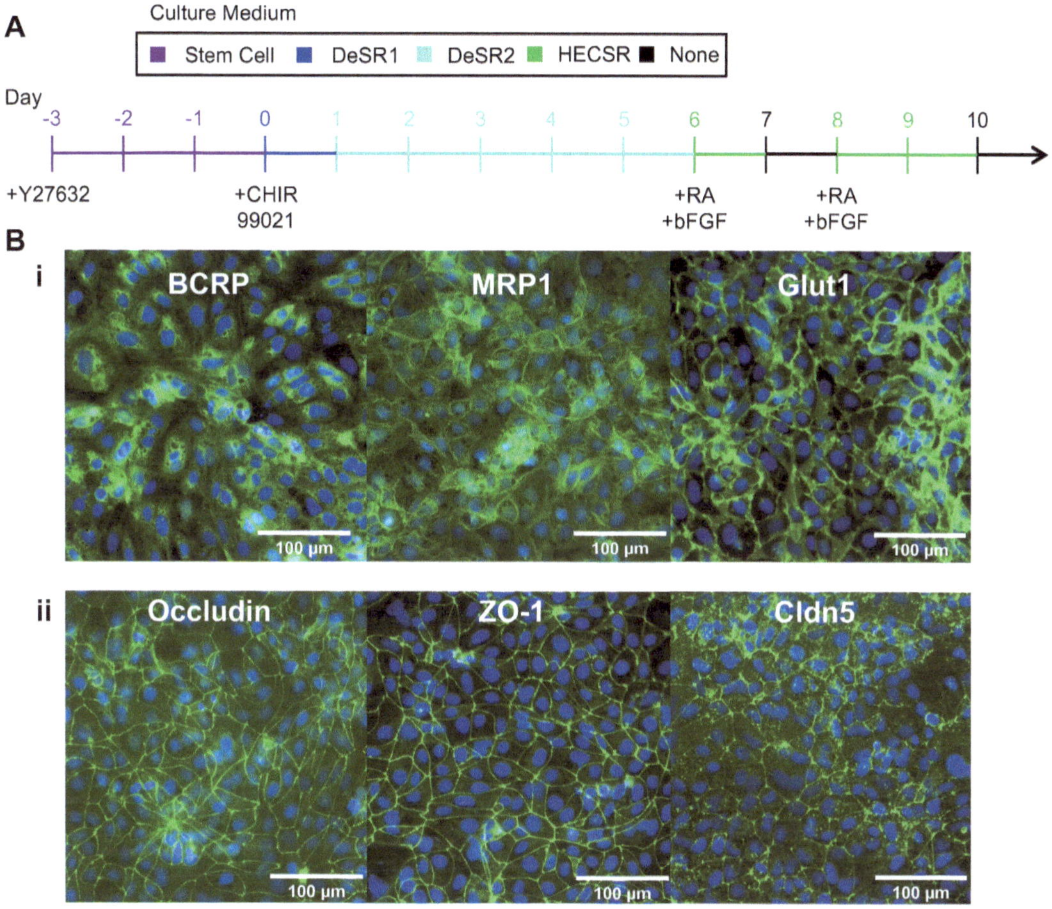

Fig. 1 Scheme for hPSC-derived BMEC-like cell differentiation and BBB marker expression. (**a**) Schematic of differentiation process of hPSCs to BMEC-like cells, color coded by medium composition. Timing and small molecule supplements are provided underneath. (**b**) Example immunofluorescence images of BMEC-like cells stained on day 10 of differentiation using IMR90-4 iPSCs. (i) Sample transporters. (ii) Sample tight junction associated proteins. Target proteins are pseudocolored green and cell nuclei are blue (Hoescht)

differentiation parameters, such as induction factor concentration, initial seeding density, subculture seeding density, and line-to-line variability [31–35] can impact the differentiation process, changing the identity and quality of the resulting cells. The sensitivity to key differentiation variables in this BMEC-like cell differentiation protocol can be observed in the included sample data.

2 Materials

Prepare all materials, unless stated otherwise, using cell culture grade sterile water and reagents in a sterile laminar flow hood with proper aseptic technique. Differentiations will require a suitable hPSC line. This method has previously been used in our lab with IMR90-4 iPSCs and H9 human embryonic stem cells (hESCs), both obtained from WiCell.

2.1 Small Molecule and Protein Aliquots

1. Y-27632 Aliquots—Y-27632 is typically provided as a sterile powder. Add water to dilute to produce a 10 mM solution. Y-27632 should solubilize quickly in solution. This is 1000× the needed concentration in the application described here. Aliquots should be 50–100 μL, dependent on rate of use. Store at −20 °C, with one working aliquot kept at 4 °C for up to 2 weeks.
2. CHIR99021 Aliquots—CHIR99021 is provided as a powder. Suspend in dimethyl sulfoxide (DMSO) to 10 mM in aliquots of 20–50 μL, dependent on use. Store aliquots at −80 °C and one working stock at 4 °C for up to 1 week.
3. Retinoic Acid Aliquots—Retinoic acid (RA) is provided as a powder. Suspend in DMSO to 10 mM and aliquots to between 50 and 200 μL, dependent on use. Store aliquots away from light at −20 °C and one working stock at 4 °C away from light for up to 1 week.
4. FGF2 Aliquots—FGF2 solution should be diluted according to manufacturer's instructions to 100 mg/mL in aliquots of about 20–50 μL, dependent on use. Store aliquots at −80 °C, avoiding freeze thaw cycles, with one working stock at 4 °C for up to 1 week.

2.2 ECM Components and Plates

1. Tissue culture-treated polystyrene 6-well plates.
2. Matrigel Aliquots—Thaw Matrigel overnight at 4 °C. Chill 1.5 mL tubes on ice. Do not thaw at room temperature. Since Matrigel has batch to batch variability in protein concentration, add sufficient thawed Matrigel solution to reach 2.5 mg per aliquot, which is sufficient to coat 5 6-well plates or about 290 cm^2 of surface area. The concentration of solution will be provided by the manufacturer and the volume is typically close to 250 μL. Once thawed, place the Matrigel on ice, and each aliquot should be stored on ice until transferred to −80 °C storage. Do not store at 4 °C.
3. Acetic Acid Solution (5 mg/mL)—Acetic acid is typically not provided sterile, so the solution should be sterile filtered. Add approximately 100 mL water to the upper chamber of a 0.4 μm sterile filter. Add 1.25 g glacial acetic acid, swirling to mix. Fill with water to the 250 mL line. Store at 4 °C.
4. Collagen Type IV Solution—Cell culture certified collagen type IV is provided as a lyophilized powder. Add sufficient preprepared acetic acid solution to reach 1 mg/mL. Collagen type IV will solubilize slowly, so prepare at least a day in advance and allow to resolubilize overnight at 4 °C. Store at 4 °C when not in use. 5 mg would require 5 mL of acetic acid solution for example. Store at 4 °C for up to 2 weeks.

5. Fibronectin Solution– Purchase at 1 mg/mL or dilute to 1 mg/mL in appropriate solution according to manufacturer's instructions. Store at 4 °C for up to 2 weeks.

2.3 Media and Differentiation Materials

1. Accutase.
2. Dulbecco's Phosphate Buffered Saline (DPBS)—Purchase without calcium or magnesium. Should be sterile and cell culture grade.
3. Dimethyl sulfoxide (DMSO).
4. Fetal Bovine Serum (FBS)—Purchase at 500 mL. Thaw overnight at 4 °C. Prepare ~40 mL aliquots in 50 mL conical tubes. Store at −20 °C. Keep one working stock at 4 °C for up to 2 weeks. If not using 40 mL of FBS in 2 weeks, create smaller 10 mL aliquots in 15 mL conical tubes and store as above.
5. Human Pluripotent Stem Cell Culture Medium—Both E8 and mTeSR1 have been used successfully to expand hPSC lines prior to BMEC-like cell differentiation.
6. DMEM/F12—Either purchase directly or mix 1:1 of Dulbecco's Modified Eagle's Media and F12.
7. DeSR1—Add ~100 mL of DMEM/F12 to the top of a sterile 0.2 μm pore size 500 mL filter. Add 5 mL of MEM Non-Essential Amino Acids (NEAA), 2.5 mL of GlutaMAX, and 3.5 μL of β-mercaptoethanol. Swirl and then fill to 500 mL with DMEM/F12 and filter through a 0.2 μm filter. Store at 4 °C for up to a month.
8. DeSR2—Add ~100 mL of DeSR1 to the top of a sterile 0.2 μm pore size 250 mL filter. Add 5 mL B-27 media supplement. Swirl and fill to 250 mL with DeSR1 and filter. Store at 4 °C for up to 2 weeks.
9. HECSR—Add ~100 mL of HESFM to the top of a sterile 0.2 μm pore size 250 mL filter. Add 5 mL B-27 media supplement. Swirl and fill to 250 mL with HESFM and filter. Store at 4 °C for up to 2 weeks.
10. HECSR +10 μM RA + 20 ng/mL FGF2—Should be made on the day of use. Do not store. If using aliquot concentrations as described in Subheading 2.2, then add 1 μL RA per 1 mL of HECSR and 0.2 μL FGF2 per 1 mL of HECSR.

2.4 Immunocytochemistry

1. PBS—Add 20 L of ultrapure water (~18 MΩ) to a 20+ L container. Add a stir bar, 4 g KCl, 4 g KH_2PO_4, 160 g NaCl, and 43.2 g Na_2HPO_4-$7H_2O$. Stir overnight to dissolve. Aliquot into 1–2 L containers for longer term storage. Keeps indefinitely.
2. Methanol.

3. 16% Paraformaldehyde solution
4. 4% Paraformaldehyde solution (PFA)—Mix 12.5 μL 16% paraformaldehyde with 37.5 μL PBS. Make fresh as needed.
5. 10% Goat Serum—Dilute 1 mL goat serum in 9 mL PBS. Store at 4 °C for up to a week.
6. Hoechst 33342—Purchase at or dilute with water to 10 mg/mL.
7. Dilute Hoechst Solution—Add 10 μL of Hoechst 33342 to 50 mL of PBS. Store protected from light at 4 °C for up to a month.
8. Fluorescent microscope capable of exciting at 488 and 352 nm.

2.5 Accumulation Assay

1. Hank's Balanced Saline Solution—Sterile.
2. Ethanol.
3. Rhodamine 123 (Rh123) Aliquots—Purchase powder, resuspend to 10 mM in ethanol. Aliquots should be 500–1000 μL and stored at −20 °C protected from light.
4. DPBS -As 2.2.
5. Cyclosporine A (CsA)—CsA is provided as a powder. Suspend in DMSO to 10 mM in aliquots of 20–50 μL, dependent on use. Store aliquots at −20 °C.
6. RIPA Buffer.
7. Plate Reader capable of exciting at 485 nm and measuring fluorescence at 530 nm.

2.6 Trans Endothelial Electrical Resistance (TEER)

1. Transwells—Corning 3640 or equivalent with 0.4 μm pore size.
2. EVOM2 or equivalent.
3. Chopstick electrodes.
4. DPBS—Cell culture and sterile without calcium and magnesium.
5. 70% Ethanol—Mix 7 mL of ethanol with 3 mL of cell culture grade sterile water
6. 70% ethanol in a spray bottle for sterilizing—Mix 70% by volume ethanol and fill to 100% volume with deionized water.

3 Methods

Unless stated otherwise, all steps should be performed in a sterile laminar flow hood with proper aseptic technique. All media should be warmed at room temperature at least 30 min before application.

3.1 Differentiation of hPSCs to BMEC-Like Cells

Differentiation of hPSCs to BMEC-like cells [31]. Equations are provided for ease of scaling.

Human pluripotent stem cells should be maintained according to previously established methods with culture at 37 °C and 5% CO_2. We suggest protocols such as Stebbins et al. [17] and maintaining cells on Matrigel-coated tissue culture-treated polystyrene 6-well plates, passaging cells between 50% and 70% confluency. Before starting a differentiation, cells should not be stressed by poor passaging, changes in feeding schedule, exposure to room temperature for prolonged periods, or storage in a non-humidified incubator. The cultures should not exhibit evidence of spontaneous differentiation upon visual inspection. Cells should also be at least 2–3 passages post thawing before initiating a differentiation.

1. A minimum of 1 h before, but up to a week before starting differentiation: Prepare a 50 mL conical tube with ~32 mL of DMEM/F12 (if using the Matrigel aliquot sizes provided in Materials).
2. Remove Matrigel aliquot from freezer and quickly suspend by pipetting 1 mL from the prepared aliquot of DMEM/F12 into the Matrigel aliquot. Pipette up and down until Matrigel is dissolved. Return DMEM/F12 + Matrigel to the DMEM/F12 aliquot.
3. Add 1 mL of DMEM/F12 and Matrigel solution per well of a tissue culture treated polystyrene 6 well plate using a 5 mL stripette. Work quickly.
4. Day −3: When hPSCs are 50–70% confluent, aspirate stem cell culture medium (*see* **Note 1**), then add 1 mL of Accutase per well. Typically, for each plate to be seeded, harvest 1 well; there will be excess cell solution (Fig. 1a).
5. Return the plate to the 37 °C incubator for 5–7 min.
6. While the cells are being incubated with Accutase, prepare a 50 mL conical containing 4 mL of DMEM/F12 per 1 mL of Accutase used to detach the cells.
7. Swirl plate gently by hand. If the cells are not detached, return to the incubator for another 2 min.
8. Once ~70% of the cells are detached, use a 1000 uL pipette to wash the well 1–5 times until cells are completely detached and singularized. Any cells still attached to the plate should have been rinsed off during this process. There should be no visible clumps of cells within the Accutase when done (*see* **Note 2**). Add cell solution to 50 mL conical containing pre-prepared DMEM/F12. Invert 3–4 times to mix before reserving 50–100 μL of cell solution to count.

9. Centrifuge cells at 200 g for 5 min. While doing this, count reserved cell concentration and calculate total cell number according to manufacturer's cell counting protocol.
10. Aspirate supernatant. There should be a visible pellet. Resuspend pellet to two million cells/mL in stem cell culture medium by gently pipetting up and down using a 5 or 10 mL stripette to generate concentrated cell solution.
11. Calculate cells needed per well (*see* **Note 3**)
12. $\frac{\text{Cells}}{\text{Well}} = \text{Optimal Seeding Density}\left(\frac{\text{cells}}{\text{cm}^2}\right) * \text{Surface Area of Well}\left(\text{cm}^2\right)$
13. Prepare a volume of stem cell medium +10 μM Y-27632 sufficient for one plus the number of wells to be seeded (*see* Table 1). Remember to account for required volume of the cell solution.
14. Add cell solution to provide enough cells for total number of wells to be seeded plus one to the prepared stem cell medium plus Y-27632. Invert three times to mix.
15. Quickly fill 3 wells with appropriate volume of the cell suspension. Invert tube to ensure the solution remains well mixed. Repeat until all wells are seeded (*see* **Note 4**).
16. Place plates into 37 °C incubator with 5% CO_2.
17. Shake plates 3 times forward and backward, then 3 times left to right to ensure even distribution of cells (*see* **Note 5**).
18. Day −2: After 24 h, aspirate spent medium, add 2 mL stem cell culture medium without Y-27632 to each well (Fig. 1a).
19. Day −1: Aspirate spent medium, add 2 mL stem cell culture medium to each well.
20. Day 0: Prepare 2 mL for each well of a 6-well plate of DeSR1 containing the optimal concentration of CHIR99021 determined via an optimization assay (Fig. 2b).
21. Aspirate medium, add 2 mL of DeSR1 + CHIR99021 to each well (*see* **Notes 6** and 7).
22. Day 1: After 22.5 to 24 h from **step 21**, aspirate spent medium and add 2 mL DeSR2 to each well (*see* **Notes 8** and **9**).
23. Day 2: Aspirate medium, add 2 mL DeSR2 to each well.
24. Day 3: Aspirate medium, add 2 mL DeSR2 to each well.
25. Day 4: Aspirate medium, add 2 mL DeSR2 to each well.
26. Day 5: Aspirate medium, add 2 mL DeSR2 to each well.
27. Day 6: Prepare a conical with sufficient HECSR +10 μM RA + 20 ng/mL FGF2.
28. Day 6: Aspirate spent medium, add 2 mL prepared HECSR +10 μM RA + 20 ng/mL FGF2 to each well (*see* **Note 10**).

Table 1
Working media volumes and coating solution volumes for wells by approximate size

Plate	Volume (mL/well)	Coating solution volume (uL/well)
6-well plate	2	1000
12-well plate	1	500
24-well plate	0.5	250
48-well plate	0.25	125
96-well plate	0.2	50
12 Well with Transwell	0.5 Apical, 1.5 Basolateral	200 (Concentrated)

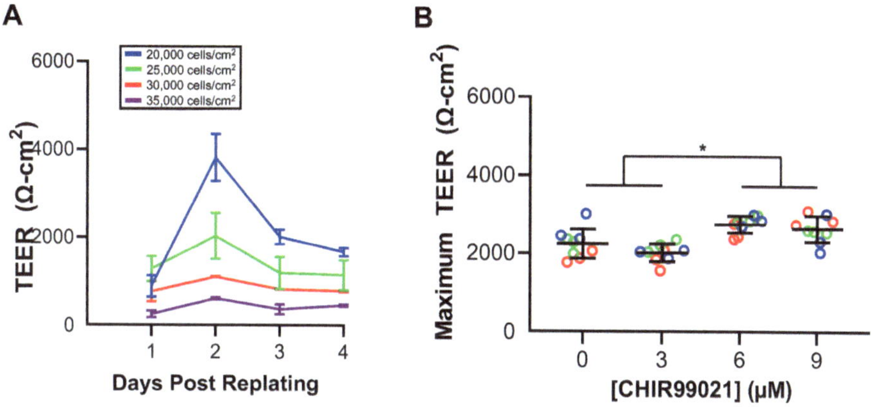

Fig. 2 Effects of culture time and CHIR concentration on BMEC-like cell barrier properties. (**a**) An example day −3 hPSC seeding density optimization of TEER. Performed with IMR90-4 iPSCs with n = 3 biological eplicates. (**b**) Maximum TEER between day 9 and day 11 from 3 BMEC-like cell differentiations of IMR90-C4 iPSCs treated with 0–9 uM CHIR99021 at day 0 for 24 h. Black line is mean with standard deviation. Circles represent biological replicates ($n = 3$), colors represent independent differentiations ($n = 3$). *$p < 0.05$ via 2 Way ANOVA with Tukey's post hoc test

29. Day 7: Prepare collagen IV and fibronectin-coated wells or Transwells. For Transwells use 100 μg/mL fibronectin and 400 μg/mL collagen IV diluted with water. For flat plates use a solution of 20 μg/mL fibronectin and 80 μg/mL collagen IV. *See* Table 1 for volumes. Incubate overnight at 37 °C and 5% CO_2. Cell medium does not need to be changed on day 7.
30. Day 8: Aspirate medium and gently wash cells with DPBS.
31. Add 1 mL Accutase per well, incubate for 20–60 min. Check every 10–15 min by gently swirling; if cells do not detach continue incubation (*see* **Note 11**).
32. During the incubation, prepare 4 mL DMEM/F12 per 1 mL of Accutase used to quench the Accutase.
33. Prepare HECSR +10 μM retinoic acid +20 ng/mL FGF2 sufficient for all wells in the experiment, and to resuspend the

cells. Typical yield is close to two to four million cells per 10 cm^2. Prepare more than needed.

34. Once 60 min of Accutase treatment have elapsed or cells are visibly detaching from the well, whichever comes first, rinse cells from each well using a 1000 μL pipette tip, triturating to singularize the cells. Filter solution through a 40 μm sterile filter to remove cell clumps into the prepared DMEM/F12 quench. Collect ~100 μL of cell solution to count.
35. Centrifuge at 200 g for 5 min. While doing this, count reserved cell concentration and calculate total number of cells according to manufacturer's cell counting protocol.
36. Aspirate supernatant, being careful not to aspirate the cell pellet. Resuspend cells to two million cells/mL.
37. Aspirate the remaining collagen and fibronectin solution from prepared plates. Seed Transwells at 500,000 cells/cm^2 (*see* **Note 12**) and plates at 100,000 cells/cm^2.
38. Place plates or Transwells into 37 °C incubator with 5% CO_2.
39. Shake plates or Transwells 3 times forward and backward, then 3 times left to right. Repeat 3 times to ensure even distribution of cells.
40. Day 9: Aspirate medium, add appropriate volume of HECSR per well. This is generally the last medium change; cells can be maintained for up to at least 6 days afterward (Fig. 3).
41. Day 10: Most cell function assays are performed on this day. A small number of 96 well seeded cells should be fixed and

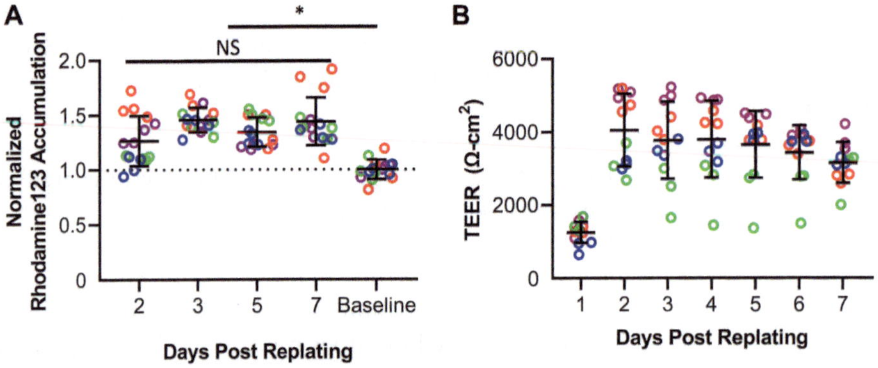

Fig. 3 Pgp activity and TEER as a function of duration post replating: H9 hESCs were differentiated to BMEC-like cells (*see* **Note 18**). (**a**) BMEC-like cells were replated onto collagen IV and fibronectin-coated 24-well plates (Rh123 Accumulation Assay) or Transwells (TEER) on day 8 and Pgp efflux transporter activity measured by Rh123 accumulation. Statistically significant increases above baseline of accumulation under CsA inhibition demonstrate continued activity through 7 days post replating (day 15 after initiation of differentiation). (**b**) BMEC-like cells were replated onto Transwell membranes on day 8 and TEER monitored daily. TEER measurements show maintained tight junction integrity through 7 days post replating (day 15 after initiation of differentiation). Black line is mean with standard deviation. Circles represent biological replicates ($n = 3$), colors represent independent differentiations ($n = 3$). *$p < 0.05$ via 2 Way ANOVA with Tukey's post hoc test

prepared for immunocytochemistry at this time point to ensure appropriate expression of proteins relevant to the properties being assessed. Examples of efflux transporters and tight junction associated proteins have been provided (Fig. 1b) (*see* **Note 13**).

3.2 Immunocytochemical Analysis

Unless stated otherwise, all steps should be performed in a sterile laminar flow hood with proper aseptic technique. Staining protocols are adapted from Stebbins et al. and Qian et al. [17, 31] and example images are provided in Fig. 1b. Typically, once a differentiation protocol is optimized, proper expression and localization of proteins of interest are assessed as quality control to ensure appropriate marker expression. For example, if assessing permeability of a hydrophilic molecule, continuous tight junction protein expression is important to validate. Table 1 lists several proteins suggested as potential quality controls along with the antibodies we have used previously.

1. Perform BMEC-like cell differentiation as described in Subheading 3.1. On day 8, cells should be seeded at 100,000 cells/cm^2 or higher to ensure a monolayer into a wells of a collagen IV and fibronectin-coated 96-well plate. (*see* **Note 14**).
2. Store 100% methanol at −20 °C or allow to chill for at least 1 h at −20 °C.
3. Day 10: Aspirate medium from each well, then wash 3 times with ~200 μL of DPBS per well.
4. Aspirate DPBS, add 50 μL methanol or 4% PFA (antibody-dependent, *see* Table 1) per well to fix the cells.
5. Incubate for 15 min at room temperature.
6. Aspirate methanol or PFA, then wash each well once with ~200 μL PBS (can switch to non-sterile PBS at this stage). Cells no longer need to be kept sterile and the remaining steps can be performed on a bench top.
7. Aspirate PBS. Add 100 μL 10% goat serum per well to block.
8. Incubate 1 h at room temperature or overnight at 4 °C.
9. During goat serum incubation, prepare primary antibody solutions (*See* Table 2 for concentrations and solutions) with 50 μL per well.
10. Aspirate blocking goat serum. Add 50 μL primary antibody solution to each well.
11. Incubate overnight at 4 °C.
12. Prepare 50 μL per well of species appropriate secondary antibody solution in 10% goat serum. Protect from light. *See* Table 2 for concentrations.

Table 2
Antibody information

Protein target	Catalog number	Manufacturer	Fixative	Concentration	Species
BCRP	MAB4155	Sigma-Aldrich	4% PFA	1:50	Mouse
MRP1	MAB4100	Sigma-Aldrich	Methanol	1:50	Mouse
Glut1	MA1-37783	Invitrogen	Methanol	1:100	Mouse
Occludin	33-1500	Invitrogen	Methanol	1:50	Mouse
Claudin5	35-2500	Invitrogen	Methanol	1:200	Mouse
ZO1	33-9100	Invitrogen	Methanol	1:200	Mouse
Mouse IgG(H + L) 488	A11001	Invitrogen	N/A	1:200	Goat

13. Aspirate primary antibody solution. Wash 3 times with ~200 μL PBS per well (*see* **Note 15**).
14. Add 50 μL per well of appropriate secondary solution.
15. Incubate, protected from light, for 1 h at room temperature or overnight at 4 °C.
16. Aspirate secondary antibody solution. Wash 2 times with ~200 μL PBS per well.
17. Aspirate PBS. Add 50 μL per well of dilute Hoechst solution. Cover and incubate for 10–20 min at room temperature.
18. Aspirate dilute Hoechst solution. Add 200 μL PBS to each well.
19. Image on a fluorescence microscope with appropriate excitation and emission filters according to secondary antibodies used.

3.3 Rhodamine 123 Accumulation Assay

Unless stated otherwise, all steps should be performed in a sterile laminar flow hood with proper aseptic technique. This protocol is adapted from Stebbins et al. [17]. This assay can be used to assess the relative activity of P-gp in the BMEC-like cells by comparing accumulation of a fluorescent substrate of P-gp, Rh123, in cells with and without P-gp inhibition. The fold change represents both the activity and the Typically, fold increases of 1.5–2.0 of CsA treated cells over control have been observed [14, 17, 31, 32].

1. Perform BMEC-like cell differentiation as described in Subheading 3.1. On day 8, BMEC-like cells should be seeded at 100,000 cells/cm^2 or higher onto a collagen IV and fibronectin-coated 24-well plate to obtain a confluent monolayer. Experiment should be performed at $n = 4$–6. The assay should include a vehicle control (DMSO), an inhibited P-gp condition (CsA), and a single blank well.

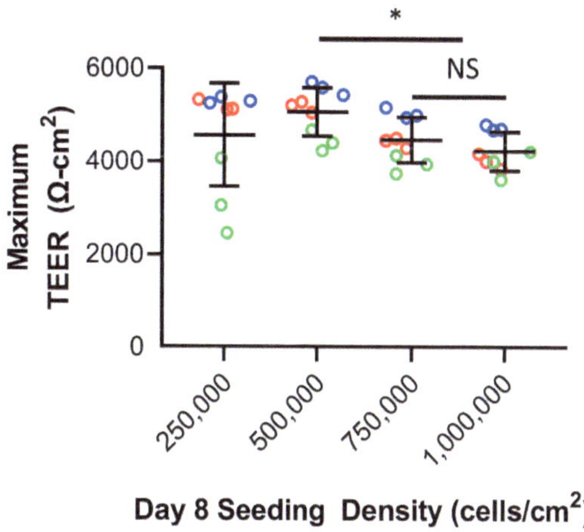

Fig. 4 Impact of BMEC-like cell Transwell seeding density on TEER. The differentiation was performed in H9 hESCs (*see* **Note 18**). On day 8, BMEC-like cells were seeded onto collagen IV and fibronectin-coated Transwells at the indicated cell density, and TEER measured every day for 4 days. Black line is mean with standard deviation. Circles represent biological replicates ($n = 3$), colors represent independent differentiations ($n = 3$). *$p < 0.05$ via 2 Way ANOVA with Tukey's post hoc test

2. On days 10–15 (*See* Fig. 4): Add 0.5 mL of HBSS per well of vehicle control +1 for the blank to a 15 mL conical. Add 1 μL DMSO per mL of HBSS. Invert 3 times to mix.
3. In another 15 mL conical tube, add 0.5 mL of HBSS per well to be treated with CsA. Add 1 μL CsA per mL of HBSS. Invert 3 times to mix. A white particulate may be visible initially but should dissolve.
4. Aspirate medium and wash each well twice with 1 mL HBSS. Add 0.5 mL of HBSS + DMSO to half the wells and HBSS + CsA to the remainder.
5. Incubate at 37 °C and 5% CO_2 for 1 h on an orbital shaker at 30 rpm.
6. While incubating, prepare HBSS +10 μM Rh123 at 0.5 mL per well. Dilution should be 1 μL of Rh123 solution per mL of HBSS if using the aliquot concentrations in Materials.
7. Split HBSS + Rh123 into two 15 mL conical tubes. One for DMSO, the other for CsA.
8. Add 1 μL per mL DMSO or CsA to the conical tubes containing HBSS + Rh123.
9. After 1 h, aspirate HBSS + DMSO/CsA, add 0.5 mL of HBSS + Rh123 + DMSO/CsA and return to shaker and incubator. Do not aspirate or change the contents of the blank control well.

10. Incubate at 37 °C and 5% CO_2 for 2 h on an orbital shaker at 30 rpm.
11. Place DPBS at 4 °C to chill.
12. After 2 h, aspirate Rh123 containing HBSS carefully. Be sure to not leave small droplets on the sides of the wells.
13. Wash each well 3 times with 1 mL chilled DPBS.
14. Aspirate DPBS. Add 200 μL RIPA buffer to each well.
15. Shake at 30 rpm for 10 min to completely lyse cells.
16. Scan each well on a plate reader with 485 excitation and 530 emission filters.
17. Freeze cell lysates at −20 °C this stage by placing the entire plate in the freezer. Alternatively, proceed to BCA assay.
18. Perform BCA assay according to manufacturer's instructions on lysate to allow proper normalization.
19. Normalize each well (i) according to the following equation:
20. $F = \frac{(Rh_i - Rh_{Blank})}{P_i}$
21. Where Rh_i is Rhodamine fluorescence from well i, P is protein concentration according to the standard curve according to the manufacturer's instructions in the BCA kit in well i, Rh_{Blank} is Rhodamine fluorescence from the blank well.
22. Normalize results to the average of all fluorescent values for the control (DMSO) wells. This will provide the fold changes typically reported.

3.4 Trans Endothelial Electrical Resistance (TEER)

All steps should be performed in a sterile laminar flow hood with proper aseptic technique. This protocol is adapted from Stebbins et al. [17]. This assay measures the resistance across a BMEC-like cell monolayer and provides an assessment of the paracellular resistance of the model. TEER measurement is nondestructive and is an important quality control assay prior to measuring molecular permeability or when interested in passage of hydrophilic molecules. Typically the BMEC-like cell monolayer should have a barrier above 1000 Ω-cm^2 [27].

1. On day 7 of the differentiation of BMEC-like cells, as described in Subheading 3.1: Coat Transwells with collagen IV and fibronectin. Prepare an additional unseeded well to determine baseline resistance of the Transwell membrane (*see* **Note 16**).
2. On day 8 of the differentiation: Seed Transwells at 500,000–1,000,000 cells/cm^2 as described in Subheading 3.1. Do not seed the unseeded control well.
3. Day 9+: Add ~10 mL of 70% ethanol to a 15 mL conical tube.
4. Place the chopstick probe in 70% ethanol for 5 min (*see* **Note 17**).

5. Prepare a second 15 mL conical containing ~15 mL DPBS.
6. Immediately before measuring TEER, remove Transwell plates from the incubator. Minimize duration that the cells are outside the incubator. Lower temperatures can cause erroneously high resistances.
7. Grasp the probe firmly with index finger and thumb.
8. Wrap remaining wire around middle and ring fingers (to avoid loose cable dragging over or touching the wells).
9. Spray 70% ethanol onto the wire.
10. Gently dip the tip of the probe into the DPBS until the TEER reading drops and stabilizes.
11. Insert probe into Transwell being careful to insert the long side into the basolateral chamber. Do not touch the cell monolayer with the probe. Contact with the monolayer will disrupt the barrier.
12. Wait until the TEER measurement is approximately steady for 30 s. There is always some volatility in the reading.
13. Record TEER.
14. Remove the probe from the well, taking care to not touch the cell monolayer.
15. Dip the probe gently in the conical containing 70% ethanol for 5 s.
16. Repeat **steps 9–12** until all wells are measured. The ethanol dips help to prevent contamination. The DPBS dips prevent ethanol from contacting the cells.
17. When all wells are measured, dip the probe in DPBS and then submerge in the 70% ethanol for 2 min.
18. Allow the probe to air dry completely before returning to storage.

3.5 BMEC-Like Cell Cryopreservation

Adapted from methods in Grifno et al. and Wilson et al. [28, 33]. Unless stated otherwise, all steps should be performed in a sterile laminar flow hood with proper aseptic technique.

1. Differentiate BMEC-like cells as described in Subheading 3.1 up through **step 29** (day 7). Do not coat plates on day 7.
2. Ensure freezing container is at room temperature and has sufficient isopropanol in the lower chamber, according to manufacturer's instructions.
3. Day 8: Aspirate medium and wash each well with 2 mL DPBS.
4. Add 1 mL Accutase per well, then incubate for 20–60 min at 37 °C at 5% CO_2. Check every 10–15 min by gently swirling. If cells do not start to detach, continue incubation (*see* **Note 11**).

5. During the incubation, prepare 4 mL DMEM/F12 quench solution per 1 mL of Accutase used to detach the cells.
6. Prepare freezing medium consisting of 10% DMSO +30% FBS + 60% HECSR volumetrically. Typically 10–20 mL is sufficient for a differentiation. Volume required will vary depending on final cell concentration. Typically, prepare enough to resuspend to the chosen concentration for approximately four million cells per 6-well plate.
7. Once 60 min of Accutase treatment have elapsed or cells are visibly detaching from the well, whichever comes first, rinse cells from each well using a 1000 μL pipette tip, triturating to singularize the cells. Filter the cell solution through a 40 μm sterile filter to remove cell clumps into the prepared DMEM/F12 quench. Reserve ~100 μL of cell solution to count.
8. Centrifuge at 200 g for 5 min. While doing this, count reserved cell concentration and calculate total cell number according to manufacturer's cell counting protocol.
9. Label an appropriate number of cryotubes. Typically BMEC-like cells, differentiation number, hPSC line, technician, number of cells per tube, and freezing date are recorded.
10. Aspirate supernatant, being careful not to aspirate the cell pellet.
11. Resuspend to one to ten million cells/mL, dependent on planned application of the cells. Pipette 1 mL into each labeled tube. Lower concentrations are more useful for plate assays, while higher concentrations are more convenient for seeding Transwells and other high cell concentration applications.
12. Place tubes into a room temperature freezing container.
13. Place freezing container into −80 °C overnight.
14. Transfer tubes to liquid nitrogen tank for long-term storage.

3.6 BMEC-Like Cell Thawing

Adapted from methods in Grifno et al. and Wilson et al. [28, 33]. Unless stated otherwise, all steps should be performed in a sterile laminar flow hood with proper aseptic technique. Cells can be expected to have similar barrier and protein expression properties following thaw as if they had not been frozen.

1. One day before thawing, prepare collagen IV and fibronectin-coated plates and Transwells as though it was **step 29** of Subheading 3.1 (day 7). Seed Transwells at 500,000 cells/cm^2 (*see* **Note 12**) and plates at 100,000 cells/cm^2.
2. Prepare sufficient HECSR +10 μM retinoic acid +20 ng/mL FGF2 for plates to be seeded. Volume is dependent on planned downstream experiments. Then add 10 μM Y-27632.
3. Retrieve vials of frozen cells from liquid nitrogen storage.

4. Thaw rapidly by holding or floating in a 37 °C water bath. Do not allow the cap to touch the water.
5. When only a small amount of ice remains in the tube, transfer the 1 mL of cell solution in the vial to a 15 mL conical tube with a 1000 μL pipetter.
6. Add 4 mL of DMEM/F12 to the conical containing the cell solution at a rate of about 2–4 mL/min.
7. Centrifuge at 200 g for 5 min.
8. Aspirate the supernatant, being careful to not disrupt the cell pellet.
9. Resuspend the cell pellet in prepared HECSR +10 μM retinoic acid +20 ng/mL FGF2 + 10 μM Y-27632 to two million cells/mL.
10. Add cell suspension to collagen and fibronectin-coated plates at 100,000 cells/cm^2. Seed Transwells at 500,000 cells/cm^2 (*see* **Note 12**) and plates at 100,000 cells/cm^2.
11. Place plates or Transwells into 37 °C incubator with 5% CO_2.
12. Shake plates or Transwells 3 times forward and backward, then 3 times left to right. Repeat 3 times to ensure even distribution of cells.
13. One day later, resume Subheading 3.1 (BMEC-like cell differentiation) at **step 40**.

4 Notes

1. Our lab has used both mTeSR1 and E8 hPSC maintenance media to expand hPSCs prior to BMEC-like cell differentiation.
2. Do not pipette too vigorously. Excess shear stress can damage the cells. Do not allow bubbles to form while triturating. This can reduce cell viability.
3. Optimal seeding density must be determined for each hPSC line via a screen. Typically a range of around 20,000–60,000 cells per cm^2 with increments of 10,000 cells/cm^2 will identify the optimal density. Typically TEER (Fig. 3) is used to screen for barrier properties in BMEC-like cells. Once the optimal density is found, further BMEC-like cell differentiations can be performed at the optimal seeding density to further ensure properties of interest (experiment dependent), such as protein expression or efflux activity, are present. *See* Fig. 2a for an example seeding screen focusing on optimizing TEER in hPSC-derived BMEC-like cells.

4. Cells will settle over time, so it is important to work quickly and mix frequently to ensure consistent seeding across wells.
5. Shaking is necessary to ensure an even distribution of cells across the plate. After cells are in the incubator, it is best to disturb the plate as little as possible. Heterogenous cell density can result in poor or failed differentiations and high batch-to-batch variability.
6. CHIR99021 concentration at day 0 should be optimized for each hPSC line used. Values typically fall around 3–6 μM. We suggest testing a range from 3–9 μM CHIR99021 at increments of 3 μM. TEER is used to screen barrier integrity in the BMEC-like cells after differentiation (Fig. 3b). Once an optimal CHIR99021 concentration is identified, subsequent experiments can be performed at the optimal CHIR99021 concentration. *See* Fig. 2b for an example of varying CHIR99021 concentration to optimize TEER in hPSC-derived BMEC-like cells.
7. Extended CHIR99021 exposure can result in toxicity. Do not expose the cells to medium containing CHIR99021 for longer than 24 h. If cells are not confluent by day 2–3, the differentiation will not produce BMEC-like cells with appreciable barrier properties.
8. Typically a medium change during BMEC-like cell differentiation is performed every 24 h (plus or minus 1 h).
9. On day 1, substantial cell death is to be expected. If noticeable cell death is visible on or after day 3, the differentiation will likely fail to produce BMEC-like cells with appreciable barrier properties.
10. HECSR can be made in advance, but RA and FGF2 should be added at the time of use.
11. If detachment takes less than 15 min, the resulting cells typically have poor barrier properties. After detachment, over four million cells per well also suggests a poor quality differentiation.
12. Previously published hPSC-derived BMEC-like cell differentiation protocols have typically used one million cells/cm^2 to ensure successful formation of a confluent monolayer [15–17, 31]. Testing has demonstrated that as low as 250,000 cells/cm^2 may be sufficient for confluent monolayer formation (Fig. 4).
13. A similar hPSC-derived BMEC-like cell differentiation protocol [20] has been shown to generate BMEC-like cells that maintain tight junction integrity for 1 week. Similarly, confluent monolayers of BMEC-like cells differentiated using the protocol described here retain TEER above 1000 Ω*cm^2 for approximately 1 week without changing medium (Fig. 3b),

and efflux transport similarly appears unaffected over that time-frame (Fig. 3a).

14. If BMEC-like cells are not confluent by day 10, tight junction visualization and localization will be especially difficult to assess.
15. This wash step is critical. If secondary antibodies are applied without appropriate washing, a low signal or nonspecific background signal could result.
16. TEER can be measured starting from day 9 onward if the BMEC-like cells form a confluent monolayer.
17. Electrodes are vulnerable to ethanol and prolonged exposure decreases the probe longevity. Do not exceed 15 min of treatment with ethanol.
18. This data is from an H9 hESC line which was found to contain a paracentric inversion in the short arm of chromosome 11.

Acknowledgments

This work was supported by NIH grants NS107461 and HL154254, and NSF grant 1703219.

References

1. Miller G (2010) Is pharma running out of brainy ideas? Science 329:502–504. https://doi.org/10.1126/science.329.5991.502
2. Gribkoff VK, Kaczmarek LK (2017) The need for new approaches in CNS drug discovery: why drugs have failed, and what can be done to improve outcomes. Neuropharmacology 120:11–19. https://doi.org/10.1016/j.neuropharm.2016.03.021
3. Pardridge WM (2005) The blood brain barrier: bottleneck in brain drug development. NeuroRx 2:12. https://doi.org/10.1602/neurorx.2.2.3
4. Abbott NJ, Patabendige AAK, Dolman DEM, Yusof SR, Begley DJ (2010) Structure and function of the blood–brain barrier. Neurobiol Dis 37:13–25. https://doi.org/10.1016/j.nbd.2009.07.030
5. de Boer AG, van der Sandt ICJ, Gaillard PJ (2003) The role of drug transporters at the blood-brain barrier. Annu Rev Pharmacol Toxicol 43:629–656. https://doi.org/10.1146/annurev.pharmtox.43.100901.140204
6. Jamieson JJ, Searson PC, Gerecht S (2017) Engineering the human blood-brain barrier in vitro. J Biol Eng 11:37. https://doi.org/10.1186/s13036-017-0076-1
7. Sabbagh MF, Nathans J (2020) A genome-wide view of the de-differentiation of central nervous system endothelial cells in culture. elife 9:e51276. https://doi.org/10.7554/eLife.51276
8. Vanlandewijck M, He L, Mäe MA, Andrae J, Ando K, Del Gaudio F, Nahar K, Lebouvier T, Laviña B, Gouveia L, Sun Y, Raschperger E, Räsänen M, Zarb Y, Mochizuki N, Keller A, Lendahl U, Betsholtz C (2018) A molecular atlas of cell types and zonation in the brain vasculature. Nature 554:475–480. https://doi.org/10.1038/nature25739
9. Nitta T, Hata M, Gotoh S, Seo Y, Sasaki H, Hashimoto N, Furuse M, Tsukita S (2003) Size-selective loosening of the blood-brain barrier in claudin-5–deficient mice. J Cell Biol 161:653–660. https://doi.org/10.1083/jcb.200302070
10. Helms HC, Abbott NJ, Burek M, Cecchelli R, Couraud P-O, Deli MA, Förster C, Galla HJ, Romero IA, Shusta EV, Stebbins MJ, Vandenhaute E, Weksler B, Brodin B (2016)

In vitro models of the blood–brain barrier: an overview of commonly used brain endothelial cell culture models and guidelines for their use. J Cereb Blood Flow Metab 36:862–890. https://doi.org/10.1177/0271678X16630991

11. Aday S, Cecchelli R, Hallier-Vanuxeem D, Dehouck MP, Ferreira L (2016) Stem cell-based human blood–brain barrier models for drug discovery and delivery. Trends Biotechnol 34:382–393. https://doi.org/10.1016/j.tibtech.2016.01.001
12. Song HW, Foreman KL, Gastfriend BD, Kuo JS, Palecek SP, Shusta EV (2020) Transcriptomic comparison of human and mouse brain microvessels. Sci Rep 10:12358. https://doi.org/10.1038/s41598-020-69096-7
13. Chagastelles PC, Nardi NB (2011) Biology of stem cells: an overview. Kidney Int Suppl 1:63–67. https://doi.org/10.1038/kisup.2011.15
14. Lippmann ES, Al-Ahmad A, Azarin SM, Palecek SP, Shusta EV (2014) A retinoic acid-enhanced, multicellular human blood-brain barrier model derived from stem cell sources. Sci Rep 4:4160. https://doi.org/10.1038/srep04160
15. Lippmann ES, Azarin SM, Kay JE, Nessler RA, Wilson HK, Al-Ahmad A, Palecek SP, Shusta EV (2012) Derivation of blood-brain barrier endothelial cells from human pluripotent stem cells. Nat Biotechnol 30:783–791. https://doi.org/10.1038/nbt.2247
16. Lippmann ES, Al-Ahmad A, Palecek SP, Shusta EV (2013) Modeling the blood–brain barrier using stem cell sources. Fluids Barriers CNS 10:2. https://doi.org/10.1186/2045-8118-10-2
17. Stebbins MJ, Wilson HK, Canfield SG, Qian T, Palecek SP, Shusta EV (2016) Differentiation and characterization of human pluripotent stem cell-derived brain microvascular endothelial cells. Methods 101:93–102. https://doi.org/10.1016/j.ymeth.2015.10.016
18. Canfield SG, Stebbins MJ, Morales BS, Asai SW, Vatine GD, Svendsen CN, Palecek SP, Shusta EV (2017) An isogenic blood-brain barrier model comprising brain endothelial cells, astrocytes, and neurons derived from human induced pluripotent stem cells. J Neurochem 140:874–888. https://doi.org/10.1111/jnc.13923
19. Vatine GD, Barrile R, Workman MJ, Sances S, Barriga BK, Rahnama M, Barthakur S, Kasendra M, Lucchesi C, Kerns J, Wen N, Spivia WR, Chen Z, Van Eyk J, Svendsen CN (2019) Human iPSC-derived blood-brain barrier chips enable disease modeling and personalized medicine applications. Cell Stem Cell 24:995–1005.e6. https://doi.org/10.1016/j.stem.2019.05.011
20. Neal EH, Marinelli NA, Shi Y, McClatchey PM, Balotin KM, Gullett DR, Hagerla KA, Bowman AB, Ess KC, Wikswo JP, Lippmann ES (2019) A simplified, fully defined differentiation scheme for producing blood-brain barrier endothelial cells from human iPSCs. Stem Cell Rep 12:1380–1388. https://doi.org/10.1016/j.stemcr.2019.05.008
21. Faley SL, Neal EH, Wang JX, Bosworth AM, Weber CM, Balotin KM, Lippmann ES, Bellan LM (2019) iPSC-derived brain endothelium exhibits stable, long-term barrier function in perfused hydrogel scaffolds. Stem Cell Rep 12:474–487. https://doi.org/10.1016/j.stemcr.2019.01.009
22. Hollmann EK, Bailey AK, Potharazu AV, Neely MD, Bowman AB, Lippmann ES (2017) Accelerated differentiation of human induced pluripotent stem cells to blood–brain barrier endothelial cells. Fluids Barriers CNS 14:9. https://doi.org/10.1186/s12987-017-0059-0
23. Workman MJ, Svendsen CV (2020) Recent advances in human iPSC-derived models of the blood–brain barrier. Fluids Barriers CNS 17(1):30. https://doi.org/10.1186/s12987-020-00191-7
24. Lu TM, Houghton S, Tarig M, Durán JGB, Minotti A, Snead A, Sproul A, Nguyen DT, Xiang J, Fine HA, Rosenwaks Z, Studer L, Rafii S, Agalliu D, Redmond D, Lis R (2021) Pluripotent stem cell-derived epithelium misidentified as brain microvascular endothelium requires ETS factors to acquire vascular fate. Significance PNAS 118(8):e2016950118. https://doi.org/10.1073/pnas.2016950118
25. Delsing L, Donnes P, Sanchez J, Clausen M, Voulgaris D, Falk A, Herland A, Brolen G, Zetterberg H, Hicks R, Synnergren J (2018) Barrier properties and transcriptome expression in human iPSC-derived models of the blood–brain barrier. Stem Cells 36(12):1816–1827. https://doi.org/10.1002/stem.2908
26. Lippmann ES, Samira AM, Palecek SP, Shusta EV (2020) Commentary on human pluripotent stem cell-based blood–brain barrier models. Fluids Barriers CNS 17(1):64. https://doi.org/10.1186/s12987-020-00222-3
27. Mantle JL, Min L, Lee KH (2016) Minimum transendothelial electrical resistance thresholds for the study of small and large molecule drug transport in a human *in vitro* blood–brain barrier model. Mol Pharm 13:4191–4198.

https://doi.org/10.1021/acs.molpharmaceut.6b00818

28. Grifno GN, Farrell AM, Linville RM, Arevalo D, Kim JH, Gu L, Searson PC (2019) Tissue-engineered blood-brain barrier models via directed differentiation of human induced pluripotent stem cells. Sci Rep 9:13957. https://doi.org/10.1038/s41598-019-50193-1
29. Katt ME, Mayo LN, Ellis SE, Mahairaki V, Rothstein JD, Cheng L, Searson PC (2019) The role of mutations associated with familial neurodegenerative disorders on blood–brain barrier function in an iPSC model. Fluids Barriers CNS 16:20. https://doi.org/10.1186/s12987-019-0139-4
30. Vatine GD, Al-Ahmad A, Barriga BK, Svendsen S, Salim A, Garcia L, Garcia VJ, Ho R, Yucer N, Qian T, Lim RG, Wu J, Thompson LM, Spivia WR, Chen Z, Van Eyk J, Palecek SP, Refetoff S, Shusta EV, Svendsen CN (2017) Modeling psychomotor retardation using iPSCs from MCT8-deficient patients indicates a prominent role for the blood-brain barrier. Cell Stem Cell 20:831–843.e5. https://doi.org/10.1016/j.stem.2017.04.002
31. Qian T, Maguire SE, Canfield SG, Bao X, Olson WR, Shusta EV, Palecek SP (2017) Directed differentiation of human pluripotent stem cells to blood-brain barrier endothelial cells. Sci Adv 3:e1701679. https://doi.org/10.1126/sciadv.1701679
32. Wilson HK, Canfield SG, Hjortness MK, Palecek SP, Shusta EV (2015) Exploring the effects of cell seeding density on the differentiation of human pluripotent stem cells to brain microvascular endothelial cells. Fluids Barriers CNS 12:13. https://doi.org/10.1186/s12987-015-0007-9
33. Wilson HK, Faubion MG, Hjortness MK, Palecek SP, Shusta EV (2016) Cryopreservation of brain endothelial cells derived from human induced pluripotent stem cells is enhanced by rho-associated coiled coil-containing kinase inhibition. Tissue Eng Part C Methods 22:1085–1094. https://doi.org/10.1089/ten.tec.2016.0345
34. Katt ME, Xu ZS, Gerecht S, Searson PC (2016) Human brain microvascular endothelial cells derived from the BC1 iPS cell line exhibit a blood-brain barrier phenotype. PLoS One 11:e0152105. https://doi.org/10.1371/journal.pone.0152105
35. Lian X, Bao X, Al-Ahmad A, Liu J, Wu Y, Dong W, Dunn KK, Shusta EV, Palecek SP (2014) Efficient differentiation of human pluripotent stem cells to endothelial progenitors via small-molecule activation of WNT signaling. Stem Cell Rep 3:804–816. https://doi.org/10.1016/j.stemcr.2014.09.005

Chapter 11

Modeling the Blood-Brain Barrier Using Human-Induced Pluripotent Stem Cells

Louise A. Mesentier-Louro, Natalie Suhy, Diede Broekaart, Michael Bula, Ana C. Pereira, and Joel W. Blanchard

Abstract

The blood-brain barrier (BBB) is a key physiological component of the brain, protecting the brain from peripheral processes and pathogens. The BBB is a dynamic structure that is heavily involved in cerebral blood flow, angiogenesis, and other neural functions. However, the BBB also creates a challenging barrier for the entry of therapeutics into the brain, blocking more than 98% of drugs from contact with the brain. Neurovascular comorbidities are common in several neurological diseases including Alzheimer's and Parkinson's Disease, suggesting that BBB dysfunction or break down likely has a causal role in neurodegeneration. However, the mechanisms by which the human BBB is formed, maintained, and degenerated in diseases remain largely unknown due to limited access to human BBB tissue. To address these limitations, we have developed an in vitro induced human BBB (iBBB) derived from pluripotent stem cells. The iBBB model can be used for discovery of disease mechanisms, drug targets, drug screening, and medicinal chemistry studies to optimize brain penetration of central nervous system therapeutics. In this chapter, we will explain the steps to differentiate the three cellular components (endothelial cells, pericytes, and astrocytes) from induced pluripotent stem cells, and how to assemble them into the iBBB.

Key words Endothelial cells, Vascularization, Permeability, Blood-brain barrier, Transwell, Induced pluripotent stem cells

1 Introduction

The blood-brain barrier (BBB) is a specialized type of brain microvasculature that is involved in cerebral blood flow, protection from peripheral pathogens, angiogenesis and other functions in the healthy brain and in neurodegenerative disease [1–3]. The BBB is composed of three primary cell types: endothelial cells, astrocytes, and pericytes. Endothelial cells are the primary capillary-forming component of the BBB, and line the outermost cell layer of the BBB, whereas both astrocytes and pericytes extend onto the brain parenchyma side of the layer. Pericytes form a thin sheath around the center of endothelial cells. Astrocytes wrap

Yu-Wen Alvin Huang and ChangHui Pak (eds.), *Stem Cell-Based Neural Model Systems for Brain Disorders*, Methods in Molecular Biology, vol. 2683, https://doi.org/10.1007/978-1-0716-3287-1_11,

around the pericyte and endothelial cells bundle in specialized processes called "astrocytic end feet," which further support the BBB [4].

The BBB is selectively permeable to certain molecules, proteins, and nutrients while restricting the passage of peripheral molecules that may be harmful to the brain tissue. Junction proteins between the endothelial cells of the BBB include tight junctions, gap junctions, and adherens junctions that enable the strict regulation of the barrier, providing protection of the brain from the periphery [5]. Methods of transport may include diffusion, receptor-mediated endocytosis, efflux transport, and more. For example, efflux pumps present in endothelial cells such as p-glycoprotein, MRP1, MRP4, MRP5, and BCRP, recognize cell-permeable solutes and transport them out of the brain and back into the bloodstream via ATP-dependent mechanisms [6]. Generally speaking, permeability is allowed only for molecules that are lipophilic, positively charged, or are smaller than 600 kilodaltons [7].

Using stem cells to create a BBB offers great advantages in testing the different cell types. Each cell type of the BBB can be differentiated from pluripotent stem cells, either embryonic or induced. Diseases may be modeled by either using patient-derived induced pluripotent stem cells or editing human pluripotent stem cells obtained from established cell lines. For example, APOE4 mutant pericytes have been used to model a BBB of a patient with Alzheimer's disease [8].

Once the stem cell-derived cells of the BBB have been differentiated, they must be assembled to create the in vitro induced BBB (iBBB). There are many ways to recapitulate the physiology of the multi-cellular BBB layer in vitro, but the most common way is to use a transwell system. A transwell insert sits inside individual wells and creates a two-layered well that enables permeability and transmigration assays (Fig. 2b). In addition, the transwell may support cell culture on either side of the insert, due to its nature of hovering above a well's bottom surface. In our iBBB model, an endothelial cell layer is cultured on the membrane (apical side) of the transwell insert and, after 24 h, the pericytes and astrocytes are added on top of the endothelial cells.

Monolayer formation (rather than a tubular formation) of the iBBB is important to reflect the human BBB. In order to measure the integrity of the iBBB's monolayer, multiple methods can be used. The first is a transendothelial electrical resistance (TEER) assay in which changes in impedance of current flow can inform about the efficiency of barrier formation. Another method that may be used is a permeability assay, with detection of small, fluorescent molecules such as FITC-Dextran crossing the barrier. Both of these techniques will be described in this chapter.

2 Materials

2.1 Cells

1. Induced pluripotent stem cells.

2.2 Media

After preparation, sterile filtered media can be stored at 4 °C for up to 1 month.

2.2.1 StemFlex™ Basal Medium

1. Basal medium with StemFlex™ supplement (StemCell Technologies, Cat. no. A3349401).
2. 1% penicillin-streptomycin.

2.2.2 Freezing Media iPSCs, NPCs, and Pericytes

1. 90% KSR (KSR, Gibco, Cat. no. 10828028).
2. 10% Dimethyl sulfoxide (DMSO, Sigma-Aldrich, Cat. no. D2650).

2.2.3 NPC (Neural Progenitor Cell) Media

1. 1:1 DMEM/F12 with GlutaMAX (Life Technologies, Cat. no.10565-042): Neurobasal media Neurobasal Medium (ThermoFisher, Cat. no. 21103049).
2. 1× N-2 Supplement (Thermo Fisher, Cat. no. 17502048).
3. 1× B-27 Serum-Free supplement (Gibco, Cat. no. 17504044).
4. 0.5× GlutaMAX Supplement (Thermo Fisher, Cat. no. 35050079).
5. 1× Eagle's Minimum Essential Medium Non-essential Amino Acid Solution (MEM-NEAA) (100×) (Sigma-Aldrich, Cat. no. M7145).
6. 1% penicillin-streptomycin.

2.2.4 Astrocyte Media

1. 500 mL basal media with supplements and penicillin-streptomycin (ScienCell, Cat. no. 1801).

2.2.5 Freezing Media for Astrocytes

1. 90% Fetal bovine serum (FBS, Gibco, Cat. no. 26140).
2. 10% DMSO.

2.2.6 DeSR1

1. DMEM/F12 with GlutaMAX.
2. 1× MEM-NEAA.
3. 1× penicillin-streptomycin.

2.2.7 DeSR2

1. DeSR1 media.
2. 1× N-2.
3. 1× B-27.

2.2.8 hECSR

1. Human Endothelial Serum-free Medium (Gibco, Cat. no. 11111044).
2. 1× MEM-NEAA.
3. 1× B-27.
4. 1% penicillin-streptomycin.

2.2.9 Freezing Media for Endothelial Cells

1. 60% KSR.
2. 30% hECSR.
3. 10% DMSO.
4. 10 μM Y27632 ROCK Inhibitor (PeproTech, Cat. no. 1293823).

2.2.10 N2B27

1. 1:1 DMEM/F12 with GlutaMAX: Neurobasal media.
2. 1× B-27.
3. 1× N-2.
4. 1× MEM-NEAA.
5. 0.5× GlutaMAX.
6. 1% penicillin-streptomycin.

2.2.11 iBBB Media

1. Astrocyte media.
2. 10 ng/mL VEGF-A (R & D Systems, Cat. no. 293-VE-050/CF). *Supplement astrocyte media with VEGF-A immediately before feeding cells.*

2.3 Cell Culture and Differentiation

1. 6-well plates.
2. Accutase (ThermoFisher, Cat. no. A1110501).
3. Activin A Protein (R & D Systems, Cat. no. 338-AC-010).
4. Automated Cell Counter or Hemocytometer.
5. BMP-4 (R & D Systems, Cat. no. 314-BP-050).
6. Centrifuge.
7. CHIR99021 (R & D Systems, Cat. no. 4423/10).
8. CO_2 Incubator, 37 °C 5% CO_2.
9. Centrifuge Tubes.
10. Disposable serological pipettes.
11. Filtered pipette tips.
12. Forskolin (R & D Systems, Cat. no. 1099/10).
13. Geltrex™ LDEV-Free hESC-qualified Reduced Growth Factor Basement (Gibco, Cat. no. A1413302).
14. Glass Bottom 48-well Culture Dishes (Mattek Corporation, Cat. no. P48G-1.5-6-F).

15. LDN193189 (Sigma-Aldrich, Cat. no. SML0559).
16. PBS.
17. Recombinant Human FGF-basic (bFGF, PeproTech, Cat. no. 100-18B-1MG).
18. SB431542 (R & D Systems, Cat. no. 1614/10).
19. Sterile Biosafety cabinet.
20. UltraPure 0.5 M EDTA (ThermoFisher, Cat. no. 15575-020).
21. Water or bead bath at 37 °C.

2.4 Transwell Assays

1. 12 or 24 well plates.
2. Inserts for 12 or 24 well plates.
3. Millicell ERS-2 Voltohmmeter (Millipore, Cat. no. MERS00002).
4. Plate reader.

2.5 Immunostaining and Imaging

1. Confocal Microscope.
2. Fluoromount-G slide mounting medium (VWR, Cat # 100502-406).
3. Hoechst 33258 (Sigma-Aldrich, Cat. no. 94403).
4. Normal Donkey Serum (Millipore-Sigma, Cat. no. S30-100 mL).
5. Paraformaldehyde (PFA).
6. Triton X-100 (Sigma-Aldrich, Cat. no.T8787-250 mL).
7. Primary antibodies:
   (a) S100β (Sigma-Aldrich, Cat. no. S2532-100 uL).
   (b) Pecam1 (Platelet endothelial cell adhesion molecule 1, R & D Systems Cat. no. AF806).
   (c) PDGFRβ (Platelet-derived growth factor receptor beta, R & D Biosystems, Cat. no. AF385).
   (d) VE-cadherin (R & D Systems Cat. no. AF938).
8. Secondary antibodies.
   (a) Alexa Fluor 488, 555, 647 secondary antibodies (all from Invitrogen).

3 Methods

3.1 Coating Cell Culture Plates

Thaw Geltrex™ Matrix overnight in a refrigerator at 2–8 °C. Dilute 0.5 mL of Geltrex™ Matrix into 49.5 mL of DMEM and transfer the diluted solution into tissue culture treated plates. Use 1–2 mL per well of 6-well plates. For coating of cell culture plates and

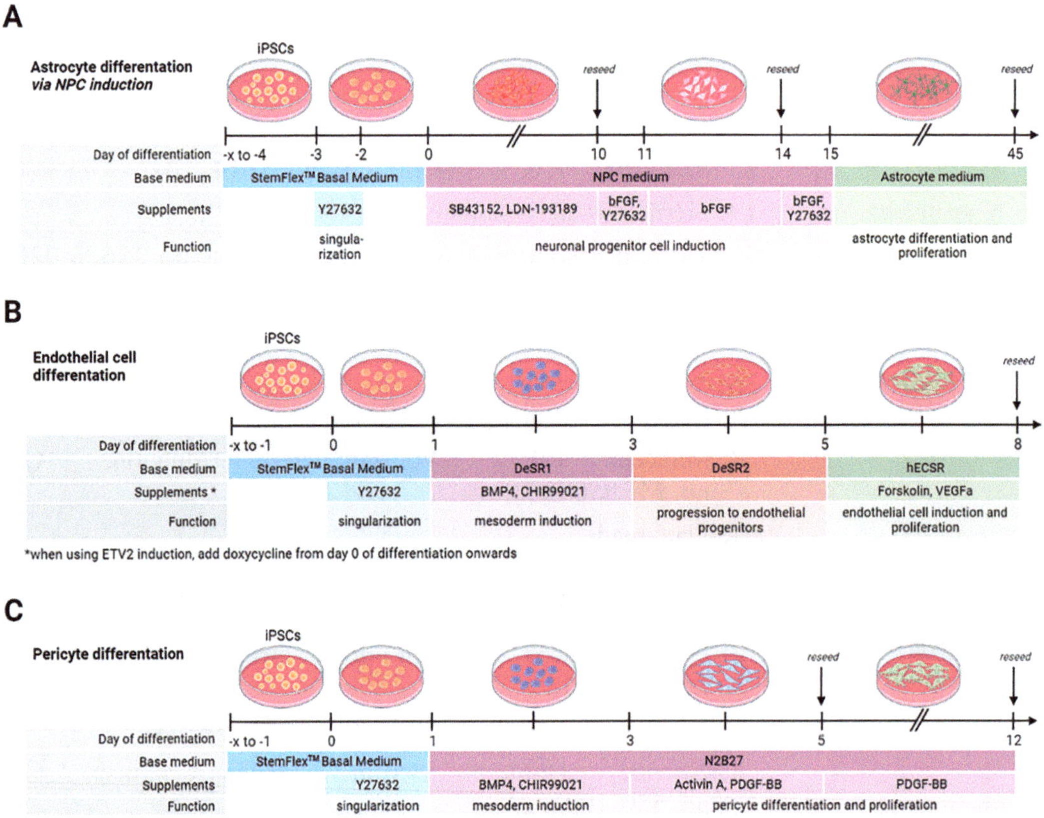

Fig. 1 Differentiation schemes vascular cell types: Schematic representation of differentiation of induced pluripotent stem cells into astrocytes via induction of (**a**) neural progenitor cells (NPC), (**b**) endothelial cells, and (**c**) pericytes. Details of medium composition can be found in Subheading 2.2. Of note, for endothelial cell differentiation using ETV2 induction, the induction antibiotic (e.g., doxycycline) must be added from differentiation day 0 onwards. Created with BioRender.com

transwells, coat for at least 20 min at 37 °C and then replace the coating solution with warmed culture medium prior to seeding of cells.

3.2 Differentiation of Human iPSCs into Astrocytes

Timing: 45–50 days (if starting from iPSCs) or 30–35 days (if starting from NPCs).

NPC differentiation was adapted from Chambers et al. [9]. Astrocytes were differentiated as described in TCW, J et al. [10] (Fig. 1a).

1. Culture iPSCs on Geltrex™-coated plates in feeder-free conditions, feeding with StemFlex™ every other day until cells reach at least 60% confluence.

   CRITICAL STEP: Ensure Geltrex™ is always stored below 8 °C to prevent unexpected polymerization of reagent.

2. Dissociate the cells with accutase for 5–10 min. Transfer dissociated cells to a falcon tube with a minimum of 1:3 accutase to media ratio, spin down cells (at 300 × *g* for 3 min at RT), resuspend the cell pellet in pre-warmed StemFlex™ supplemented with 10 μM Y27632, and plate the cells at 100,000 cells/cm^2 onto Geltrex™-coated plates.
3. Feed cells with StemFlex™. Continue feeding cells every other day until they reach >95% confluence (approximately 3–4 days, depending on the cell line).
4. (Day 0) Once cells reach confluence, switch to NPC medium supplemented with 10 μM SB43152 and 100 nM LDN193189.
5. (Day 1–9) Feed cells daily with NPC media plus 10 μM SB43152 and 100 nM LDN193189.
6. (Day 10) Split NPCs 1:3 with accutase and re-seed onto fresh Geltrex™-coated plates. Feed with NPC media supplemented with 20 ng/mL bFGF and 10 μM Y27632.
7. (Day 11–13) Feed NPCs daily with NPC media plus 20 ng/mL bFGF.
8. (Day 14) Split NPCs 1:3 with accutase and re-seed onto fresh Geltrex™-coated plates. Feed with NPC media plus 20 ng/mL bFGF and supplement with 10 μM Y27632.
9. (Day 15) NPCs can be frozen for stocking. This is *day 0 of astrocyte differentiation:* feed cells with Astrocyte Medium (AM).
10. (Day 1–30) Feed cells every 2 days with AM, passage using accutase (*See* Troubleshooting #1) and plate cells at 15,000 cells/cm^2 once cells reach 90% confluence. NPCs should be fully differentiated into astrocytes in 30 days.

3.3 Differentiation of Human iPSC into Brain Microvascular Endothelial Cells

Timing: 8 days.

Brain endothelial cell differentiation was adapted from Blanchard et al. and Qian et al. [8, 11] (Fig. 1b). For a higher yield of endothelial cells, we recommend concomitant activation of the transcription factor ETV2 (*See* Troubleshooting #2) [12].

3.3.1 Without ETV2-Inducible Activation

1. Culture iPSCs on Geltrex™-coated 6-well plates in feeder-free conditions, feeding every other day with StemFlex™ until the cells reach ~80% confluence.
2. (Day 0) Dissociate the cells with accutase for 5 min. Transfer dissociated cells to a falcon tube with a minimum of 1:3 accutase to media ratio, spin down cells (at 300 × *g* for 3 min at RT), resuspend the cell pellet in warmed StemFlex™ supplemented with 10 μM Y27632, and plate the cells at 20,800 cells/cm^2 onto Geltrex™-coated plates (or 200,000 for each well of a 6-well plate).

3. (Day 1) Replace medium with DeSR1 supplemented with 10 ng/mL BMP4 and 6 μM CHIR99021.
4. (Day 3) Replace medium with DeSR2 medium.
5. (Day 5) Replace medium with hECSR medium supplemented with 50 ng/mL VEGF-A, and 2 μM Forskolin.
6. (Day 7) Replace medium with hECSR medium supplemented with 50 ng/mL VEGF-A, and 2 μM Forskolin.
7. (Day 8) Split endothelial cells 1:3 with Accutase and re-seed onto fresh Geltrex™-coated plates, or directly onto Geltrex™-coated transwell inserts at 0.5 to 1 × 10^6 cells/cm^2. Feed with hECSR supplemented with 50 ng/mL VEGF-A.
8. (Day 9) Feed cells every 2–3 days with hECSR media supplemented with 50 ng/mL VEGF-A until ready for iBBB co-culture.

3.3.2 With ETV2-Inducible Activation

CRITICAL: For this protocol, iPSCs need to be transfected with an inducible system for expression of ETV2. We used a PiggyBac inducible system from AddGene (PB_iETV2_P2A_GFP_Puro, Catalog #168805) and transfected the cells using Lipofectamine (Lipofectamine™ Stem Transfection Reagent, #STEM00001). This system has a selection cassette that confers resistance to puromycin, and the expression of ETV2 is induced by adding doxycycline.

1. Culture iPSCs on Geltrex™-coated 6-well plates in feeder-free conditions, feeding every other day with StemFlex™ until the cells reach ~80% confluence. *If the inducible system contains a selection cassette, we suggest that iPSCs are maintained with the selection antibiotic* (e.g., *puromycin) to increase purity (See* **Note 1**, *See Troubleshooting #3).*
2. (Day 0) Dissociate the cells with accutase for 5 min. Transfer dissociated cells to a falcon tube with a minimum of 1:3 accutase to media ratio, spin down cells (at 300 × *g* for 3 min at RT), resuspend the cell pellet in warmed StemFlex™ supplemented with 10 μM Y27632, and plate the cells at 20,800 cells/cm^2 onto Geltrex™-coated plates (or 200,000 for each well of a 6-well plate).
3. (Day 1) Replace medium with DeSR1 supplemented with 10 ng/mL BMP4, 6 μM CHIR99021, and 5 μg/mL doxycycline.
4. (Day 3) Replace medium with DeSR2 medium with 5 μg/mL doxycycline.

5. (Day 5) Replace medium with hECSR medium supplemented with 50 ng/mL VEGF-A, 2 μM Forskolin, and 5 μg/mL doxycycline.
6. (Day 7) Replace medium with hECSR medium supplemented with 50 ng/mL VEGF-A, and 2 μM Forskolin, and 5 μg/mL doxycycline.
7. (Day 8) Dissociate the cells using Accutase and re-seed onto fresh Geltrex™-coated plates, or directly onto Geltrex™--coated transwell inserts at 0.5 to 1 × 10^6 cells/cm^2. Feed with hECSR supplemented with 50 ng/mL VEGF-A and 5 μg/mL doxycycline.
8. (Day 9) Feed cells every 2–3 days with hECSR media supplemented with 50 ng/mL VEGF-A and 5 μg/mL doxycycline until ready for iBBB co-culture.

3.4 Differentiation of Human iPSCs into Pericytes

Timing: 6 days.

Pericyte differentiation was adapted from Patsch et al. [13] (Fig. 1c).

1. Culture iPSCs on Geltrex™-coated plates in feeder free conditions, feeding every other day with StemFlex™ until the plate reaches at least 60% confluence.
2. (Day 0) Dissociate the cells with accutase for 5 min. Transfer dissociated cells to a falcon tube with a minimum of 1:3 accutase to media ratio, spin down cells (at 300 × *g* for 3 min at RT), resuspend the cell pellet in warmed StemFlex™ supplemented with 10 μM Y27632, and plate the cells at 37,000–40,000 cells/cm^{-2} onto Geltrex™-coated plates.
3. (Day 1) Replace media with N2B27 media supplemented with 25 ng/mL BMP4 and 8 μM CHIR99021.
4. (Day 3 and Day 4) Feed the cells daily with N2B27 media supplemented with 2 ng/mL Activin A and 10 ng/mL PDGF-BB.
5. (Day 5) Dissociate pericytes with accutase and re-seed onto fresh 0.1% gelatin-coated plates at 35,000 cells/cm^2 and expand for another 5–7 days in N2B27 supplemented with 10 ng/mL PDGF-BB. Maintain in N2B27 with PDGF-BB until cells are ready for iBBB (*see* **Note 2**, *See* Troubleshooting #4).

3.5 Barrier Model

1. Following endothelial cell differentiation (B), dissociate endothelial cells with accutase for 5 min. Transfer dissociated cells to a falcon tube with a minimum of 1:3 accutase to media ratio, spin down cells (at 300 × *g* for 3 min at RT).

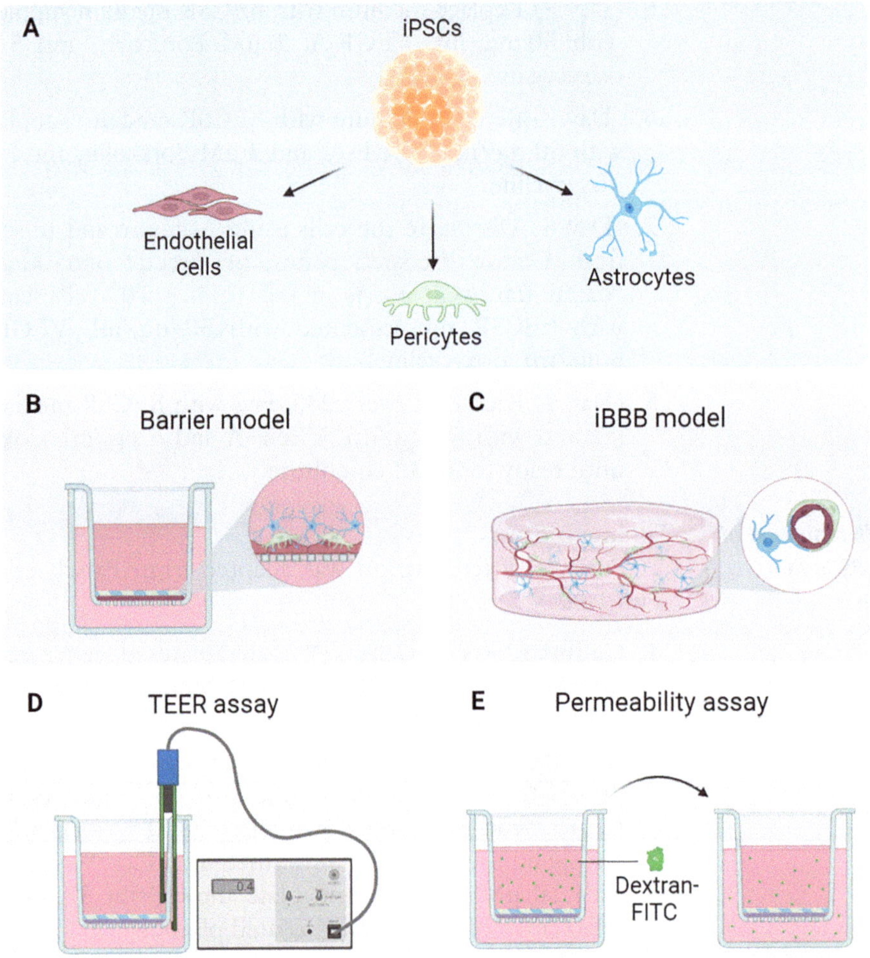

Fig. 2 In vitro blood-brain barrier models: schematic representations of in vitro blood-brain barrier (BBB) models. (**a**) After differentiating induced pluripotent stem cells (iPSCs) to endothelial cells, pericytes, and astrocytes, cell types can be combined in two models mimicking the BBB. (**b**) Endothelial cells can be seeded onto Geltrex™-coated transwell membrane cell culture inserts, followed by seeding of astrocytes and pericytes 24 h later to form a barrier. (**c**) Alternatively, cells can be combined in a 5:1:1 endothelial cell to pericytes to astrocytes ratio in Geltrex™ to form the iBBB model. (**d**) Transendothelial electrical resistance (TEER) can be read using chopstick electrodes measuring the current flow impedance over the barrier. (**e**) Additionally, transwell membrane cell culture inserts allow for the investigation of barrier function through the measurement of FITC-labeled dextran through the barrier. Created with BioRender.com

2. (Day 0) Resuspend cell pellet in hECSR supplemented with 10 μM Y27632 and seed onto 24-well Geltrex™-coated transwell membrane cell culture inserts at a density of 0.5 to 1×10^6 cells/cm^2 (*see* **Notes 3** and **4**; *see* Troubleshooting #5).
3. (Day 1) Dissociate differentiated astrocytes (3.2) and pericytes (3.4) with accutase and seed each cell type on top of the endothelial cell layer in the transwell insert at a density of 50,000 cells/cm^2 for each cell type (Fig. 2a, b).

4. (Day 2–8+) Feed cells every 2 days with iBBB media. Measure TEER recordings via chopstick probes (Fig. 2d). Reserve 3 wells of the Geltrex™-coated transwell plate as a cell-free blank control for TEER recordings. Perform permeability assays to follow barrier formation over time. Typically, peak TEER values are reached in 4 to 6 days, but different iPSC lines might require different formation times.

3.6 Transwell Permeability Assay

Timing: 2–3 h (Fig. 3e)

1. Prior to performing permeability assays, run a standard curve titrating concentrations of FITC-labeled dextran to determine the linear working range. Be sure to also include one Geltrex™-coated well without cells to compare relative

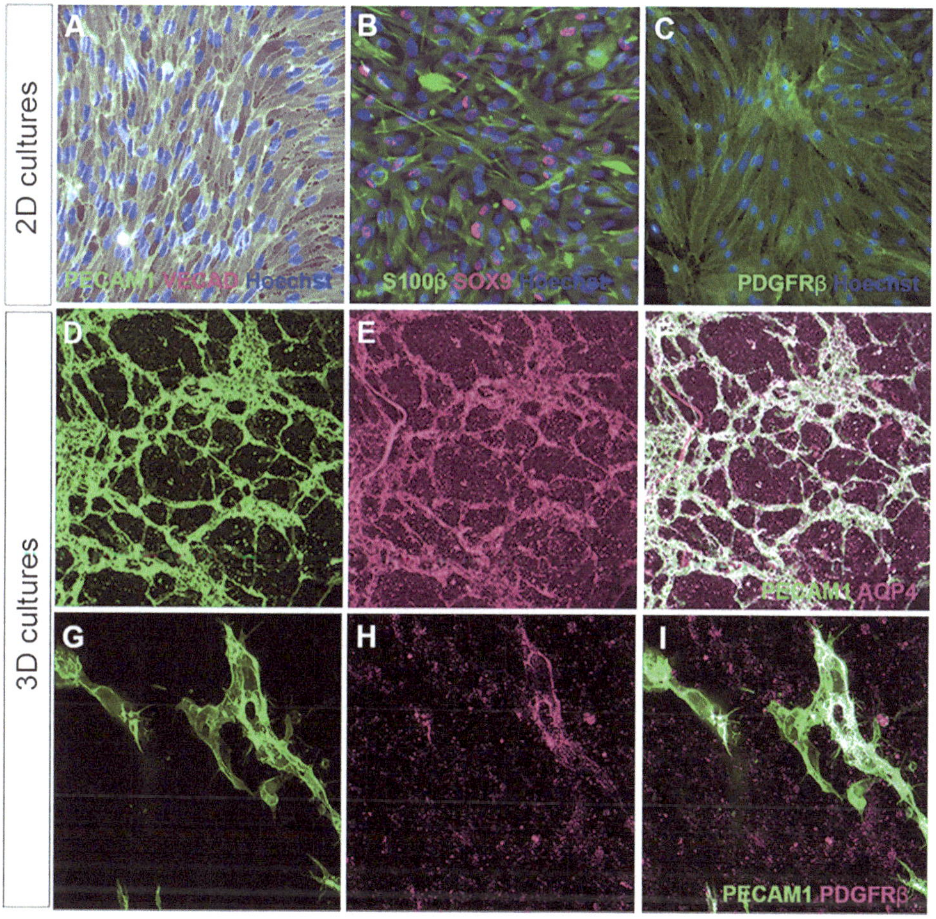

Fig. 3 Expression of cell-specific markers in 2D and 3D cultures: Immunofluorescence for cell-specific markers of (**a**) endothelial cells, (**b**) astrocytes, (**c**) and pericytes in 2D cultures. After assembly of 3D cultures, a vascular network composed of (**d**) PECAM1+ endothelial cells, (**e**) Aquaporin-4+ astrocyte endfeet, and (**h**) PDGFRβ+ pericytes is observed

permeability. We suggest a concentration of 250 μg/mL dextran molecules for use.

2. For a 24-well plate with inserts, add 100 μL of FITC-labeled dextran or other labeled compound of interest diluted in media to the apical side of the transwell system and replace the basolateral side with 500–600 μL of fresh medium (Fig. 2e).
3. Incubate the transwell plate at 37 °C with 5% CO_2 for 1 h.
4. Transfer 100 μL of the bottom chamber to a 96-well plate. Afterwards, remove the media from both the apical and basolateral sides and replace with fresh media.
5. Measure fluorescence intensity of the 96-well plate containing assay media via plate reader (e.g., EnSpire, Perkin Elmer). When using FITC-dextran, set the excitation to 490 nm and the emission to 520 nm.
6. Calculate barrier permeability using the following formula:

$$\text{Papp} = \frac{(dQ/dt)}{C_0 \times A}$$

Where Papp is the permeability coefficient, dQ/dt is the concentration of compound in the apical compartment as a function of time, C_0 is the initial concentration at time zero, and A is the area of the transwell in cm^2. The permeability coefficient can be normalized to an individual sample and averaged to obtain a relative permeability.

3.7 iBBB Culture

Timing: 15–30 days.

1. Thaw the appropriate volume of Geltrex™ overnight at 2–8 °C (calculate 50 μL per iBBB; *See* Troubleshooting).
2. Prepare and pre-warm media necessary for astrocytes (AM), pericytes (N2B27), brain endothelial cells (hECSR), and iBBB cultures.
3. Dissociate differentiated astrocytes, endothelial cells, and pericytes with accutase at 37 °C for 5–10 min. Transfer each dissociated cell type to a separate falcon tube with a minimum of 1:3 accutase to media ratio, spin down cells (at 300 × *g* for 3 min at RT) and resuspend pellets in each cell type's corresponding media.
4. Using a hemocytometer, count each corresponding cell type. We recommend diluting in media for working concentrations of 1×10^6 cells/mL for ease of use. During this step, ensure the cells counted via hemocytometer are in single cell suspension.
5. Calculate the volume of cells needed to set up iBBB cultures. Each iBBB culture requires 2.5×10^5 endothelial cells, 5×10^4 astrocytes, and 5×10^4 pericytes. Combine the calculated number of cells from each cell type together into a 15 mL centrifuge tube. It is recommended to calculate cell numbers

to reach 10% additional to the originally planned count as some of the Geltrex™ and cells will be lost due to pipetting. Spin down the cells (at 300 × *g* for 3 min at RT).

6. Aspirate the media carefully, leaving the cell pellet undisturbed, and put the cells on ice.
7. Resuspend cell pellet in Geltrex™ with enough volume for 50 μL of this solution for each iBBB culture. Carefully pipette the mixture to homogeneously distribute the cells throughout the Geltrex™ mixture (Fig. 2c), avoiding air bubbles.

   CRITICAL STEP: The cell suspension must remain on ice before seeding onto glass bottom plates. Failure to do so will result in premature Geltrex polymerization and inability to properly seed iBBBs.
8. Carefully pipette 50 μL Geltrex™ and encapsulated cell mixture onto the glass-bottom well of a 48-well MatTek plate. It is recommended to use a pipette tip as a brush to carefully stroke the cell mixture solution around to evenly distribute the Geltrex™ and cells throughout the bottom of the dish (*See* Troubleshooting #6).
9. Transfer the iBBB cultures into a 37 °C incubator for 30–45 min to allow the Geltrex™ to polymerize. Check the cultures underneath a light microscope to ensure single cell suspension of cells in Geltrex™.
10. After the Geltrex™ has been polymerized, add 250 μL of warmed iBBB medium to each MatTek well of a 48-well plate, ensuring complete submersion of the culture in media.
11. (Day 1–14) Feed iBBB cultures every 2 days with iBBB media (*see* Troubleshooting #7).
12. (Day 15–21) Feed iBBB cultures every 2 days with astrocyte media.
13. (Day 21+) Feed cultures with a doubled volume of astrocyte media (500 μL) every 2 days. If medium culture appears yellow within 24 h, start changing media every day.
14. iBBB cultures should be ready for downstream assays or fixation between week 2 and 4. Cultures can be used for experiments, and subsequently fixed, stained, and imaged.

3.8 Fixation and Staining of Cultures

Timing: 4–7 days.

1. Aspirate the medium from the iBBB cultures and quickly wash with PBS. After aspirating the PBS fix the cultures with 4% paraformaldehyde solution at 4 °C overnight.
2. Remove the cultures from 4 °C and rinse the cultures with PBS twice. For extremely dense cultures additional washing may be required. The iBBBs can be stored in PBS at 4 °C until they are ready to be stained.

3. Permeabilize the samples with PBST solution (PBS with 0.3% Triton X-100) at room temperature for 1 h with gentle shaking.
4. Incubate the cultures in blocking solution (5% Normal Donkey Serum) at room temperature for 4 h with gentle shaking, or overnight at 4 °C.
5. Dilute primary antibodies to the recommended concentration in PBST.
6. Aspirate blocking solution from the iBBB cultures, being careful to remove as much of the solution as possible without disturbing the tissue. Add 100–150 μL of primary antibody solution for each well, ensuring the iBBB matrix is completely submerged. Incubate overnight at 4 °C with gentle shaking.
7. Wash cultures thoroughly with PBS three times, incubating for 10–20 min each wash.
8. Dilute secondary antibody and nuclear stain (e.g., Hoechst) at 1:1000 or according to manufacturer's instructions in PBST. Aspirate the solution on the iBBB and add 100–150 μL of secondary antibody solution, ensuring the iBBB is completely submerged. Cover in aluminum foil, and incubate at room temperature for 2 h with gentle shaking, or overnight at 4 °C.
9. Wash cultures with PBST three times, incubating 10 and 20 min between washes. Store in PBS or anti-fading mounting media at 4 °C. The samples are ready to be imaged.

CRITICAL

Mounting detached iBBBs (*see* Troubleshooting #8):

Vascular cells may promote tissue folding and detachment from the bottom of the well. In that case, we suggest that you mount the iBBB as described below:

1. Carefully remove each iBBB from the 48-well plate, using forceps, and place them on a 35 mm MatTek dish.
2. Carefully pipette ~50–100 μL fluoromount solution onto the well containing the iBBB, dispensing slowly to prevent bubbles from forming.
3. Gently place a coverslip over the fluoromount. Lightly press down on the coverslip and wipe off any excess fluoromount solution from the sides of the coverslip.
4. Afterward, paint the edges with nail polish to create an airtight seal. Store the plates protected from light, at 4 °C.
5. Image the cultures on an inverted confocal microscope.

3.9 Anticipated Results

Each cell type should express key markers according to its own corresponding cell type (Fig. 3). TEER values for the brain endothelial cells should increase for co-cultures with astrocytes and pericytes. Additionally, iBBB co-cultures should be less permeable to small molecules compared to endothelial cells alone. Three-dimensional iBBB cultures generated using this protocol should develop self-assembled vascular networks, with pericytes organizing themselves to the proximal locations of the vascular endothelial cells after 2 weeks. Astrocytes should project their "end-feet" to the vasculature and express key transport molecules such as aquaporin 4 (AQP4), creating a polarized expression of AQP4 around the vasculature (Fig. 3).

4 Notes

1. Endothelial cell differentiation: if using an inducible system with a selection antibiotic (e.g., puromycin), you may want to determine whether it is beneficial to keep the selection antibody throughout the differentiation to improve purity. This step may vary according to the cell line and transfection efficiency.
2. Pericyte maintenance: after differentiation, cells can be kept in N2B27 with PDGF-BB for the first passage. After that, they can be maintained in N2B27 without PDGF-BB.
3. The membrane material of the transwell insert may affect attachment, with available materials being polycarbonate, polyester (PET), or polytetrafluoroethylene (PTFE). Coating the transwell in an appropriate extracellular matrix increases the probability of strong attachment. Consider comparing Geltrex, Vitronectin, Matrigel, Laminin, and Collagen if monolayer troubleshooting is needed.
4. Additionally, pore size of the transwell insert matters when optimizing attachment. Transwells with pore sizes of 0.4, 3, and 8 microns may be used. If you desire to build a BBB for usage in cell transmigration assays, 3 or 8 microns are the classic pore sizes used. One thing to consider when using different pore sizes, however, is that smaller pore sizes confer higher pore density on the transwell insert, while larger pores will have lower pore density. In general, greater numbers of pores increase the number of attachment points for the seeded cells, therefore 0.4 microns is the best option to use when optimizing attachment (but not for use in transmigration assays).

5 Troubleshooting

#	Technique	Problem	Possible reason	Solution
1	Astrocyte cultures	Astrocytes do not detach	Strong adherence to coating	Use TrypLE select (Gibco™ 12,563,029) instead of Accutase to accelerate detachment
2	Endothelial cell differentiation	Low yield of endothelial cells	Intrinsic effects of cell line	Use activation of the transcription factor ETV2
3	Endothelial cell differentiation using ETV induction	Cell death during differentiation	Keeping selection antibiotic during differentiation	Lower or omit puromycin starting from day 0 of differentiation
4	Pericyte cultures	Low growth in later passages	Intrinsic effects of cell line	Use Y27632 when passaging cultures. Refresh to medium without Y27632 after 24 h
5	Monolayer formation	Poor cell attachment	Surface or coating	Compare different surface materials (e.g., PET, polycarbonate) and/or coating with ECM compositions (e.g., Geltrex, Matrigel, laminin, collagen)
6	iBBB assembly	Geltrex gelatinizes before plating the cells	High temperature	Work rapidly and keep geltrex solutions on ice in between pipetting
7	iBBB maintenance	iBBB peels off bottom of MatTek dish	Vascular cells remodeling the ECM	Various options: (a) Add extra Geltex on top of existing cells after 7–10 days to provide additional ECM for vascular outgrowth (b) When assembling the iBBB, seed a smaller volume of cell suspension per well (25 μL instead of 50 μL), and let it expand horizontally across the well over time (c) Feed the detached cultures with care not to aspirate them. Image them after mounting on a 35 mm dish
8	Imaging iBBBs	Cannot focus	iBBB is detaching from the bottom of the plate and leading to an inaccurate working distance for the objective lens	Mount the detached iBBB as described in 3.8.

Acknowledgments

We thank Camille Goldman and Rikki Rooklin for critically reading the chapter, Tatyana Kareva for help with cell transfections, Stephen Fitzsimons and all Blanchard Lab members for helpful discussions, the Neuroscience Microscopy core at Icahn School of Medicine at Mount Sinai for imaging advice and support. Licenses has been granted to use the BioRender content under agreement numbers OU25APQJH4 and OF25APQP9P. This work was supported by Cure Alzheimer's Foundation, Brain Injury of America Association, NIH grant UH3-NS115064, R01NS114239.

References

1. Persidsky Y, Ramirez SH, Haorah J et al (2006) Blood-brain barrier: structural components and function under physiologic and pathologic conditions. J Neuroimmune Pharmacol 1: 223–236
2. Pardridge WM (2012) Drug transport across the blood-brain barrier. J Cereb Blood Flow Metab 32:1959–1972
3. Sweeney MD, Kisler K, Montagne A et al (2018) The role of brain vasculature in neurodegenerative disorders. Nat Neurosci 21: 1318–1331
4. Kadry H, Noorani B, Cucullo L (2020) A blood-brain barrier overview on structure, function, impairment, and biomarkers of integrity. Fluids Barriers CNS 17:69
5. Stamatovic SM, Johnson AM, Keep RF et al (2016) Junctional proteins of the blood-brain barrier: new insights into function and dysfunction. Tissue Barriers 4:e1154641
6. Qosa H, Miller DS, Pasinelli P et al (2015) Regulation of ABC efflux transporters at blood-brain barrier in health and neurological disorders. Brain Res 1628:298–316
7. Bellettato CM, Scarpa M (2018) Possible strategies to cross the blood-brain barrier. Ital J Pediatr 44:131
8. Blanchard JW, Bula M, Davila-Velderrain J et al (2020) Reconstruction of the human blood-brain barrier in vitro reveals a pathogenic mechanism of APOE4 in pericytes. Nat Med 26:952–963
9. Chambers SM, Fasano CA, Papapetrou EP et al (2009) Highly efficient neural conversion of human ES and iPS cells by dual inhibition of SMAD signaling. Nat Biotechnol 27:275–280
10. Tcw J, Wang M, Pimenova AA et al (2017) An efficient platform for astrocyte differentiation from human induced pluripotent stem cells. Stem Cell Rep 9:600–614
11. Qian T, Maguire SE, Canfield SG et al (2017) Directed differentiation of human pluripotent stem cells to blood-brain barrier endothelial cells. Sci Adv 3:e1701679
12. Wang K, Lin RZ, Hong X et al (2020) Robust differentiation of human pluripotent stem cells into endothelial cells via temporal modulation of ETV2 with modified mRNA. Sci Adv 6: eaba7606
13. Patsch C, Challet-Meylan L, Thoma EC et al (2015) Generation of vascular endothelial and smooth muscle cells from human pluripotent stem cells. Nat Cell Biol 17:994–1003

Chapter 12

A Three-Dimensional Primary Cortical Culture System Compatible with Transgenic Disease Models, Virally Mediated Fluorescence, and Live Microscopy

Sophie Brown, Elaina Atherton, and David A. Borton

Abstract

In vitro cell culture models can offer high-resolution and high-throughput experimentation of cellular behaviors. However, in vitro culture approaches often fail to fully recapitulate complex cell processes involving synergistic interactions between heterogeneous neural cell populations and the surrounding neural microenvironment. Here, we describe the formation of a three-dimensional primary cortical cell culture system compatible with live confocal microscopy.

Key words Three-dimensional cell culture, Live imaging, Primary cell culture, In vitro

1 Introduction

Parkinson's disease, Alzheimer's disease, and other dementias, multiple sclerosis, and epilepsy are among the large number of debilitating neurological disorders, many of which lack effective treatments. The paths toward successful therapeutic intervention for such disorders are often obstructed by a limited understanding of disease etiology and complex underlying mechanisms. Further, current neuroimaging techniques used to study the human brain are typically non-invasive with limited spatial and temporal resolution, which hinders direct study of cellular mechanisms of disease. While in vivo animal models can provide a path to understanding neurological disorders, they are often defined by a slow iterative process and study conclusions are limited to only those pathways that are well conserved across species. Alternatively, in vitro modeling of the brain can provide repeatable, well-controlled, and environmentally isolated experimentation for effective molecular and cellular investigations of neural function in health and disease.

Yu-Wen Alvin Huang and ChangHui Pak (eds.), *Stem Cell-Based Neural Model Systems for Brain Disorders*, Methods in Molecular Biology, vol. 2683, https://doi.org/10.1007/978-1-0716-3287-1_12,

To date, myriad sources are used to form in vitro neural cell cultures including immortalized cell lines, induced-pluripotent stem cells, and human patient derived or animal primary tissue [1–5]. In the case of primary rodent cultures, a tissue sample is collected from the host and then dissociated into a cellular suspension by means of enzymatic and mechanical manipulation. Embryonic or neonatal cells, having gone through fewer differentiated states, are more likely to survive in culture [6]. For primary monolayer cultures, tissues are dissociated into a single cell suspension and seeded onto flat culture plates where they attach to the two-dimensional surface. However, when dissociated cells are left in suspension, they aggregate and reassemble into three-dimensional structures [7]. Compared to two-dimensional cultures which physically constrain cell mechanics and morphologies to the culture surface, three-dimensional cell cultures provide increased physiological relevance with cytoarchitecture and cell-cell interactions that are more akin to those found in the living brain [6, 8–12]. Without specific selection for a unique cell type, dissociated cells will form heterogeneous neural cell cultures including both glial and neuronal subtypes with evidence of neuronal remodeling and formation of axonal networks with fully developed synapses and microcircuitry [13–20]. Additionally, in vitro primary cultures can be formed from an array of transgenic animal models, which allows for study of disease phenotypes in an engineered tissue environment, with control over exact cell type concentrations and the ability to supplement or deplete specific cell types.

We have previously reported on the use of a three-dimensional, self-assembled, primary cortical cell culture platform for investigative studies of neuroinflammatory effects on functional neural network structures [19, 20]. Briefly, we derived three-dimensional cortical cultures from primary rodent tissue utilizing a protocol adapted from the Laboratory of Dr. Hoffman-Kim whose previous work demonstrated neural cell subtype diversity and functionality of intra-cellular interactions in primary rodent cortical spheroids [11, 21]. We further optimized the protocol to seed cortical cells into an equibiaxial tension "trampoline" shaped microtissue, previously described by Schell et al., to create a culture geometry compatible with longitudinal microscopy studies [22]. The trampoline shape provides a stable ~300 μm thick microtissue, consisting of a self-assembled extracellular matrix, and heterogeneous cell population [19] (Fig. 1). Development of a custom injection mold allowed us to seed microtissues within 350 μm of the bottom of a glass bottom multi-well culture plate. The injection mold was devised in order to bring the microtissue within working distance of high-power objectives of an inverted confocal microscope. Concurrently we optimized methods to induce fluorescence in several key neural cell types within the 3D microtissues via

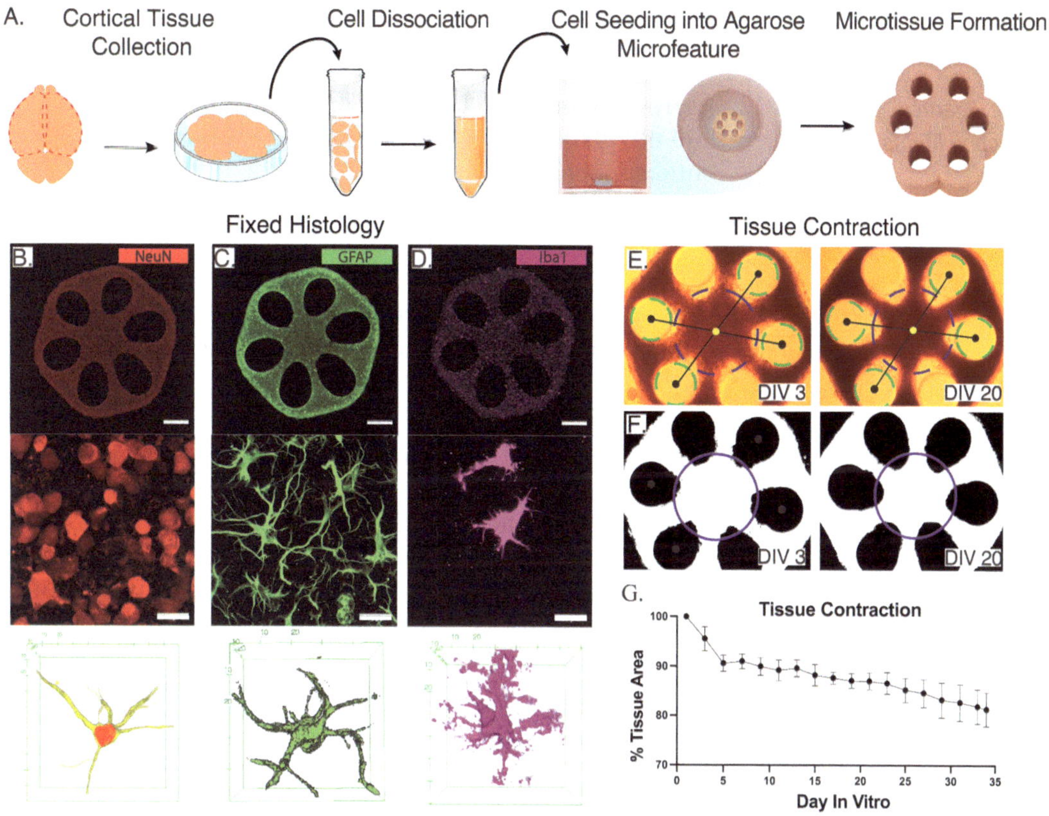

Fig. 1 Formation of three-dimensional primary corticasl microtissue: primary rat cortical tissue is collected at postnatal day 1–2 and dissociated by enzymatic means to form a single cell suspension prior to seeding into agarose-molded microfeature, (**a**) at 24 h post-seeding 3D microtissue forms and immunolabeling of (**b**) neurons (NeuN, red), (**c**) astrocytes (GFAP, green), and (**d**) microglia (Iba1, magenta) demonstrates heterogeneous cell population with intricate 3D morphologies. (**e**) Microtissues show macromorphology changes characterized by tissue contraction over weeks. (**f**) Contraction of the center of the microtissue was measured with custom MATLAB code by placing an ROI in the center of the tissue (purple), binarizing the image so that the tissue is white, and then measuring the percent of white pixels within the ROI. (**g**) Microtissues contracted the most within the first 14 days in vitro (89.63% tissue area) and continued to contract through day 34 (81.34% tissue area). (Reproduced from Atherton et al. [20] with permission from *Scientific Reports*)

adeno-associated viral (AAV) vectors. Combined with our engineered plate set-up and unique microtissue geometry, we were able to use AAV-mediated fluorescence to perform high-resolution microscopy over several weeks, capturing longitudinal evolution of intricate cellular morphologies and functions at the single cell level. As this multimodal culture and live-imaging platform allows us to monitor the direct and indirect influence of cellular dynamics in and on the surrounding neural microenvironment, we highlight its application to study complex multicellular processes in both healthy and diseased states.

2 Materials

2.1 Agarose Injection Molds

1. Agarose solution: 1 g Ultra-Pure agarose powder, autoclaved in 500 mL glass bottle, 50 mL sterile 1× Dulbecco's phosphate-buffered saline (PBS).
2. Stainless Steel Injection Mold: machined stainless-steel mold with 6 concentric peg holes of 600 μm diameter at one end of a 24.3 mm long hollow cylinder with a 3 mm diameter circular base. Design for the injection mold piece can be found in: https://github.com/neuromotion/3D_PrimaryCellCulture_Methods.
3. Custom 3D-printed PLA-lift piece: 3D-printed piece to "lift" the injection mold approximately 200 μm from the bottom of the glass culture plate. The lift is a circular riser that press fits temporarily into the inner diameter of a single well on a 24-well glass-bottom plate (*see* **Note 1**). Design of the 3D-printed PLA-lift piece can be found in: https://github.com/neuromotion/3D_PrimaryCellCulture_Methods.
4. 24 well culture plate with #1.5H coverslip glass bottom (0.17 ± 0.005 mm) (CellVis).
5. 5 cc sterile syringe.
6. 11.43 cm long and 2 mm diameter stainless-steel rod.
7. 0.5 cc sterile syringe with 1/2" 29 gauge needle.
8. Ice or ice pack.

2.2 Tissue Dissociation

1. Timed pregnant Sprague Dawley female rats were purchased from Charles River laboratory. Rats arrive between embryonic day 15 and 17 and they are subsequently monitored every day thereafter until pups are born. Pups are collected no later than postnatal day 2.
2. Hibernate® A-B27 Solution: 48.9 mL Hibernate® A media (Transnetyx Tissue Inc), 0.125 mL GlutaMAX™, 1 mL 1× B27 Neuronal supplement. Solution can be made in advance but recommended that B-27 supplement is added on the day of dissociation.
3. Papain Solution: 6 mg of lyophilized papain powder (Transnetyx Tissue Inc), 3 mL Hibernate® A minus Calcium (Transnetyx Tissue Inc). Add 3 mL of cold Hibernate® A minus Ca to lyophilized papain, mix well by inversion. Wait until solution becomes clear then place vial into 30 °C water bath 10 min prior to use.

4. Complete Cortical Media (CCM): 48.4 mL Neurobasal™-A medium, minus phenol red, 0.125 mL GlutaMAX™, 0.5 mL Penicillin Streptomycin, 1 mL 1× B27 Neuronal supplement.
5. Tissue Dissection kit: sharp-pointed delicate dissection scissors, two curved precision tip forceps, stainless steel laboratory spatula with scoop, sterile gauze pads, sterile razor blades.
6. Tissue Processing Supplies: sterile 40 μm filters, sterile 50 mL and 15 mL conical tubes, hemocytometer, sterile 35 mm Petri dish, serological pipettes, p1000 and p200 pipettes and tips, trypan blue dye.

2.3 Adeno-Associated Viral Transduction

1. Adeno-associated viral particles are stored in −80 °C in small-volume aliquots. For day-of transduction, virus is thawed on ice and diluted into CCM to reach the final desired viral genome per milliliter concentration.
2. Ready to use AAV1 particles produced from pAAV-hSyn1-mRuby2-GSG-P2A-GCaMP6s-WPRE-pA (Addgene).
3. Ice.

2.4 Live Confocal Microscopy

1. Olympus FV3000 confocal laser scanning microscope.
2. Stage Top Incubator (Tokai Hit) (*see* **Note 2**).

2.5 Immunohistochemistry

1. Fixative: 4 g sucrose, 45 mL 1× PBS, 5 mL formalin to create an 8% sucrose, 4% paraformaldehyde solution.
2. Blocking solution: 0.4 g bovine serum albumin, 8.9 mL 1× PBS, 1 mL normal goat serum, 100 μL TritonX-100.
3. PBS with 0.2% TritonX −100 (1× PBT) solution: 500 mL of 1× phosphate buffered saline, 100 μL of TritonX-100.
4. Primary Antibodies: mouse anti-NeuN, chicken anti-GFAP, Rabbit anti-Iba. Dilute antibodies to desired concentration in blocking solution (*see* **Note 3**).
5. Secondary Antibodies: anti-mouse CY3, anti-chicken AlexaFluor 488, anti-rabbit AlexaFluor 635, DAPI. Dilute antibodies to desired concentration in blocking solution (*see* **Note 3**).

2.6 Clearing Solutions

1. 50% formamide: 1 mL 100% formamide solution, 1 mL PBS.
2. 10% Polyethylene glycol (PEG)/25% Formamide: 1 mL of 20% PEG solution, 1 mL 50% formamide.
3. 20% PEG/50% formamide: 1 mL of 40% PEG solution, 1 mL of 100% formamide.

3 Methods

3.1 Agarose-Molded Multi-well Plate (Fig. 2)

1. Place sterilized stainless steel injection mold into a single well of a glass-bottom 24-well plate. The injection mold will be suspended in place approximately ~200 μm from the bottom of the plate by the custom 3D-printed PLA-lift piece (*see* **Note 1**).
2. Add 1 g of Ultra-Pure agarose powder to a 500 mL autoclavable glass bottle and sterilize with high heat and pressure (121 °C min for 30 min).
3. Add 50 mL of sterile PBS to cooled sterile agarose powder to create a 2% agarose solution. Add the PBS to agarose powder and gently swirl. Place bottle in laboratory microwave and heat in 30 s increments until agarose goes into solution and reaches a rolling boil (approximately 1.5 min).
4. Place the hot agarose into a biosafety cabinet for preparation of the agarose-molded multi-well plates. Using a sterile 5 cc syringe, inject 1.2 mL of hot agarose through the stainless-steel injection mold placed atop 1 well of the 24-well glass-bottom plate. Carefully remove the syringe, leaving the injection mold and PLA-lift in place. Place the plate on an ice pack for a minimum of 7 min to allow agarose to cool and solidify.
5. Carefully place the 2 mm diameter stainless-steel rod into the stainless-steel injection mold and scrape excess agarose away from the 6 peg holes. Steadily remove the injection mold and PLA-lift piece simultaneously, leaving behind the agarose-molded microfeature at the bottom of the well. To clean the injection mold before the next molding, use the plunger from a 0.5 cc syringe to force residual agarose out of the hollow injection mold.
6. Add 1 mL of CCM to keep agarose molds from dehydrating and place in 37 °C incubator until needed for cell-seeding (*see* **Note 4**).

3.2 Neonatal Cortical Tissue Collection

1. A whole litter of postnatal day 0–2 rat pups are collected and sexed. An equal number of male and female pups are pooled for use in a single dissociation experiment.
2. Following approved Institutional Animal Care and Use Committee (IACUC) protocols for non-CO_2 euthanasia of neonate rat pups, whole brains are then dissected away using sterile small scissors and scooped spatula and placed into 35 mm Petri dish containing 4 mL of HA-27 media (*see* **Note 5**). Carefully remove the thin, vascularized dura layer before dissecting away the cortex from the right and left hemispheres.

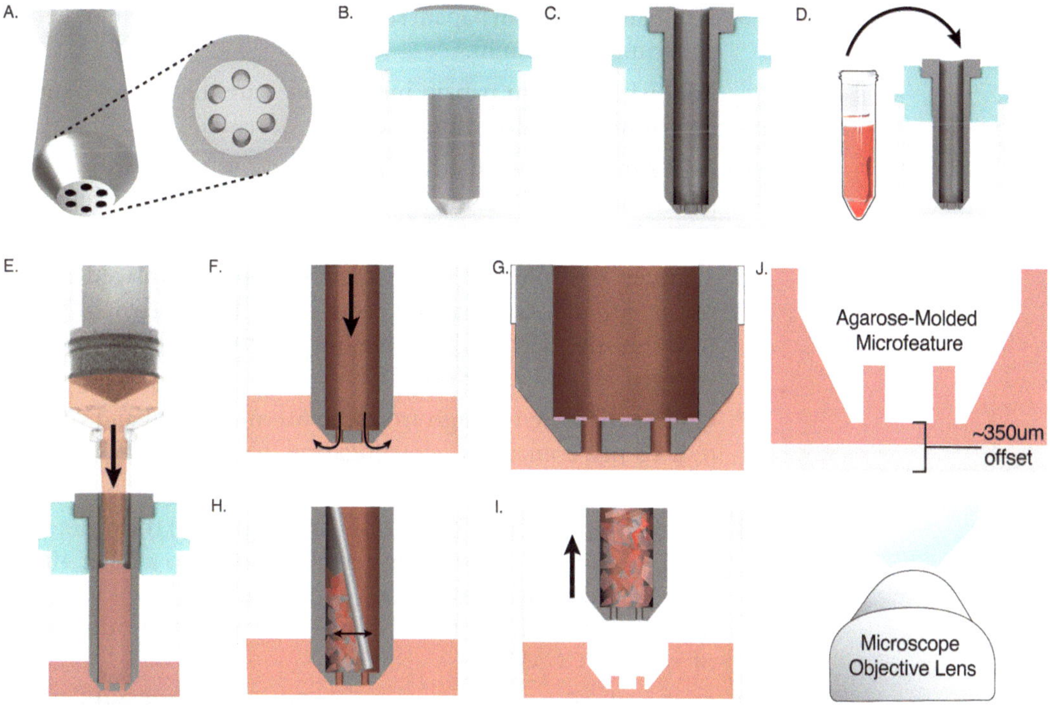

Fig. 2 Agarose-molded microfeatures on multi-well plate. (**a**-**c**) Custom machined stainless steel injection mold is placed atop a single well of a 24-well plate using a 3D-printed PLA-lift piece. (**d**, **e**) Hot agarose is pushed through the injection mold with a luer lock syringe. (**f**) Agarose moves from the injection mold into the well through the peg holes and is allowed to cool until agarose solidifies. (**g**, **h**) To remove the mold from cooled solid agarose, the pegs need to be separated from the agarose inside the injection column using a sterile stainless-steel rod. (**i**, **j**) Once separated, the mold is carefully removed (**i**) leaving behind the agarose-molded microfeature at the bottom of the well. (Reproduced from Atherton et al. [20] with permission from *Scientific Reports*)

Using forceps, gently transfer cortices into a new 35 mm sterile Petri dish containing fresh HA-B7 media.

3. Once all cortices have been collected and pooled together, forceps are used to cut each cortex piece into smaller pieces. Each piece of cortex is cut into approximately one third of its original size.

3.3 Tissue Dissociation and Plate Seeding

1. Prepare the papain solution by adding Hibernate A [minus calcium] to the sterile lyophilized papain powder, bringing final concentration to 2 mg/mL. Solution will be cloudy at first. Mix by gentle inversion, and once papain goes fully into solution, transfer the entire volume to sterile 15 mL conical tube. Place the 15 mL conical in 30 °C water bath for 10 min.
2. Gently place each piece of cortex into the papain vial, being careful to add as little of the HA-B27 media as possible (*see* **Note 6**).

3. Incubate cortices in papain solution at 30 °C for 30 min, gently agitating every 5 min.
4. Following 30 min papain incubation, carefully remove supernatant without disturbing the pellet of tissue.
5. Add 2 mL of prewarmed (37 °C) HA-B7 media to the cortices pellet and triturate using a P1000 pipette tip. Be careful not to overmix.
6. Transfer the solution to a sterile 40 μm filter placed over a sterile 50 mL conical tube and allow for passive diffusion through the filter before washing with another 3 mL of warmed HA-B7 (*see* **Note 7**).
7. Collect the flow-through from the 50 mL conical and transfer to a 15 mL conical for centrifugation. Centrifuge for 5 min at 150 × *g*.
8. Carefully remove the supernatant and resuspend the pellet with 2 mL of warmed CCM. Transfer the solution to a clean 40 μm filter placed over a new 50 mL conical tube. Allow for passive diffusion through the filter before washing with another 3 mL of warmed CCM.
9. Collect the flow-through from the 50 mL conical and transfer to a 15 mL conical for centrifugation. Centrifuge for 5 min at 150 × *g*.
10. Repeat the filter, centrifuge step once more for a total of 3 filters. Following the final centrifugation step, remove the supernatant and add 5 mL of warmed CCM and resuspend the pellet for cell counting.
11. Take a small aliquot of cell suspension, add Trypan Blue at a 1: 2 dilution to count cells by hemocytometer and assess live and dead cell populations to determine cell viability. If viability is lower than 90%, it is not recommended to use for seeding and microtissue formation (*see* **Note 8**).
12. Calculate total number of cells (typically average of 5e6 cells per pup).
13. Centrifuge for 5 min at 150 × *g*. Resuspend pellet in CCM to final concentration of 2.7e8 cells/mL (2.7e6 cells in 10 μL of CCM per microtissue).
14. Remove agarose-molded plates from the incubator and remove all CCM from each well, including the liquid around the pegs while being careful to not touch the bottom of the agarose with the pipette tip.
15. Carefully add 10 μL of cell suspension, dropwise, into the agarose microfeature.
16. Place the plate back into the incubator for 25–30 min to allow cells to settle.

17. After 30 min, remove the plate from the incubator and add 1 mL of CCM dropwise to each well, being careful to not disturb the cell pellet at the bottom of the microfeature. Place the plate back in the incubator at 37 °C 5% CO_2.
18. At 24 h post-seeding (DIV1), the cells will self-assemble into the trampoline shape, and begin to contract around the pegs. Change media on DIV1, and then every other day thereafter (*see* **Note 9**).

3.4 Adeno-Associated Viral Transduction (Fig. 3)

1. Adeno-associated viral (AAV) transduction of microtissues can be performed at various experimental timepoints. For the purpose of this protocol, we describe AAV transduction at DIV1 for delivery of a genetically encoded calcium indicator (GCaMP) and imaging of spontaneous calcium transients and DIV14, 16, and 18.
2. AAV1 particles should be removed from the −80 °C freezer and thawed on ice immediately prior to use. Dilute thawed viral particles in warmed CCM to a final concentration of 1e10 vg/mL.
3. At DIV1, remove microtissue plates from the incubator and replace the 1 mL of CCM in each well with 500 μL of freshly prepared viral media.
4. Following a 3 day viral media incubation, a media change is performed at DIV4.

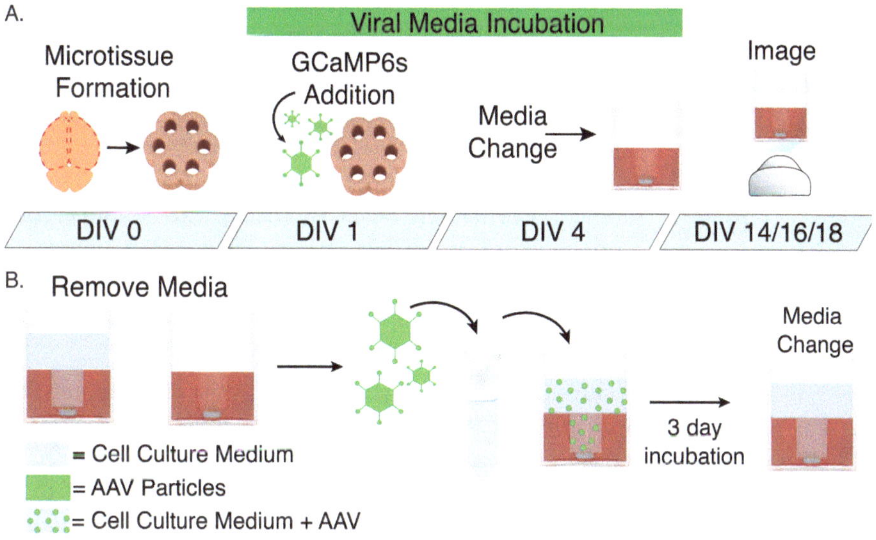

Fig. 3 Adeno-associated viral transduction of three-dimensional cortical microtissues. (**a**) Three-dimensional cortical microtissues are transduced at DIV1 with AAV viral particles to induce fluorescent labeling of neural cell populations and subsequent longitudinal live imaging studies. (**b**) AAV particles are diluted into cell culture media and allowed to incubate over microtissues for 3 days before viral media is removed and replaced with regular cell culture media

5. Microtissues are cultured until live imaging sessions begin at DIV14, with normal media changes every other day (*see* **Note 9**).

3.5 Live Confocal Imaging

1. On imaging day, the microscope room was prepped by first turning on the stage top incubator at least 15 min prior to imaging to warm to 37 °C and pre-equilibrate with pre-mixed 5% $C0_2$. Additionally, sterile h_2O was carefully added by syringe to the humidifying chamber surrounding the plate holder.
2. Microtissue plates were transferred from the cell culture room to the microscopy room for live imaging sessions. Plates were transported in a heated and cushioned container, note this may not be necessary if the transport distance is minimal.
3. For 4 min calcium video acquisition, the microtissue plate was placed on the stage top incubator with the multi-well plate adaptor. Using the 10× objective and an aperture setting of 800 μm, the 488 nm laser and EGFP detector were used to visualize fluorescence in individual microtissues. The objective was centered over the microtissue and brought within the focal plane, maximizing the number of fluorescent cells in the field of view. Time-lapse videos were recorded at 15 frames per second.
4. For high-resolution volumetric imaging, the microtissue was centered over the 30× oil immersion objective. Using both the 488 nm and the 560 nm laser, fluorescent signal from the GCaMP and the mRuby reporters were captured in a single microtissue. Due to the inability to clear the live tissues, a z-stack of approximately 80 μm was taken from the middle of the ~150 μm thick microtissue.

3.6 Immunohistochemistry

1. At the termination of the experiment, remove microtissue plates from the incubator, transfer to a certified fume hood, and aspirate culture media from each well, leaving behind only the agarose mold and the microtissue. Carefully add 1.5 mL of a 4% paraformaldehyde 8% sucrose fixative solution to each well and incubate at room temperature for 2 h before removing fixative solution and replacing with 1.5 mL of 1× PBS. Perform 3 separate 2 h washes with 1× PBS prior to removing microtissue plate from the fume hood.
2. Remove the agarose gels, including the microtissues, from each well and transfer to a clear surface (*see* **Note 10**). Using a clean razor blade, carefully cut away the excess agarose from around the microtissue, taking care not to cut into the microtissue. Once the surrounding agarose has been cut away as much as possible, transfer the microtissue in the remaining agarose to a

glass scintillation vial filled with warmed PBS. Ensure that the agarose and microtissue are fully submerged in the PBS and place the vial on a hotplate at 200 °C for approximately 10 min. Gently agitate the vial every minute until the agarose has melted away and the microtissue is floating free. Once the fixed microtissues are floating freely, transfer to a 35 mm Petri dish and wash several times with warmed PBS to ensure any residual agarose is melted away from the microtissue (*see* **Note 11**).

3. Once microtissue have been harvested from the agarose molds, transfer to a 48-well tissue culture plate and wash with a final 500 μL of room temperature PBS.
4. Remove PBS solution from each well and replace with 300 μL of Blocking Solution. Place plate on shaker with gentle agitation and incubate for 2 h in blocking solution.
5. Remove blocking solution and replace with 300 μL of primary antibody cocktail. Parafilm the plate and return to shaker for overnight incubation, again with gentle agitation.
6. On the following day, remove the primary antibody solution and replace with 500 μL of 1× PBT solution and return the plate to the shaker for a 2 h wash. Repeat 2 h PBT wash step once more before removing PBT and adding 300 μL of blocking solution to each well and incubating for additional 2 h on the shaker. After 2 h blocking, replace each well with 300 μL of secondary antibody cocktail. Parafilm the plate and return to shaker for overnight incubation.
7. On the following day, remove secondary antibody solution and replace with 500 μL of 1× PBT solution to each well. Perform 4 separate 1 h washed with 500 μL of PBT to each well. Following the washes, add 300 μL of DAPI solution to each well and place on the shaker for a 1 h incubation. Remove DAPI solution and replace each well with 500 μL of PBS. Store plate at 4 °C until needed for clearing and imaging.
8. On the day of histology imaging, remove the plate from 4 °C refrigeration and remove PBS from each well. Transfer microtissues to 1.5 mL microcentrifuge tubes and add a minimum of 500 μL of clearing solution #1 and place on the shaker for 15 min (*see* **Note 12**).
9. Remove clearing solution #1 and replace with a minimum of 500 μL of clearing solution #2, place on the shaker for 5 min incubation.
10. Remove 500 μL of clearing solution #2 and replace with a fresh 500 μL clearing solution #2 and place back on the shaker for 30–60 min (microtissue should no longer be opaque).

11. Microtissues can then be imaged for immunohistochemical labeling of cell type-specific antigens of interest.

4 Notes

1. We chose to 3D print the lift piece in PLA, however, we note that other common printing filaments would be sufficient as well. The piece was designed using AutoDesk Fusion360 software. The well dimensions (inner well diameter and inner well depth) were taken from the plate manufacturer and used to determine the diameter of the lift piece as well as the lift height however this must be optimized if using different plate manufacturers or different printing filaments. Several iterations of the print were necessary due to the limited printing resolution of the extrusion printer used (Prusa i3mk3). Before agarose molds were made, the PLA-lift pieces were lightly sprayed with 70% ethanol, allowed to air dry in the biosafety cabinet, and then placed under UV light for 20 min to decontaminate before use.
2. The Tokai Hit STX stage top incubator was selected due to its compatibility with the Olympus FV3000 Confocal microscope. Depending on the microscope manufacturer, a different environmental chamber can be used for live imaging. However, it is recommended that for any imaging session lasting 30 min or longer, temperature and CO_2 should be provided to the plate to ensure microtissue health. We observed distinct differences in whole tissue firing rates of microtissues that were imaged at room temperature and ambient oxygen levels for periods longer than 30 min.
3. Depending on the specific antibodies selected for immunolabeling, we suggest doubling the concentration recommended by the vendor for 2D immunofluorescent labeling. We provide specific concentrations used for a triple-antibody primary cocktail used to label neuronal nuclei, astrocytes, and microglia (*see* Table 1).
4. Media should be changed a minimum of 2 times prior to seeding the cells into the agarose molds. Once the agarose molds have been made, the plate needs to be stored in a warm, humid environment (preferably a sterile autoclave). We have found that evaporation of the media causes a precipitate to form on the agarose mold thus we recommend storing in the incubator until cell seeding.
5. Moving as quickly as possible when collecting the cortical tissue is important. We have found the most efficient workflow is to collect whole brains from the entire litter into 35 mm Petri

Table 1
The antibodies used for immunostaining for confocal microscopy

Primary antibody	Vendor	Concentration
Mouse Anti-NeuN	Millipore, MAB377	1:500
Chicken Anti-GFAP	Millipore, AB5541	1:200
Rabbit Anti-Iba1	FUJIFILM, 01919741	1:500
Secondary antibody	**Vendor**	**Concentration**
Anti-Mouse CY3	Jackson ImmunoResearch, 115-165-068	1:800
Anti-Chicken AF488	Thermo Fisher Scientific, A11039	1:800
Anti-Rabbit AF635	Thermo Fisher Scientific, A31577	1:800

dishes (2 brains per 35 mm dish with 4 mL of HA-B27 media), then move on to dissect away the cortices. It is best to use the scissors for midline incision along the scalp and skull before scooping out the whole brain with the scooped spatula. To remove the dura, pinch with the forceps to make a gentle incision into cortical tissue. Then using the reverse dull side of the forceps, gently tease away the dura, followed by dissection of the cortex from each hemisphere.

6. Prepare the Papain solution after all the whole brains have been collected into 35 mm Petri dishes and before the cortical dissection begins. Be sure to wait for the Papain to fully go into solution before transferring to a sterile 15 mL conical tube and placing in the 30 °C water bath. We typically use 6 mg of papain for 8–10 sets of cortices (whole brains).

7. If performing a large dissociation (10 or more pups) it is suggested to use two filters for the first filtration step as there will be more cell debris to clog the filter. It is fine to use more than 5 mL of HA-B27 to wash the filters during this first step. At this point, now that the cells in a single cell suspension, we recommend using only a p1000 tip for transfer of liquids and cells as the wider-bore opening of the tip will minimize shearing stress to the cells and help to maintain viability.

8. To perform cell count, we recommend a 1:4 dilution of cell suspension into CCM, then an additional 1:2 dilution into trypan blue. The total dilution factor for cell counting is then 1:8.

9. To perform a media change, remove all media from the top of agarose and replace. Do not remove any media that remains within the microfeature inner well. We recommend feeding microtissues every other day. However, allowing the microtissues to sit an extra day (e.g., over the weekend) and be put on a MWF feeding schedule has not proven harmful to the microtissue health

and overall viability. Additionally, it is recommended that microtissues are not fed immediately before a live imaging session. If microtissues need to be fed on an imaging day, we suggest feeding after the imaging session is complete.

10. Using curved forceps, carefully pierce the sides of the agarose, taking care not to disturb the microtissue at the center, and using a "corkscrew" motion gently lift upward to release the agarose mold from the bottom of the well. Transfer to a clean surface to cut away the excess agarose prior to melting.
11. For transfer of free-floating microtissues, use a clean razor blade to cut the end of a p1000 pipette tip, then use the pipette with cut tip as normal. The opening should be just larger than the microtissue diameter.
12. Due to the high viscosity of the clearing solutions, it is difficult to achieve adequate movement of the microtissue in the solution while on the plate shaker. To mitigate this, we have found it best to use 1.5 mL centrifuge tubes and to tape them together in a round sphere shape, packed with foil. This foil ball, containing the microcentrifuge tubes at the center, will be free to roll around the plate shaker tray and allow the clearing solution to adequately slosh around the microtissues in each tube.

Acknowledgments

This work was supported in part by Brown University internal research support (D.A.B.) and by the Center for Neurorestoration and Neurotechnology (N2864-C) from the United States (U.S.) Department of Veterans Affairs, Rehabilitation Research and Development Service, Providence, RI. The contents of this manuscript do not represent the views of VA or the United States Government. This work was also supported by NIA award R21AG077697 (D.A.B.) and the Brown University Carney Institute for Brain Science Zimmerman Innovation Award in Brain Science.

References

1. Goshi N, Morgan RK, Lein PJ et al (2022) Correction to: a primary neural cell culture model to study neuron, astrocyte, and microglia interactions in neuroinflammation. J Neuroinflammation 19:49
2. Anderl JL, Redpath S, Ball AJ (2009) A neuronal and astrocyte co-culture assay for high content analysis of neurotoxicity. J Vis Exp
3. Lancaster MA, Renner M, Martin CA et al (2013) Cerebral organoids model human brain development and microcephaly. Nature 501:373–379
4. Clevers H (2016) Modeling development and disease with organoids. Cell 165:1586–1597
5. Zhao J, Fu Y, Yamazaki Y et al (2020) APOE4 exacerbates synapse loss and neurodegeneration in Alzheimer's disease patient

iPSC-derived cerebral organoids. Nat Commun 11:5540
6. Juurlink BHJ, Walz W (1999) Neural cell culture techniques. In: Boulton AA, Baker GB, Bateson AN (eds) Cell neurobiology techniques. Humana Press, Totowa, NJ, pp 53–102
7. Delong GR, Sidman RL (1970) Alignment defect of reaggregating cells in cultures of developing brains of reeler mutant mice. Dev Biol 22:584–600
8. Irons HR, Cullen DK, Shapiro NP et al (2008) Three-dimensional neural constructs: a novel platform for neurophysiological investigation. J Neural Eng 5:333–341
9. Zhuang P, Sun AX, An J et al (2018) 3D neural tissue models: from spheroids to bioprinting. Biomaterials 154:113–133
10. Fennema E, Rivron N, Rouwkema J et al (2013) Spheroid culture as a tool for creating 3D complex tissues. Trends Biotechnol 31: 108–115
11. Dingle YT, Boutin ME, Chirila AM et al (2015) Three-dimensional neural spheroid culture: An in vitro model for cortical studies. Tissue Eng Part C Methods 21:1274–1283
12. Xiang Y, Tanaka Y, Patterson B et al (2017) Fusion of regionally specified hPSC-derived organoids models human brain development and interneuron migration. Cell Stem Cell 21: 383–398 e387
13. Seeds NW (1973) Differentiation of aggregating brain cell cultures. In: Sato G (ed) Tissue culture of the nervous system. Springer, Boston, MA, pp 35–53
14. Peng L, Juurlink BHJ, Hertz L (1991) Differences in transmitter release, morphology, and ischemia-induced cell injury between cerebellar granule cell cultures developing in the presence and in the absence of a depolarizing potassium concentration. Dev Brain Res 63:1–12
15. Trujillo CA, Gao R, Negraes PD et al (2019) Complex oscillatory waves emerging from cortical organoids model early human brain network development. Cell Stem Cell 25:558–569 e557
16. Ulloa Severino FP, Ban J, Song Q et al (2016) The role of dimensionality in neuronal network dynamics. Sci Rep 6:29640
17. Quadrato G, Nguyen T, Macosko EZ et al (2017) Cell diversity and network dynamics in photosensitive human brain organoids. Nature 545:48–53
18. Lam D, Enright HA, Cadena J et al (2019) Tissue-specific extracellular matrix accelerates the formation of neural networks and communities in a neuron-glia co-culture on a multielectrode array. Sci Rep 9:4159
19. Atherton E, Brown S, Papiez E et al (2021) Lipopolysaccharide-induced neuroinflammation disrupts functional connectivity and community structure in primary cortical microtissues. Sci Rep 11:22303
20. Atherton E, Hu Y, Brown S et al (2022) A 3Din vitromodel of the device-tissue interface: functional and structural symptoms of innate neuroinflammation are mitigated by antioxidant ceria nanoparticles. J Neural Eng 19: 036004
21. Boutin ME, Kramer LL, Livi LL et al (2018) A three-dimensional neural spheroid model for capillary-like network formation. J Neurosci Methods 299:55–63
22. Schell JY, Wilks BT, Patel M et al (2016) Harnessing cellular-derived forces in self-assembled microtissues to control the synthesis and alignment of ECM. Biomaterials 77:120–129

Chapter 13

Method to Generate Dorsal Forebrain Brain Organoids from Human Pluripotent Stem Cells

Rebecca Sebastian, Narciso S. Pavon, Yoonjae Song, Karmen T. Diep, and ChangHui Pak

Abstract

Region-specific brain organoids, such as dorsal forebrain brain organoid, have become increasingly useful to model early brain development. Importantly, these organoids provide an avenue to investigate mechanisms underlying neurodevelopmental disorders, as they undergo developmental milestones resembling early neocortical formation. These milestones include the generation of neural precursors which transition into intermediate cell types and subsequently to neurons and astrocytes, as well as the fulfillment of key neuronal maturation events such as synapse formation and pruning. Here we describe how to generate free-floating dorsal forebrain brain organoids from human pluripotent stem cells (hPSCs). We also describe validation of the organoids via cryosectioning and immunostaining. Additionally, we include an optimized protocol that allows high-quality dissociation of the brain organoids to live single cells, a critical step for downstream single-cell assays.

Key words Dorsal forebrain organoids, Corticogenesis, Neurodevelopment, Human embryonic stem cells (hESCs), Induced-pluripotent stem cells (iPSCs)

1 Introduction

To date, neuronal cells derived from hPSCs have provided a robust platform for studying various neurological diseases. These cultures are primarily 2D and are often guided toward a specific neuronal fate through the use of small molecules. For example, inducing dual-SMAD inhibition by inhibiting transforming growth factor-β (TGF-β) and bone morphogenic protein (BMP) pathways via SB431542 and dorsomorphin drives the differentiation of hPSCs toward a neural progenitor fate, which then can be further patterned into specific neuronal types through subsequent addition of appropriate small molecules [1]. hPSC-derived neurons have been widely accepted and have amassed a large library of comparative research. However, these 2D models lack spatial architecture

Yu-Wen Alvin Huang and ChangHui Pak (eds.), *Stem Cell-Based Neural Model Systems for Brain Disorders*, Methods in Molecular Biology, vol. 2683, https://doi.org/10.1007/978-1-0716-3287-1_13,

and cellular diversity and thus, do not accurately represent the in vivo cellular environment. Still, the paradigms and methodologies that were established for 2D cell culture have been instrumental for the innovation of 3D cell culture [2].

Recently, cell aggregates have been used to form 3D structures called organoids which serve as an alternative model for tissue formation in vitro. Compared to 2D differentiated cells, 3D cultures hold the advantages of better representing human physiology, displaying unique organ morphologies, and retaining cellular diversity, all of which contribute to recapitulating the intricacies of in vivo brain development. Similar to 2D cells, depending on the patterning factors applied, brain organoids can develop into cerebral organoids (unguided) or region-specific brain organoids (guided) [3–5]. The unguided cerebral organoids forego patterning factors and instead harness the intrinsic differentiation capabilities of hPSCs to produce a heterogeneous cerebral organoid containing tissue representative of various brain regions and cell types [6]. Region-specific organoids arise from guided differentiation through the addition of relevant morphogens to direct organoid development toward a specific fate that homogeneously represents a particular region of the brain i.e., optic cup, thalamus, hindbrain, or cortical organoids [2, 3, 5, 7]. Improved reproducibility is possible with the use of patterning factors to consistently guide brain organoid development to a specific regional fate [2].

Here, we introduce a protocol that generates dorsal forebrain organoids from hPSCs using guided differentiation based on previous publication (Fig. 1) [8]. This protocol offers the benefit of using free-floating cultures without the need for matrigel embedding. Using this specific protocol, our group has validated cell type diversity and functionality of developing dorsal forebrain organoids using immunostaining, single-cell gene expression, and calcium imaging [9]. At early developmental timepoints (~ day 21), we see strong expression of neural progenitor markers such as SOX2 and PAX6, that decline overtime as the brain organoids mature. As the brain organoids reach intermediate stages of development (~ day 50 and onward), EOMES+ intermediate progenitor cells, as well as SATB2+ upper and BCL11B+ deeper layer cortical neurons start to appear. At ~day 100, early astrogenesis and active neuronal maturation are detected by the presence of S100B+ and NEUN+ markers, respectively. Beyond day 100, we observe SYN+ presynaptic and HOMER+ postsynaptic markers along MAP2+ dendrites, indicative of ongoing synaptogenesis within the dorsal forebrain organoids (*see* Table 1). Ultimately, this protocol will generate a high yield of dorsal forebrain organoids displaying

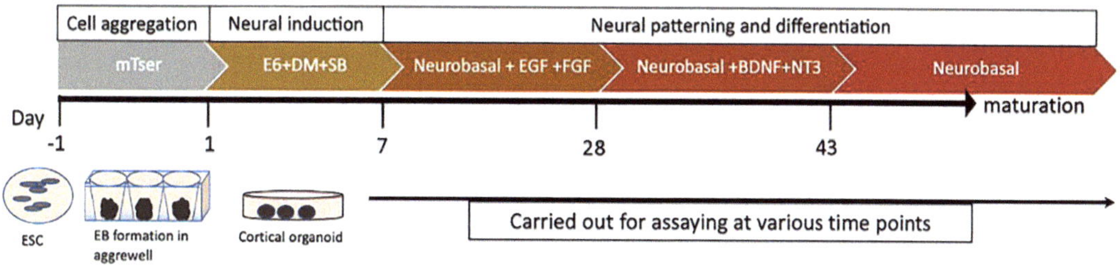

Fig. 1 Timeline: schematic showing protocol utilized for organoid generation, including the media and the respective media components at different time points

Table 1
Antibodies for various stages of organoid development: shows various antibodies that are expected to express at various time points during organoid development

Marker (antibody)	Tested temporal expression
PAX6	~day 21 onwards
KI67	~day 21 onwards
NESTIN	~day 21 onwards
SOX2	~day 21 onwards
FOXG1	~day 21 onwards
HOPX	~day 100 onwards
TBR1	~day 100 onwards
TBR2	~day 100 onwards
SATB2	~day 100 onwards
CTIP2	~day 100 onwards
S100B	~day 100 onwards
GFAP	~day 100 onwards
MAP2	~day 100 onwards
SYNAPSIN	~day 100 onwards
PSD95	~day 100 onwards
HOMER	~day 100 onwards

relatively homogeneous cell populations and prototypical developmental trajectories representing neocortical formation. Thus, the features exemplified by this brain organoid protocol will be useful for teasing apart the molecular underpinnings of neurodevelopmental disorders and for the integration of high throughput pharmacological assays.

2 Materials

2.1 Cell Lines

Human embryonic stem cells (hESCs): H1 (WA01) cells.

Human-induced pluripotent stem cells (iPSCs).

2.2 Reagents Necessary for hPSC Culture

Reagents for hPSC culturing	
Reagent/solution	**Stock concentration**
mTeSR+ Basal Medium	N/A
mTeSR+ 5× Supplement	N/A
Dulbecco's Phosphate-Buffered Saline; no calcium, no magnesium (DPBS)	N/A
ReLeSR	N/A
Y-27632 Dihydrochloride (Y Compound)	10 mM
Matrigel	8–12 mg/mL

2.3 Brain Organoid Generation

Reagents for organoid generation		
Reagents	**Stock conc**	**Final conc.**
Dorsomorphin (DM)	2.5 mM (1 ug/mL)	2.5 uM (1 ng/mL)
SB431542 (SB)	10 mM	10 uM
Human-EGF	20 ug/mL	20 ng/mL
Human-FGF2	20 ug/mL	20 ng/mL
Human BDNF	20 ug/mL	20 ng/mL
Human-NT3	20 ug/mL	20 ng/mL
Penicillin-streptomycin	100×	1% (vol/vol)
B27 Supplement without vitamin A (B27)	50×	2% (vol/vol)
Essential 6	N/A	N/A
Neurobasal	N/A	N/A

2.3.1 Neural Induction Medium

1. To Essential 6 medium (E6), add 1% penicillin-streptomycin, 2.5 uM DM, and 10 uM SB.

2.3.2 Neural Differentiation Medium

1. To Neurobasal medium, add 2% B27 without vitamin A, 1% Glutamax, 1% penicillin-streptomycin, 20 ng/mL EGF, and 20 ng/mL FGF. Antifungal can be added if desired.

2.3.3 Neural Maturation Medium

1. To Neurobasal medium, add 2% B27 without vitamin A, 1% Glutamax, 1% penicillin-streptomycin, 20 ng/mL BDNF, and 20 ng/mL NT3.
2. For maturation medium without growth factors: to Neurobasal medium, add 2% B27 without vitamin A, 1% Glutamax, 1% penicillin-streptomycin.

2.4 Cryosectioning

Materials for cryosectioning	
Consumables	**Quantity**
15 mL and/or 50 mL conical tubes	Varies
Wide bore tips	Varies
Cryomolds	Varies
Ice blocks	1
Dry ice	Varies
200 proof ethanol	~50–100 mL
Styrofoam cooler	1
Microscope glass slides	Varies
Heat-warming plate	N/A
Reagents	Quantity
1× DPBS solution or 1× PBS	Varies
32% PFA	3 mL
Sucrose	15 g
Gelatin from porcine skin	7.5 g

2.5 Antigen Retrieval

Materials for antigen retrieval	
Consumables	**Quantity**
Coplin jar	Varies
Steamer	1
pH meter	1
Reagents	Quantity
Trisodium citrate (dihydrate)	2.94 g
Tween® 20	0.5 mL
Triton-X	1 mL
1× PBS	499 mL

2.6 Immunohistochemistry

Materials for immunohistochemistry	
Consumables	**Quantity**
Coplin jar	Varies
PAP pen (hydrophobic pen)	1
Humidified chamber	1
Rectangle glass coverslips #1.5	Varies
Reagents	Quantity
Triton-X	1 mL
1× PBS	499 mL
Normal Goat Serum (NGS)	Varies
DAPI Fluoromount	Varies

3 Methods

3.1 hPSC Culture

1. Ensure cells have reached 70–100% confluence.
2. Coat a 6-well plate with Matrigel and incubate it for 30 min at 37 °C, 5% CO_2.
3. Warm mTeSR media in a water bath.
4. In a 15 mL conical, add 12 μL of Y compound to 12 mL of mTeSR media.
5. Vacuum out old media.
6. Add 2 mL of DPBS into a single well of cells.
7. Vacuum out the DPBS.
8. Add 1 mL of ReLeSR to the same well.
9. Immediately vacuum out the ReLeSR.
10. Incubate the cells for 5 min at 37 °C, 5% CO_2.
11. Add 1 mL of mTeSR + Y compound into the well and transfer the media and cells solution into a 15 mL conical.
12. Add 11 mL of mTeSR + Y compound into the 15 mL conical.
13. Remove plate coated with matrigel from the incubator.
14. Vacuum out the matrigel from the 6-well plate.
15. Distribute 2 mL of solution from the 15 mL conical into each of the wells.
16. Gently swirl the solution within the plate to spread the cells evenly.
17. Incubate overnight at 37 °C, 5% CO_2.

3.2 EB Generation

Option 1: Aggrewell method.

(Day −1)

Prepare Aggrewell Plate (24-well format)

1. Add 500 uL Anti-Adherence Rinsing Solution (StemCell) to each well of the 24-well Aggrewell 800 plate that you are using.
2. Spin 5 min at 200–300 × *g*, then check under a microscope to make sure there are no air bubbles, then aspirate solution.
3. Next, rinse each well with 1 mL of mTeSR and aspirate. Proceed to preparing cells for seeding on aggrewell.

Prepare Cells

1. Wash each well with DPBS, vacuum.
2. Add 1 mL accutase per well. Incubate at room temperature for 5 min.
3. Remove from the incubator and add 1 mL mTeSR+Y compound (per well in a 6-well plate) to lift cells from wells and then transfer to a 15 mL conical (*see* **Note 1**).
4. Spin, 5 min at 200–300 × *g*. Remove supernatant and resuspend pellet in 1 mL of mTeSR+Y compound.
5. Count the total number of cells in resuspension and then calculate the volume needed for 3×10^6 cells.
6. Transfer volume needed for 3×10^6 cells to a new conical and bring up total volume to 2 mL with mTeSR+Y compound.
7. Transfer all contents of 15 mL conical from previous step to one of the wells of the Aggrewell plate prepared earlier.
8. Centrifuge Aggrewell plate at 100 × *g* for 3 min, then incubate for 24 h at 37 °C (*see* **Note 2**).

(Day 0)

1. Once proper embryoid bodies (EB) have formed, using a cut P1000 tip or a wide bore P1000 tip, gently pipet media in each Aggrewell up and down and transfer EBs to a labeled conical.
2. Rinse well with additional 1 mL of Neural Induction media to collect any EBs left behind.
3. Once all EBs have settled to the bottom of conical, carefully aspirate media in conical and replace with 2 mL E6 media.
4. Then transfer EBs to two 10 cm ultra-low attachment dishes with 10 mL of Neural Induction media for each dish, dividing equally (*see* **Note 3**).

(Day 1)

1. No media change. EBs are incubated for another 24 h.

Option 2: Ultra-Low Attachment 96-Well U-bottom Plate Method.

(Day −1)

1. Wash each well with DPBS, vacuum.
2. Add 1 mL Accutase per well. Incubate at room temperature for 5 min.
3. Remove from the incubator and add 1 mL mTeSR+Y compound (per well) to lift cells from wells and then transfer to a 15 mL conical.
4. Spin, 5 min at 200–300 × *g*. Remove supernatant and resuspend pellet in 1 mL of mTeSR+Y compound. (The goal is to seed 10,000 cells in 100 uL of mTSeR+Y compound per U-bottom well).
5. Count the total number of cells in resuspension and then calculate the volume needed for 10,000 cells per U-bottom well, i.e.: 10,000 × 96 wells = 96,000 cells (*see* **Note 3**).
6. Transfer volume needed for 10,000 cells per U-bottom well to a new conical and bring up total volume in conical to make 100 uL per well, i.e.: For 100 wells, total volume should be 10 mL (*see* **Note 4**).
7. Transfer all contents of conical to reservoir. Then using a multichannel pipette, dispense cells into 96-well U-bottom plate (*see* **Note 5**).
8. Allow cells to aggregate overnight in mTeSR+Y compound, then proceed to Day 0.

(Day 0)

1. After cells have aggregated, add 100 uL of fresh Neural Induction media.

(Day 1)

1. No media change. EBs are incubated for another 24 h.

3.3 Neural Induction (Vontinued from Day 0)

(Day 2 to Day 5)

1. Carefully aspirate most of the media and replace it with Neural Induction media (*see* **Note 6**).
2. Media change every day up to day 5, as mentioned above with Neural Induction media.

3.4 Patterning and Differentiation

(Day 6 to Day 24)

1. Replace media with Neural Differentiation medium. Replace media daily up to day 15. If desired, start shaking on an orbital shaker on this day (*see* **Note 7**).
2. Replace media every other day from day 16 to day 24.

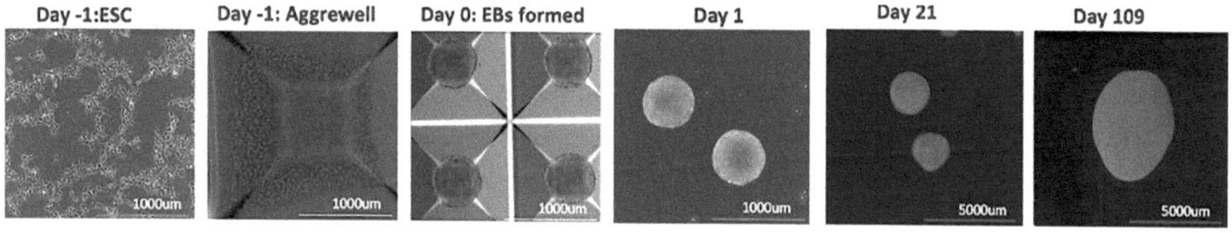

Fig. 2 Anticipated results: figure shows progression of forebrain organoids from hPSC stage to mature organoid

For 96-well U-bottom plate method:

1. Use cut P200 or wide bore P200 tip to transfer organoids from U-bottom wells to 10 cm ultra-low attachment dishes with 10 mL of Neural Differentiation media per dish. Here onward, media changes are done in the 10 cm ultra-low attachment dishes. Divide organoids equally between dishes. Might require pipetting up and down several times to release organoid from wells.

3.5 Maturation

(Day 25–43)

1. On day 25, replace media with Neural Maturation medium. Continue replacing with Neural Maturation medium up to day 43.
2. Day 43 onwards replace media without growth factors. Media changes can be performed every 3–4 days until desired day of collection (Fig. 2) (*see* **Notes 8** and **9**).

3.6 Cryosectioning

Day 1: Fixation.

Preparation

1. Make fresh 4% PFA/DPBS solution by adding 3 mL of 32% PFA to 21 mL of DPBS to make 24 mL of 4% PFA/DPBS.
2. Make fresh 30% sucrose solution by dissolving 15 g of sucrose in 50 mL of DPBS.
3. Make fresh gelatin solution by dissolving 10 g of sucrose in 100 mL of D-PBS. Then add 7.5 g of gelatin from porcine skin to the solution. Mix thoroughly (*see* **Note 10**).

Fixing Organoids

1. Collect organoids at the desired time point after culturing and transfer 5–6 organoids to each conical tube depending on organoid size using a wide bore tip or a cut p1000 tip (*see* Table 2).
2. Remove excess media and wash 3× with DPBS.
3. Then, fix organoids by adding 0.1–0.2 mL of 4% PFA/DPBS solution per organoid in conical, depending on organoid size (*see* Table 2). Then incubate overnight at 4 °C.

Table 2
Suggested volumes for fixative and cryopreservation solutions: shows different time points (often serves as an indicator for organoid size) and the appropriate volume of solutions and number of organoids that can be cryopreserved in a 15 mL conical tube

Day	Volume for fixation and cryopreservation	# of Organoids/tube
21	~0.1–0.2 mL/organoid	5–10/15 mL conical
50	~0.5–1 mL/organoid	5–10/15 mL or 50 mL conical
100+	~0.5–1 mL/organoid	5–10/50 mL conical

Day 2: Cryoprotection.

1. For cryoprotection the next day, remove the fixation solution and wash 3× with DPBS.
2. Add 0.1–0.2 mL of 30% sucrose solution per organoid in each conical depending on organoid size (*see* Table 2). Then incubate at 4 °C for 24–48 h or until organoids have settled to the bottom of conical.

Day 3: Gelatin Embedding.

1. Once organoids have sunk to the bottom of conical, use freshly made gelatin solution or prewarm gelatin solution in a 37 °C water bath if stored in 4 °C.
2. Label cryomolds and place them on an ice block. Add 1 mL of the warmed gelatin to the cryomold and allow the gelatin to solidify. Then add another 1 mL of gelatin to the conical tube with organoids. Using a cut tip, transfer all organoids in the conical tube to a labeled cryomold. Number of organoids per cryomold can vary based on preference.
3. Then add an additional 1 mL to fill the remainder of the mold.
4. Make dry ice/ethanol slurry by adding dry ice to 100% ethanol in a styrofoam cooler. When ethanol stops boiling excessively, add the molds to the dry ice/ethanol slurry until they are completely frozen.
5. Store the frozen molds at −80 °C until ready for cryosectioning.

Cryosectioning.

1. Before cryosectioning, let the organoids in cryomold blocks acclimate in the cryostat for 30 min to an hour at least.
2. Collect two organoid sections per microscope glass slide. Place all collected sections onto a heat-warming plate to help evaporate remaining gelatin residue.
3. Sectioning to desired thickness. Label slides and store in −20 °C or proceed to immunostaining immediately.

3.7 Immunohistochemistry

Preparation

1. Make 500 mL of 0.02% PBS/T solution by adding 1 mL of Triton-X to 499 mL of PBS.
2. Make 10 mL of 10% NGS blocking solution by diluting 1 mL of normal goat serum (NGS) in 9 mL of 0.02% PBS/T from **step 1**.
3. Make 10 mL of 5% NGS blocking solution by diluting 0.5 mL of normal goat serum (NGS) in 9.5 mL of 0.02% PBS/T from **step 1**.

Day 1: Blocking and primary antibody staining (overnight).

1. Obtain slides from storage and thaw at room temperature for about 10 min.
2. Next, using a PAP pen, draw a rectangle around the sections on the slide (1 rectangle per 2 sections on each slide). Let it dry at room temperature for 15 min.
3. If antigen retrieval is needed (*see* **Note 11**), proceed to antigen retrieval steps (*see* Subheading 3.8). If antigen retrieval is not necessary, proceed to the next step.
4. Wash the slides with warm 0.02%PBS/T for 10 min to remove extra gelatin. Using about 500 uL of 0.02% PBS/T for each slide. This step may be repeated if excess gelatin remains on organoid sections.
5. In a humidified chamber, pipet 10% NGS in 0.02% PBS/T onto slides. Using about 500 uL of blocking solution per slide. Incubate for 1 h at room temperature. Incubation can be done with light agitation/shaking using an orbital shaker or belly dancer.
6. After incubation, remove the blocking solution and rinse with 0.02% PBS/T.
7. Prepare primary antibody mix by diluting primary antibodies to desired concentrations in 5% NGS in 0.02% PBS/T.
8. In a humidified chamber, pipet primary antibody mix onto slides. Using 300–500 uL of primary antibody mix per slide.
9. Cover chamber and incubate in 4 °C overnight.

Day 2: Secondary antibody staining and mounting with coverslips.

1. Remove the primary antibody mix and wash slides 3 times with 0.02% PBS/T. 5 min each wash.
2. Prepare secondary antibody mix by diluting primary antibodies and DAPI to desired concentrations in 0.02% PBS/T.

3. In a humidified chamber, pipet secondary antibody mix onto slides. Using about 300–500 uL of secondary antibody mix per slide.
4. Cover and protect from light. Incubate at room temperature for 2 h.
5. After 2 h, remove the secondary antibody mix and wash the slides 3 times with 0.02% PBS/T. 5 min each wash.
6. Remove excess liquid around sections, but do not completely dry it out.
7. Dispense enough mounting media to properly cover organoid sections and gently put a coverslip on. Be careful not to create any bubble on organoid sections.
8. Dry slides and store for imaging.

3.8 Antigen Retrieval

Preparation

1. Make 1 L Citrate buffer by adding 2.94 g of Trisodium citrate into 950 mL of DI water. Use a pH meter to adjust pH to 6.0. Then add 0.5 mL of Tween® 20 to the solution. Add DI water to get the volume of the solution up to 1 L.
2. Make 500 mL of 0.02% PBS/T solution by adding 1 mL of Triton-X to 499 mL of PBS.

Antigen Retrieval

1. Fill Coplin jar with citrate buffer and immerse slides with organoid sections into the jar.
2. Place the jars inside a steamer and steam for 20 min.
3. While waiting, prepare another set of Coplin jars and fill with PBS/T.
4. After 20 min, carefully remove Coplin jars from the steamer and transfer slides to Coplin jars with 0.02% PBS/T.
5. Wash the slides in 0.02% PBS/T 3 times. 5 min each wash. A belly dancer can be used for wash steps.
6. Proceed to primary antibody staining.

3.9 Live Single-Cell Dissociation for scRNAseq

Preparation

1. Make 2 mL of fresh digestion solution by adding 1.9 mL of 1× HBSS, 50 uL papain, 11 uL 0.5 M EDTA, and 7 mg L-Cysteine. Then mix the solution and warm it in 37 °C for 15 min until no longer cloudy. Finally, sterilize filter with a 0.22 um filter.
2. Prepare DNAse I solution by reconstituting DNAse in 500 uL 1× HBSS. Gently pipet so as to not shear the DNAse. Use 50 uL of DNAse I per 1.9 mL of digestion solution.

Dissociation of organoids into single cells

1. Rinse organoids 3 times with 1× HBSS (10× HBS salt, 1 M HEPES, and 0.004 M NaHCO3 diluted to 1×).
2. Mince organoids into small pieces using a sharp sterile razor and transfer to a 15 mL conical with digestion mixture. Incubate for 15 min in 37 °C.
3. After incubation, triturate digestion mixture containing organoids with DNase I solution from above. Then incubate again for another 10 min at 37 °C.
4. Next filter with 70 uM filter followed by a 30 uM filter.
5. After filtering, centrifuge and pellet cell mixture. Then resuspend in culture media. Before sequencing, cells are filtered with 40 μM FlowMi pipet to remove debris.

4 Notes

1. EBs should be formed using hPSCs that have been passaged 3–4 times (i.e., stable passages will increase the chances of better EB formation).
2. Some lines may need to incubate longer (up to 48 h) for proper EB formation. (In this case, if needed media can be replaced by gently aspirating and replacing without disturbing cells in the Aggrewell).
3. As EBs grow larger, they should be divided further into dishes to avoid fusion and malnourishment.
4. Account for pipetting error by preparing for at least 4–5 more extra wells.
5. Remember to pipette up and down every time before dispensing. This ensures an even distribution of cells (cells can settle at the bottom of the reservoir).
6. Only half media change is performed in 96-well U-bottom plate (i.e., until organoids are transferred to 10 cm ultra-low attachment dishes on day 6).
7. Once EBs look healthy and stable (usually between day 6 and above), EBs can be shaken on an orbital shaker. Shaking can be done at 55 rpm for 10 cm dishes as this reduces fusion. Shaking speed might need to be adjusted according to the diameter of the dish. 55–60 rpm appears to be ideal for the 10 cm dishes.
8. If the medium in dishes turns yellow too quickly (sometimes with visible increase in cell debris), it could mean that there are too many organoids in the dish. Try to keep 15–20 organoids per dish. Less if organoids are larger.

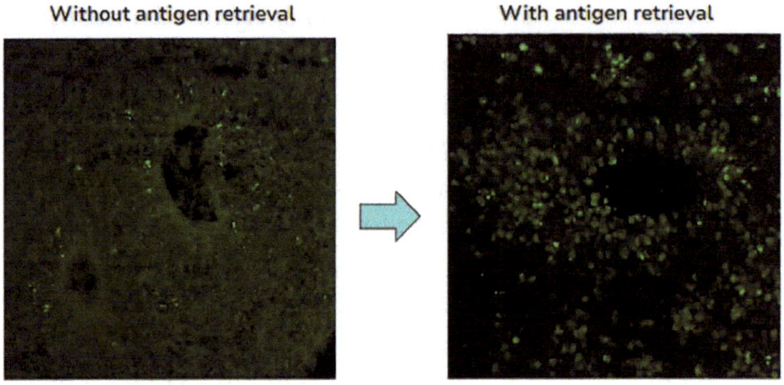

Fig. 3 IHC image with and without antigen retrieval: representative images showing brain organoid sections immunostained for BrdU (Bromodeoxyuridine) antibody, without (left) and with (right) antigen retrieval

9. Volume of media can be increased up to 20 mL in 10 cm dishes to ensure organoids are not undernourished. However, accumulation of cell debris is also an indicator of stressed, starved organoids, or unhealthy organoids.
10. Warming helps with dissolving. Gelatin solidifies at colder temperatures.
11. Antigen retrieval is not necessary for all antibodies. Check antibody details or figure out experimentally (Fig. 3).

Acknowledgments

We thank past and present members of the Pak lab for the various experimental and troubleshooting assistance. This work was supported by NIMH to Pak lab (R01 MH122519 and R21 MH130843) and NIGMS T32 BTP training program (T32 GM135096 to N.P.)

References

1. Chambers SM, Fasano CA, Papapetrou EP, Tomishima M, Sadelain M, Studer L (2009) Highly efficient neural conversion of human ES and iPS cells by dual inhibition of SMAD signaling. Nat Biotechnol 27:275–280. https://doi.org/10.1038/nbt.1529
2. Paşca AM, Sloan SA, Clarke LE, Tian Y, Makinson CD, Huber N, Kim CH, Park J-Y, O'Rourke NA, Nguyen KD, Smith SJ, Huguenard JR, Geschwind DH, Barres BA, Paşca SP (2015) Functional cortical neurons and astrocytes from human pluripotent stem cells in 3D culture. Nat Methods 12:671–678. https://doi.org/10.1038/nmeth.3415
3. Sidhaye J, Knoblich JA (2021) Brain organoids: an ensemble of bioassays to investigate human neurodevelopment and disease. Cell Death Differ 28:52–67. https://doi.org/10.1038/s41418-020-0566-4
4. Renner M, Lancaster MA, Bian S, Choi H, Ku T, Peer A, Chung K, Knoblich JA (2017) Self-organized developmental patterning and differentiation in cerebral organoids. EMBO J 36:1316–1329. https://doi.org/10.15252/embj.201694700
5. Muguruma K, Nishiyama A, Kawakami H, Hashimoto K, Sasai Y (2015) Self-organization of polarized cerebellar tissue in 3D culture of

human pluripotent stem cells. Cell Rep 10:537–550. https://doi.org/10.1016/j.celrep.2014.12.051

6. Lancaster MA, Renner M, Martin C-A, Wenzel D, Bicknell LS, Hurles ME, Homfray T, Penninger JM, Jackson AP, Knoblich JA (2013) Cerebral organoids model human brain development and microcephaly. Nature 501:373–379. https://doi.org/10.1038/nature12517
7. Del Dosso A, Urenda J-P, Nguyen T, Quadrato G (2020) Upgrading the physiological relevance of human brain organoids. Neuron 107:1014–1028. https://doi.org/10.1016/j.neuron.2020.08.029
8. Sloan SA, Andersen J, Paşca AM, Birey F, Paşca SP (2018) Generation and assembly of human brain region–specific three-dimensional cultures. Nat Protoc 13:2062–2085. https://doi.org/10.1038/s41596-018-0032-7
9. Sebastian R, Jin K, Pavon N, Bansal R, Potter A, Song Y, Babu J, Gabriel R, Sun Y, Aronow B, Pak C (2022) Single cell transcriptomic profiling of human brain organoids reveals developmental timing- and cell-type-specific vulnerabilities induced by NRXN1 CNVs in schizophrenia. 2022.08.24.505165

Chapter 14

A 3D Bioengineered Neural Tissue Model Generated from Human iPSC-Derived Neural Precursor Cells

Selene Lomoio and Giuseppina Tesco

Abstract

Available models to study neuropathological diseases include cell cultures and animal models. Brain pathologies, however, are often poorly recapitulated in animal models. 2D cell culture systems are well established and have been used since the early 1900s to grow cells on flat dishes. However, conventional 2D neural culture systems, which lack key features of the brain's 3D microenvironment, often inaccurately represent the diversity and maturation of multiple cell types and their interaction under physiological and pathological conditions.

To improve CNS modeling, we have designed a 3D bioengineered neural tissue model generated from human iPSC-derived neural precursor cells (NPCs). This NPC-derived biomaterial scaffold, composed of silk fibroin with an intercalated hydrogel, matches the mechanical properties of native brain tissue and supports the long-term differentiation of neural cells in a donut-shaped sponge within an optically clear central window. This chapter describes integrating iPSC-derived NPCs in these silk-collagen scaffolds and differentiating them into neural cells over time.

Key words Neural Precursor Cells, Human Neural Tissue, 3D Bioengineered Model, Cell culture

1 Introduction

Available models to study neuropathological diseases include cell cultures and animal models. Brain pathologies, however, are often poorly recapitulated in animal models [5]. Advances in stem cell technology and genome editing have allowed reprogramming primary cells from human subjects into induced pluripotent stem cells (iPSCs) and their differentiation in neurons, astrocytes, and microglia.

2D cell culture systems are well established and have been used since the early 1900s to grow cells on flat dishes. However, 2D systems aren't entirely representative of physiological environments. Growing on a flat surface isn't an ideal approach to understanding how cells grow and function in the human body, where they are typically surrounded by other cells in three dimensions. So,

Yu-Wen Alvin Huang and ChangHui Pak (eds.), *Stem Cell-Based Neural Model Systems for Brain Disorders*, Methods in Molecular Biology, vol. 2683, https://doi.org/10.1007/978-1-0716-3287-1_14,

conventional 2D neural culture systems, which lack key features of the brain's 3D microenvironment [10], often inaccurately represent the diversity and maturation of multiple cell types and their interaction under physiological and pathological conditions [4, 8].

To address these limitations, investigators are designing in vitro models assembled using appropriate cell types, biomimetic extracellular matrix, and microenvironmental cues to ideally recapitulate the standard brain physiology [10].

In this context, we have designed, in collaboration with Dr. David Kaplan at Tufts University, a 3D bioengineered neural tissue model generated from human iPSC-derived neural precursor cells (NPCs), differentiated according to the StemCell Technologies method, using an embryoid body protocol without SMADi, [3]. This NPC-derived biomaterial scaffold [6, 9], composed of silk fibroin with an intercalated hydrogel, was adapted from a previous design generated using primary embryonic rodent neural cells [11] and human-induced neural stem cells [1, 2] to match the mechanical properties of native brain tissue. Our improved protocol offers several advantages: NPCs can be expanded, frozen, banked, and subsequently differentiated. Having NPCs stocks derived from multiple subjects allows synchronization of their differentiation and minimizes experimental variability. The highly innovative model favors the growth of the differentiating neural cells in a donut-shaped porous silk sponge within an optically clear collagen-filled central window and supports axonal connectivity and synapse formation studies (e.g., cell-based electrophysiology, trafficking, synaptic functionality). These scaffolds can sustain compartmentalized, 3D neural-like tissues for over a year without necrosis [10]. Under these experimental conditions, neural cells develop networks and action potentials typical of fully mature neurons [3]. The architecture of the scaffolds was optimized to meet the metabolic demand of high-density cultures in terms of free diffusion of nutrients and oxygen, a fundamental requisite for long-term studies. Each element of this model is tunable and amenable to experimental modification, from the pore size of the scaffold biomaterial [7] to the hydrogel composition and incorporated cell types [10].

This chapter describes integrating iPSC-derived NPCs in these silk-collagen scaffolds and differentiating them into neural cells over time.

2 Materials

2.1 NPC Thawing

1. 6-well Tissue Culture Plate.
2. 15 mL Conical Tube.

3. Matrigel hESC-Qualified Matrix (Corning).
4. DMEM/F-12.
5. Complete STEMdiff Neural Progenitor Medium (StemCell Tech): STEMdiff Neural Progenitor Basal Medium, Neural Progenitor Supplement A (50X), and Supplement B (1000X).

2.2 NPC Passaging and Expansion

1. 6-well Tissue Culture Plate.
2. 15 mL Conical Tube.
3. Matrigel hESC-Qualified Matrix (Corning).
4. DMEM/F-12.
5. Complete STEMdiff Neural Progenitor Medium (StemCell Tech).
6. Accutase.
7. Trypan Blue.

2.3 Scaffold Coating

1. Silk Scaffolds (5 mm outer diam. × 2 mm central hole diam. × 1 mm ht.).
2. Autoclaved tweezers.
3. 15 mL Conical Tubes.
4. Coating medium: 20 ug/mL poly-L-ornithine (Sigma) and 10 ug/mL laminin (Sigma) in DMEM/F-12.

2.4 Scaffold Seeding

1. Coated scaffolds.
2. Sterile Cell Strainer 40 um Nylon Mesh.
3. 96-well Cell Culture-Treated Flat-Bottom Microplate.
4. Complete STEMdiff Neural Progenitor Medium (StemCell Tech).
5. Accutase.
6. Trypan Blue.
7. Autoclaved tweezers.

2.5 Collagen Filling

1. 96-well Cell Culture-Treated Flat-Bottom Microplate.
2. 24-well Tissue Culture Plate.
3. Complete STEMdiff Neural Progenitor Medium (StemCell Tech).
4. Collagen Solution: Collagen Type-1 Rat Tail, 10X PBS pH 7.4, 1N NaOH—ratio 88:10:2.
5. Autoclaved tweezers.

2.6 Neural Induction and Long-Term Maintenance

1. 24-well Tissue Culture Plate.
2. BrainPhys Neuronal Medium (StemCell Tech).
3. NeuroCult SM1 Neuronal Supplement (StemCell Tech).
4. N2 Supplement (ThermoFisher).

3 Methods

3.1 NPC Thawing

1. Coat one 6-well tissue culture-treated plate with Matrigel for at least 1 h (*see* **Note 1**).
2. Warm (37 °C) sufficient DMEM/F-12 and complete STEMdiff Neural Progenitor medium (*see* **Note 2**).
3. Add 5 mL of warm DMEM/F-12 to a 15 mL conical tube.
4. Quickly thaw cells in a 37 °C water bath by gently shaking the cryovial until only a small frozen pellet remains.
5. Gently transfer cells from the cryovial to the tube containing DMEM/F-12.
6. Count viable cells using Trypan Blue and a hemocytometer.
7. Centrifuge cells at 300 × *g* for 5 min.
8. Aspirate medium without disturbing the pellet. Gently resuspend the cell pellet in complete STEMdiff Neural Progenitor medium to achieve a final concentration of 1×10^6 cells/mL.
9. Remove the Matrigel solution from the 6-well plate (prepared in **step 1**) using a serological pipette or by aspiration and add 1 mL per well of complete STEMdiff Neural Progenitor medium. Pay attention not to scratch the coated surface.
10. Add 350/500k cells (350/500 uL) to one well of the coated 6-well plate (*see* **Note 3**).
11. Place the plate in a 37 °C incubator. Move the plate in several quick, short, back-and-forth, and side-to-side motions to distribute the NPCs across the surface of the wells.

3.2 NPC Passaging and Expansion

1. Coat 4 or 5 6-well tissue culture-treated plates with Matrigel for at least 1 h (*see* **Note 4**).
2. Thaw Accutase (*see* **Note 5**).
3. Warm (37 °C) sufficient DMEM/F-12 and complete STEMdiff Neural Progenitor medium.
4. Aspirate medium and add 1 mL of Accutase.
5. Incubate at 37 °C for 5–10 min.
6. Using a 1 mL pipette, pipette the cell suspension up and down to dislodge the remaining attached cells.

7. Add 3 mL of DMEM/F-12 to the well and transfer the NPC suspension to a 15 mL conical tube.
8. Count viable cells using Trypan Blue and a hemocytometer.
9. Centrifuge cells at 300 × *g* for 5 min.
10. Aspirate medium without disturbing the pellet. Gently resuspend the cell pellet in complete STEMdiff Neural Progenitor medium to achieve a final concentration of 1×10^6 cells/mL.
11. Remove the Matrigel solution from the 6-well plate (prepared in **step 1**) using a serological pipette or by aspiration and add 1 mL per well of complete STEMdiff Neural Progenitor medium. Pay attention not to scratch the coated surface.
12. Add 350/500k cells (350/500 uL) to one well of the coated 6-well plate.
13. Place the plate in a 37 °C incubator. Move the plate in several quick, short, back-and-forth, and side-to-side motions to distribute the NPCs across the surface of the wells.

3.3 Scaffold Coating

1. Collect the autoclaved scaffolds in a 15 mL conical tube using autoclaved tweezers and remove the excess water with a 1 mL pipette (*see* **Note 6**).
2. Add the coating medium to the 15 mL conical tube containing the scaffold. Make sure that the volume of the coating matrix is sufficient to cover the scaffolds completely.
3. Incubate overnight at 4 °C.

3.4 Scaffold Seeding

1. Using autoclaved tweezers, transfer the scaffold from the 15 mL conical tube containing the coating matrix to a 96-well plate (one scaffold per well). Ensure that the scaffolds are lying flat at the bottom of the well.
2. Use a 200 uL pipette directly on the donut to carefully aspirate the excess coating matrix (*see* **Note 7**).
3. Warm (37 °C) sufficient DMEM/F-12 and complete STEMdiff Neural Progenitor medium.
4. Aspirate medium from NPC plates and add 1 mL of Accutase.
5. Incubate at 37 °C for 5–10 min.
6. Using a 1 mL pipette, pipette the cell suspension up and down to dislodge the remaining attached cells.
7. Add 3 mL of DMEM/F-12 to the well.
8. Before transferring to a 15 mL conical tube, strain the NPC suspension using a sterile cell strainer 40 um Nylon mesh.
9. Count viable cells using Trypan Blue and a hemocytometer.
10. Centrifuge cells at 300 × *g* for 5 min.

11. Aspirate medium without disturbing the pellet. Gently resuspend the cell pellet in complete STEMdiff Neural Progenitor medium to achieve a final concentration of 2×10^6 cells in 100 uL.
12. Add 100 uL of cell resuspension to each scaffold avoiding bubbles, and incubate at 37 °C until the following day.
13. Perform daily media change with 200 uL of complete STEMdiff Neural Progenitor medium for 5 days following seeding (*see* **Note 8**).

3.5 Collagen Filling

1. Move scaffolds to a clean and dry well of a 96-well plate using autoclaved tweezers (one scaffold per well). Ensure that the scaffolds are lying flat at the bottom of the well.
2. On ice, prepare 100 uL of collagen solution per scaffold. Keep everything on ice and use it immediately after preparation (*see* **Note 9**).
3. Use a 200 uL pipette directly on the donut to gently aspirate the excess media.
4. Infuse each scaffold with a 100 uL cold collagen solution prepared in **step 2**. Avoid bubbles.
5. Incubate the collagen-infused scaffolds at 37 °C until the gelation occurs—30 min to 1 h (*see* **Note 10**).
6. Once solidified, transfer the scaffolds to a 24-well plate (one scaffold per well) and flood them with 1.5 mL of complete STEMdiff Neural Progenitor medium.
7. Perform daily media change with 1.5 mL of complete STEMdiff Neural Progenitor medium for 5 days following collagen filling.

3.6 Neural Induction and Long-Term Maintenance

1. On day 6, following collagen filling, prepare BrainPhys medium supplemented with SM1 (50X) and N2 (100X). (*See* **Note 11**).
2. Carefully aspirate spent STEMdiff Neural Progenitor medium and flood each well with 1.5 mL of BrainPhys/SM1/N2 medium.
3. Perform partial (500 uL) media changes every 4 days until desired differentiation is achieved.

4 Notes

All the steps described above must be performed in a sterile environment, such as a biosafety cabinet. Turn on the UV light, and let it run for 30 min before using the biosafety cabinet. Wipe down all surfaces, tubes, pipettes, and reagents with 70% ethanol.

1. Prepare Matrigel solution using cold DMEM/F-12 according to manufacturer instructions. Thaw Matrigel on ice. Do not allow Matrigel aliquots to warm to room temperature. You should coat the plates for at least 1 h at room temperature. Alternatively, if you are not using Matrigel-coated dishes on the same day as coating, wrap them in parafilm, and store them at 4 °C for up to 3 days. Make sure wells do not dry out. Allow stored coated plates to come to room temperature for 1 h before plating cells.
2. Prepare complete STEMdiff Neural Progenitor Medium according to manufacturer instructions. Prepare 50 mL aliquots and store the medium at 4 °C for up to 2 weeks. Do not freeze the complete medium.
3. The appropriate number of cells per well depends on the iPS line the NPCs are derived from. You will have to determine it empirically. A good rule of thumb is to seed a number of cells per well that will reach 90% confluency in 4–6 days.
4. In our hands, NPCs have an ideal splitting ratio of around 1:4/1:5. So the estimated number of plates to coat for this step is based on our experience and the lines we work with. You may have to adjust it accordingly.
5. Accutase should not be defrosted in a warm water bath since it is sensitive to temperatures above 37 °C and will be inactivated after 45 min at 37 °C. We usually defrost a new Accutase bottle overnight at 4 °C and then aliquot it. We store the aliquots at − 20 °C and thaw them at room temperature before use. Always remember to mix Accutase properly after defrosting. As the solution defrosts, the ingredients will not go into solution evenly.
6. The scaffolds can be stored long-term in ultrapure sterile water at 4 °C. Before coating, scaffolds must be autoclaved in ultrapure water using a 120 °C, 20-min liquid cycle. Ensure they are thoroughly cooled down before proceeding with the coating step.
7. The step of aspirating the coating matrix is highly critical. Excessive residual coating matrix will interfere with cell attachment/adherence. Scaffolds must be thoroughly dried before cell seeding. When the coating matrix is removed, the scaffold color will turn pink to white.
8. For media change following NPCsNeural progenitor cells (NPCs) seeding, scaffolds must be moved daily to a clean well of a 96-well plate using autoclaved tweezers. This will guarantee that non-attached/dead cells will be left behind and won't impact overall cell survival.

9. It is highly critical always to keep all the components on ice until use. Prepare at least 125 uL of collagen solution per scaffold to account for pipetting loss.
10. To monitor the gelation process, observe the leftover collagen solution. Release a 10 uL pipette tip in the tube with the excess solution. Gelation is completed when the tip is firmly embedded in the collagen solution and cannot be displaced by inverting the tube. The collagen solution will have a milky appearance once gelation has occurred.
11. Brainphys complete medium can be prepared and stored in 50 mL aliquots at 4 °C for up to spiepr IndexRangeEnd ID="ITerm10"1 montspiepr IndexRangeEnd ID="ITerm8"h.

Acknowledgments

This work was supported by awards from the National Institutes of Health: 1R21AG065792 and 5R01AG061838 (to GT). We thank the Kaplan Lab (Department of Biomedical Engineering at Tufts University, Boston) for providing the silk sponge scaffolds and their methodological support. We also thank our colleagues Beatrice Menicacci, Isabel Paine, and Arlo Simmerman for their feedback and comments.

References

1. Cairns DM, Chwalek K, Moore YE et al (2016) Expandable and rapidly differentiating human induced neural stem cell lines for multiple tissue engineering applications. Stem Cell Rep 7: 557–570
2. Cairns DM, Rouleau N, Parker RN et al (2020) A 3D human brain-like tissue model of herpes-induced Alzheimer's disease. Sci Adv 6: eaay8828
3. Cantley W, Du C, Lomoio S et al (2018) Functional and sustainable 3D human neural network models from pluripotent stem cells. ACS Biomater Sci Eng 4:4278. Article ASAP. https://doi.org/10.1021/acsbiomaterials.8b00622. Publication Date (Web): October 1, 2018
4. Cenini G, Hebisch M, Iefremova V et al (2021) Dissecting Alzheimer's disease pathogenesis in human 2D and 3D models. Mol Cell Neurosci 110:103568
5. D'avanzo C, Aronson J, Kim YH et al (2015) Alzheimer's in 3D culture: challenges and perspectives. BioEssays 37:1139–1148
6. Dingle YL, Bonzanni M, Liaudanskaya V et al (2021) Integrated functional neuronal network analysis of 3D silk-collagen scaffold-based mouse cortical culture. STAR Protoc 2: 100292
7. Nazarov R, Jin HJ, Kaplan DL (2004) Porous 3-D scaffolds from regenerated silk fibroin. Biomacromolecules 5:718–726
8. Penney J, Ralvenius WT, Tsai LH (2020) Modeling Alzheimer's disease with iPSC-derived brain cells. Mol Psychiatry 25:148–167
9. Rockwood DN, Preda RC, Yucel T et al (2011) Materials fabrication from Bombyx mori silk fibroin. Nat Protoc 6:1612–1631
10. Rouleau N, Cantley WL, Liaudanskaya V et al (2020) A long-living bioengineered neural tissue platform to study neurodegeneration. Macromol Biosci 20:e2000004
11. Tang-Schomer MD, White JD, Tien LW et al (2014) Bioengineered functional brain-like cortical tissue. Proc Natl Acad Sci U S A 111: 13811–13816

Chapter 15

FACS-Based Sequencing Approach to Evaluate Cell Type to Genotype Associations Using Cerebral Organoids

Liam Murray, Meagan N. Olson, Nathaniel Barton, Pepper Dawes, Yingleong Chan, and Elaine T. Lim

Abstract

Recent technological developments have led to widespread applications of large-scale transcriptomics-based sequencing methods to identify genotype-to-cell type associations. Here we describe a fluorescence-activated cell sorting (FACS)-based sequencing method to utilize CRISPR/Cas9 edited mosaic cerebral organoids to identify or validate genotype-to-cell type associations. Our approach is high-throughput and quantitative and uses internal controls to enable comparisons of the results across different antibody markers and experiments.

Key words oFlowSeq, Cerebral organoids, FACS, Sequencing, Genotype to cell type

1 Introduction

Recent advances in high-throughput single-cell transcriptomics sequencing-based technologies such as single-cell or single-nucleus RNA sequencing [1–3] and spatial RNA sequencing [4–6], as well as data integration methods [7], have enabled widespread applications in investigation of disease-associated cell types and cell type specific transcriptomic perturbations.

To provide an orthogonal high-throughput sequencing-based approach for independently validating cell type-to-genotype discoveries from transcriptomics-based approaches, we developed a method called oFlowSeq which utilizes CRISPR-Cas9 gene editing, fluorescence-activated cell sorting (FACS), and targeted sequencing to associate genetic variants with cell types in a high-throughput manner [8]. oFlowSeq was inspired by the FlowSeq method that has been previously demonstrated using bacteria and yeast [9–11]. The oFlowSeq method has been adapted to include CRISPR-Cas9 editing of genetically mosaic cerebral organoids, lending the name "organoid FlowSeq", to allow researchers to

Yu-Wen Alvin Huang and ChangHui Pak (eds.), *Stem Cell-Based Neural Model Systems for Brain Disorders*, Methods in Molecular Biology, vol. 2683, https://doi.org/10.1007/978-1-0716-3287-1_15,

examine the effects of synthetic mutations on cell type proportions within cerebral organoids. The oFlowSeq method is also broadly applicable to other organoid and multicellular systems.

The oFlowSeq method controls for inherent technical variability as unedited wildtype cells and edited mutant cells are differentiated within the same cerebral organoids. In addition to normalizing the numbers of edited reads by the numbers of unedited reads, we use the distribution of long mutations (defined as mutations with >10-bp insertions or deletions) for normalizing the distribution of short mutations (defined as mutations with ≤10-bp insertions or deletions), thus enabling the comparison of normalized distributions of mutations across different antibody markers and across different experiments.

2 Materials

Prepare and store all reagents at 4 °C (unless indicated otherwise). Diligently follow all waste disposal regulations when disposing waste materials.

2.1 iPSC Cell Culture

1. StemFlex.
2. Y-27632.
3. SMC4 cocktail.
4. Accutase.
5. Matrigel Basement Membrane Matrix.
6. DMEM/F-12, HEPES.
7. DPBS, no calcium, no magnesium.

2.2 Nucleofection

1. P3 Primary Cell 4D-Nucleofector X Kit L.
2. TrueCut Cas9 protein v2.
3. Multi-guide Gene Knockout Kit v2.

2.3 Cell Counter

1. NanoEnTek EVE cell counting slides.
2. 0.4% Trypan Blue.

2.4 Cerebral Organoid Dissociation

1. 0.25% Trypsin-EDTA.
2. 30 μm cell strainers.

2.5 Cell Staining

1. Paraformaldehyde (PFA).
2. Permeabilization buffer (DPBS, 0.02% sodium azide, 2% FBS, and 0.1% saponin).

3. Alexa Fluor 488-conjugated mouse IgG2A antibody (IC003G).
4. Alexa Fluor 488-conjugated NeuN antibody (NBP-92693AF488).
5. Alexa Fluor 488-conjugated Nestin antibody (IC1259G).
6. Alexa Fluor 488-conjugated TRA-1-60 antibody (NB100-730F488).

2.6 DNA Extraction and Library Preparation

1. DNA FFPE Tissue Kit.
2. Q5 Hot Start Master Mix.
3. Monarch PCR & DNA Cleanup Kit.
4. KAPA Library Quantification Kit.

3 Methods

Carry out all procedures at room temperature unless otherwise specified. Cerebral organoids were differentiated using a previously reported protocol for spontaneously differentiated cerebral organoids [12].

3.1 Nucleofection of Cas9 and Multi-guide RNA

1. Prepare 6-well tissue culture plates coated with 0.5 mg of Matrigel in 6 mL of DMEM for each plate (1 mL per well). Passage iPSCs on the Matrigel-coated plates until the cells are ~75–90% confluent. Aspirate the cell culture media and wash the iPSCs twice with 1 mL/well of 1 × DPBS.
2. Add 1 mL of Accutase into each well and incubate in 37 °C for 10 min to dissociate the cells. Dilute the well with 9 mL of StemFlex. Transfer the media with dissociated cells into a 15 mL conical tube and spin down at 300 × *g* for 5 min to pellet the cells. Wash the cell pellet once with 10 mL of StemFlex and spin down to pellet the cells.
3. Aspirate the supernatant and resuspend the cells in 1 mL of StemFlex. Mix 10 μL of cells with 10 μL of Trypan Blue and load 10 μL into one side of a NanoEnTek EVE cell counting slide. Count the concentration of live cells and aliquot out 500,000 cells into a 1.5 mL tube. Spin down at 300 × *g* for 5 min to pellet the cells and remove the supernatant.
4. Incubate 3 μL of 100 μM multi-guide RNA with 2 μL of 20 μM Cas9 nuclease for 10 min. Add the 5 μL of Cas9-guide RNA complex to 82 μL nucleofector solution with 18 μL supplement and resuspend the iPSCs. Pipette the solution gently into the cuvette supplied.

5. Place the cuvette into the nucleofector and start the CA-137 program on the 4D nucleofector. Upon completion, use a Pasteur pipet to add prewarmed iPSC cell culture media (StemFlex with 10 μM Y-27632 and 1:100 SMC4) to the cuvette and remove the solution to one well of a 24-well plate coated with Matrigel. After 24 h, expand the cells onto 6-well plates for another 6 days.
6. Differentiate cerebral organoids using a published protocol [12] for at least 3 months.

3.2 Cerebral Organoid Dissociation

1. Using a cut P200 pipette tip, pick up 20 cerebral organoids into a single 1.5 mL tube. Centrifuge at 300 × *g* for 5 min and remove excess media.
2. Wash the cerebral organoids twice in 600 μL of ice-cold 1 × DPBS and centrifuge at 300 × *g* at 4 °C for 10 min to pellet the cells.
3. Add 500 μL of Trypsin-EDTA to each tube and incubate in 37 °C for 30 min on a shaking heat block at 300 RPM. Pipette to manually dissociate clumps of cells.
4. Add 200 μL of StemFlex to each tube and centrifuge at 300 × *g* to pellet the cells. Wash the cells once with 700 μL of 1 × DPBS and centrifuge at 300 × *g* to pellet the cells.
5. Resuspend the cell pellet in 700 μL of 1 × DPBS and filter through 30 μm strainers to remove any remaining Matrigel or clumps of cells.

3.3 Staining and FACS

1. Fix and permeabilize the cells (*see* **Note 1**) with 300 μL of 4% PFA and 300 μL of permeabilization buffer for 45 min at 4 °C. Centrifuge at 1000 × *g* to pellet the cells.
2. Resuspend the cell pellet in 245 μL of permeabilization buffer and 5 μL of antibody (1:50 ratio) and incubate on an orbital shaker for 1 h. Centrifuge at 1000 × *g* to pellet the cells. Wash the cells twice with 700 μL of 1 × DPBS.
3. Resuspend the cells in 300 μL of 1 × DPBS for each tube.
4. For each antibody, sort one million cells on a Sony SH800 FACS sorter using a 100 μm sorting chip and sorting speed of 4000 events per second. Sort the cells into 15 mL tubes containing 3 mL of 1 × DPBS. Add additional 1 × DPBS to top up each tube to 10 mL in total and centrifuge at 300 × *g* to pellet the cells.
5. Resuspend the cell pellet in 700 μL of 1 × DPBS and transfer to 1.5 mL tubes. Centrifuge at 300 × *g* to pellet the cells.

3.4 DNA Extraction and Library Preparation for Sequencing

1. Perform DNA extraction of the sorted cells using the standard protocol for the DNA FFPE Tissue Kit, and quantify the extracted DNA using Nanodrop.
2. Perform PCR reactions using the standard protocol for Q5 Hot Start Master Mix using the extracted DNA and including DNA from unedited iPSCs, with the locus-specific primers.
3. Clean up the PCR product using the standard protocol for Monarch PCR & DNA Cleanup Kit.
4. Perform a second round of PCR reactions using the standard protocol for Q5 Hot Start Master Mix using the locus-specific PCR products with Illumina barcode adapters.
5. Clean up the PCR product using the standard protocol for Monarch PCR & DNA Cleanup Kit.
6. Quantify the concentration of the DNA using the KAPA library quantification kit and pool in equal concentrations for Illumina paired-end sequencing with at least 15% PhiX spike-in control or human whole-genome libraries (*see* **Note 2**).

3.5 Sequence Analyses

1. Filter away sequences with <0.1% reads compared to the total number of reads and sequences that did not match the primers used for the samples, as these sequences are likely to be background noise or sequencing artifacts. In addition, filter away mutant sequences that were present in unedited iPSCs.
2. For each sample, calculate the total number of reference unedited reads. For each mutant sequence within the sample, calculate $\text{Ratio}_{\text{mutation}}$, where $\text{Ratio}_{\text{mutation}} = \frac{\text{Num\_reads}_{\text{mutation}}}{\text{Num\_reads}_{\text{reference}}}$.
3. Calculate the mean of the ratios using only long mutations ($\text{Mean\_Ratio}_{\text{long\_mutations}}$), defined as mutations resulting in >10-bp insertions or deletions. $\text{Mean\_Ratio}_{\text{long\_mutations}}$ is used to normalize the distribution of $Ratio_{mutation}$, to account for experimental biases such as biases in fluorophore intensities and to serve as an internal control for comparing results across different samples and experiments.
4. For each mutant sequence within the sample, calculate a normalized ratio, where $\text{Normalized\_Ratio}_{\text{mutation}} = \frac{\text{Ratio}_{\text{mutation}}}{\text{Mean\_ratio}_{\text{long\_mutations}}}$.
5. Calculate an odds ratio for each antibody marker of interest by comparing the normalized ratios of each mutant sequence in the positively versus negatively sorted cells for the marker, where

$$\text{Marker}(\text{Odds\_Ratio}_{\text{mutation}}) = \frac{\text{Positive\_bin}\left(\text{Normalized\_Ratio}_{\text{mutation}}\right)}{\text{Negative\_bin}\left(\text{Normalized\_Ratio}_{\text{mutation}}\right)}.$$

An odds ratio that is significantly greater than 1 suggests that the mutant sequence is enriched in the positively sorted cells versus the negatively sorted cells. An odds ratio that is significantly less than 1 suggests that the mutant sequence is depleted in the positively sorted cells versus the negatively sorted cells.

4 Notes

1. We typically use at least three million cells for each antibody staining to ensure that we can obtain at least 200,000 cells post FACS sorting.
2. We aim to sequence each pool of cells at a minimum of 100,000 reads and the number of sorted cells should be greater than 100,000 to ensure that we are likely to detect sequences from unique individual cells.

Acknowledgments

We thank George Church, Mohammed Uddin, ChangHui Pak, and members of their labs, as well as members in the Program in Bioinformatics & Integrative Biology at UMass Chan Medical School, for their expertise, advice, and suggestions in developing the oFlowSeq method. This study was supported by the National Institutes of Health grants (NHGRI RM1HG008525 to George Church; NIMH R01MH113279 to George Church), Robert Wood Johnson Foundation grant (74178 to George Church), and startup funds by UMass Chan Medical School (Elaine Lim and Yingleong Chan).

References

1. Macosko EZ, Basu A, Satija R et al (2015) Highly parallel genome-wide expression profiling of individual cells using nanoliter droplets. Cell 161:1202–1214
2. Klein AM, Mazutis L, Akartuna I et al (2015) Droplet barcoding for single-cell transcriptomics applied to embryonic stem cells. Cell 161:1187–1201
3. Zheng GX, Terry JM, Belgrader P et al (2017) Massively parallel digital transcriptional profiling of single cells. Nat Commun 8:14049
4. Stahl PL, Salmen F, Vickovic S et al (2016) Visualization and analysis of gene expression in tissue sections by spatial transcriptomics. Science 353:78–82
5. Rodriques SG, Stickels RR, Goeva A et al (2019) Slide-seq: a scalable technology for measuring genome-wide expression at high spatial resolution. Science 363:1463–1467
6. Chen A, Liao S, Cheng M et al Large field of view-spatially resolved transcriptomics at nanoscale resolution. bioRxiv 2021: 2021.2001.2017.427004
7. Lim ET, Chan Y, Dawes P et al (2022) Orgo-Seq integrates single-cell and bulk transcriptomic data to identify cell type specific-driver genes associated with autism spectrum disorder. Nat Commun 13:3243
8. Dawes P, Murray LF, Olson MN et al (2023) oFlowSeq: a quantitative approach to identify protein coding mutations affecting cell type

enrichment using mosaic CRISPR-Cas9 edited cerebral organoids. Hum Genet. https://doi.org/10.1007/s00439-023-02534-4

9. Kosuri S, Goodman DB, Cambray G et al (2013) Composability of regulatory sequences controlling transcription and translation in Escherichia coli. Proc Natl Acad Sci U S A 110:14024–14029
10. Raveh-Sadka T, Levo M, Shabi U et al (2012) Manipulating nucleosome disfavoring sequences allows fine-tune regulation of gene expression in yeast. Nat Genet 44:743–750
11. Sharon E, Kalma Y, Sharp A et al (2012) Inferring gene regulatory logic from high-throughput measurements of thousands of systematically designed promoters. Nat Biotechnol 30:521–530
12. Lancaster MA, Knoblich JA (2014) Generation of cerebral organoids from human pluripotent stem cells. Nat Protoc 9:2329–2340

Chapter 16

Dynamic Measurement of Endosome-Lysosome Fusion in Neurons Using High-Content Imaging

Qing Ouyang, Michael Schmidt, and Eric M. Morrow

Abstract

Endocytosis is a dynamic cellular process that actively transports particles into a cell. Late endosome fusion with the lysosome is a crucial step in the delivery of newly synthesized lysosomal proteins and endocytosed cargo for degradation. Disturbing this step in neurons is associated with neurological disorders. Thus, studying endosome-lysosome fusion in neurons will provide new insight into the mechanisms of these diseases and open new possibilities for therapeutic treatment. However, measuring endosome-lysosome fusion is challenging and time consuming, which limits the research in this area. Here we developed a high throughput method using pH-insensitive dye-conjugated dextrans and the Opera Phenix® High Content Screening System. By using this method, we successfully separated endosomes and lysosomes in neurons, and time-lapse images were collected to capture endosome-lysosome fusion events in hundreds of cells. Both assay set-up and analysis can be completed in an expeditious and efficient manner.

Key words Endosome-lysosome fusion, Opera Phenix®, High throughput assay, Time-lapse images, Dye-conjugated dextran, Neurological disorders

1 Introduction

The endosome-lysosome (endolysosomal) system of eukaryotic cells is compartmentalized and efficient. The endocytic pathway consists of distinct membrane compartments which allow cells to bring in exogenous material. It also works to allow membrane protein and lipid delivery and recycling, organelle removal, and signaling [1–3]. The process of endocytosis starts with collecting cargo proteins at the site of the plasma membrane. In turn, an endocytotic vesicle, now known as cargo, and fission is formed. The newly endocytic vesicles deliver their contents and membrane to the early endosome (EE), where efficient sorting occurs. Some cargo may get recycled back to the plasma membrane by way of recycling endosomes (RE). Other cargos move towards the perinuclear region. Early endosomes change their make-up, structure/morphology, and luminal pH as they mature into late endosomes

Yu-Wen Alvin Huang and ChangHui Pak (eds.), *Stem Cell-Based Neural Model Systems for Brain Disorders*, Methods in Molecular Biology, vol. 2683, https://doi.org/10.1007/978-1-0716-3287-1_16,

(LE). Finally, the late endosome (LE) will fuse with lysosomes to form endolysosomes. Throughout this time, intraluminal vesicles are moved into the lysosomal lumen, resulting in the degradation of contents [4, 5].

The endolysosomal system is critical for neuronal function. Many neurodegenerative diseases have been linked to the dysfunction of the endolysosomal system, including Alzheimer's disease, Parkinson's disease, Huntington's disease, amyotrophic lateral sclerosis (ALS), and frontotemporal lobar dementia (FTLD) [6–9]. Gene mutations that affect endocytic trafficking and fusion or impair lysosomal function have been implicated in these neurological disorders [10–12]. In some situations, the specific step of endosome to lysosome fusion in neurons is impaired [11, 13]. Thus, studying endosome-lysosome fusion in neurons will provide insight into the mechanisms of these diseases and open new possibilities for therapeutic treatment. However, measuring endosome-lysosome fusion is challenging and time consuming. Previous methods either use a quantitative fluorescence imaging assay based on the green fluorescence enhancement of 4,4-difluoro-5,7-dimethyl-4- bora-3a,4a-diaza-s-indacene (Bodipy)-avidin [14, 15] or Oregon Green–avidin [12] upon biotin binding or use a co-labeling scheme by using Alexa 488-dextran to mark lysosomes and Alexa 594-dextran to mark endosomes [16]. These methods have their own defects: Bodipy-avidin is discontinued and the enhancement of green fluorescence could be eliminated by photobleaching with long time exposure; Oregon Green-avidin is pH sensitive which is not suitable to use in acidic vesicles; Moreover, these methods were all using conventional confocal microscopy which limits the throughput and the sample size, thereby, limiting the statistical rigor of the dataset.

Here we developed a high throughput and high content imaging method, combining pH insensitive dye-conjugated dextrans and the Opera Phenix® High Content Screening System. This system has temperature and CO_2 control for live-cell imaging; spinning disk confocal modes to vastly improve the speed of image acquisition, which is suitable for imaging of fast dynamic processes like endosome fusion and considerably reduces photo damage for live cells. It takes high-throughput 2-D and 3-D imaging of fluorescent samples in microplates automatically which largely increase the sample size. Moreover, the Harmony® high-content analysis software provides ready-made image analysis building blocks that allow the researcher to create, configure, and customize high-content image analysis protocols. By using this method, we successfully separated endosomes and lysosomes in neurons, and time-lapse images were collected to capture endosome-lysosome fusion events in hundreds of cells [11]. Both assay set-up and analysis can be completed in an expeditious and efficient manner.

2 Materials

1. CellCarrier-96 Ultra microplates, black (PerkinElmer cat# 6055302).
2. Poly-D-lysine hydrobromide (MW70,000–150,000) (Sigma cat# P6407-5MG).
3. Gibco Neurobasal®-A medium (cat# 10-888-022) supplemented with 2% B-27™ Supplement (cat# 17504044) and 1% GlutaMAX™ Supplement (cat# 35050061).
4. Papain Dissociation System: Worthington Biochemical Corporation (cat# LK003150).
5. 1XPBS, pH 7.2 (Gibco, cat# 20-012-027).
6. Tetramethylrhodamine (TMR)-Dextran (ThermoFisher Scientific, cat# D1817).
7. AlexaFluor-647-dextran (ThermoFisher Scientific, cat# D22914).
8. Hoechst 33342 (Thermo Scientific™, cat# 62249).
9. Gibco Neurobasal®-A medium-minus phenol red (cat# 12349015). supplemented with 2% B-27™ Supplement (cat# 17504044) and 1% GlutaMAX™ Supplement (cat# 35050061).
10. Bafilomycin A1 (Sigma, cat# B1793).
11. Opera Phenix® High-Content Screening System (PerkinElmer).

3 Methods

3.1 Cell Culture

For mouse primary hippocampal neuron cultures, hippocampi are dissected from P0-P1 mice, dissociated with papain (20 units/mL) in Earle's Balanced Salt Solution (EBSS) at 37 °C water bath for 2 × 15 min, and triturated to dissociate cells with a 1 mL pipette. Hippocampal neurons are seeded on black CellCarrier-96 Ultra microplates pre-coated with 100 ug/mL poly-D-lysine (*see* **Note 1**) at a cell density of 28,000 cells/well in Neurobasal®-A medium supplemented with 2% B-27™ and 1% GlutaMAX™. Primary neuronal cultures are maintained at 37 °C incubator with a humidified atmosphere of 95% air and 5% CO_2.

3.2 Cell Treatment

3.2.1 One Day Before Imaging: Lysosome Labeling

On day in vitro (DIV) 4, cells are incubated with 0.25 mg/mL of TMR-dextran for 2 h at 37 °C. Cells are then washed twice with 1× PBS and chased overnight with supplemented Neurobasal-A media to label lysosomes (*see* **Note 2**).

3.2.2 The Day of Imaging: Endosome Labeling and Bafilomycin A1 Treatment

On DIV 5, cells are incubated with 0.5 mg/mL AlexaFluor-647-dextran and Hoechst-33342 (1:5000) for 10 min at 37 °C, washed twice with 1× PBS, and imaged immediately with supplemented Neurobasal-A media-minus phenol red (*see* **Note 3**). For bafilomycin A1 experiments, cells are treated with 100 nM bafilomycin A1 while incubating with AlexaFluor-647-dextran and Hoechst-33342 and imaged with supplemented Neurobasal-A media-minus phenol red with 100 nM bafilomycin A1 (*see* **Note 4**).

3.3 Time-Lapse Imaging Using Opera Phenix® High-Content Screening System

3.3.1 Define the Global Experiment Settings

In the Global Control section, define the global experiment settings: Plate Type-96 PerkinElmer CellCarrier- Ultra microplates; Autofocus Methods-Two Peak (default); Objective- 63xwater, NA1.15; Optical Mode: confocal; Binning:2 (*see* **Note 5**).

3.3.2 Select the Wells, Fields, and Planes to Be Measured

In the Navigation section, mark the wells that are included in imaging in the Plate panel; select the desired fields of one well in the Well panel (6 fields/well in our experiments). In the Layout Selection-Stack panel, defines the first plane (bottom) at 1 μm, distance (Vertical distance between the stack planes) to 0.5 μm and total number of planes (Z-stack range) to 5, which will set thc last plane (top) at 3 μm.

3.3.3 Select Channels

In the Channel Selection Pancl, click "+" button to load an existing channel from the database and adds it to the Channel Selection. In our experiment, we chose TRITC for TMR-dextran, DAPI for Hoechst 33342, and Alexa647 for AlexaFluor-647-dextran. Enter the desired Exposure Time, Power (power setting of the excitation laser or transmission LED used for this channel), and Height (Focus height above the plate bottom for snapshot) for each channel (*see* **Note 6**). We used 1000 ms, 100% power, 1.4 μm for TMR -dextran channel; 700 ms, 30% power, 1.4 μm for DAPI channel; 1000 ms, 50% power, 1.4 μm for Alexa 647-dextran channel in our experiments. Select one field and take snapshots to check each channel's setting result and adjust these parameters accordingly to make sure signals are not oversaturated or too dim.

3.3.4 Set Time Series

After all the previous setting, the experiment definition for a single time point is now complete. The maximum duration of imaging for a single time point (Max. Duration) is calculated and automatically displayed in the Global Control section and in the Time Series panel (called Shortest Interval here). In the Time Series panel, set the number of sequences to 1. In Sequence 1, enter the Number of Timepoints to 7 and set Fixed Interval to 20 min (*see* **Note 7**). This will allow you to collect time-lapse images every 20 min over the span of 2 h for all the fields we choose. The time series is visualized in the Navigation – Define Layout panel. The Timepoints panel is used to visualize the current time series configuration.

3.3.5 Save the Experiment for Future Use or Modification

3.3.6 Run Experiments

The Run Experiment tab is used for running a plate measurement with the current experiment settings. Use the current experiment defined on the Setup tab (default). Enter the desired plate name in the Plate Name text box. Click "Start" to start the plate measurement with the current experiment settings. Check if the experiment can be executed and no error message is shown in the message panel before you leave the system to run automatically.

3.4 Imaging Analysis: Build Up a Late Endosome-Lysosome Fusion Analysis Protocol

The Harmony software contains a set of analysis building blocks for each step to allow you to easily create customized analysis sequences. The building blocks are processed in top-down order. Intermediate results of the selected (and previous) building blocks are immediately displayed. Each building block offers multiple parameters (the first-level inputs) to adapt it to the researcher's specific application. Fine-tuning can be done using the second and third-level inputs (Fig. 1).

1. Click "New" to *initiate a new analysis protocol*: choose a Time-point (for example, T0), one representative image (or field), and select all the stack planes of that image for testing your parameters of each building block.
2. *Input Image block* is the first building block of each analysis sequence (Fig. 1a). The original image is displayed in the Content Area. In Flatfield Correction, choose "Basic"; in Stack Processing, choose "Maximum Projection".
3. Add *Find Image Region block* (Fig. 1b): In Channel, choose "Alexa 647"; ROI, choose "None"; Method (algorithm), choose "Whole image region"; Output Population is named "Whole Image", Output Region is named "Whole Image Region".
4. Add *Find Nuclei block* (Fig. 1c): This is a building block for detecting nuclei in our Hoechst-33,342 stained image. DAPI channel is used to find Nuclei. In Channel, choose "DAPI"; ROI, choose "Whole Image" and "Whole Image Region"; Method (algorithm), choose C (*see* **Note 8**). The second and third-level parameters in Method C contain "common threshold," "area," "splitting coefficient," "individual threshold," "contrast" for fine tuning the method so you can get the best result for nuclei finding. The Output Population of this step is named "All Nuclei".

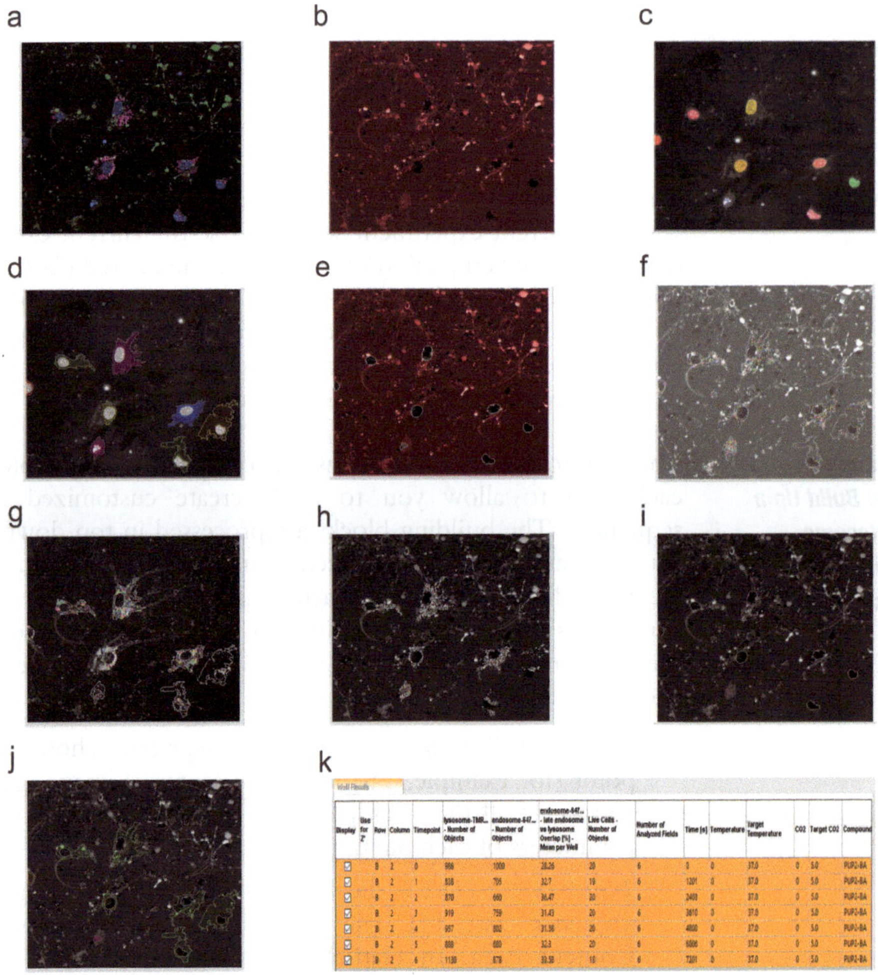

Fig. 1 Image Analysis Pipeline: (**a**) input image, (**b**) find image region, (**c**) find nuclei, (**d**) find cytoplasm, (**e**) select region, (**f**) find spots (Alexa647-dextran, endosome), (**g**) find spots (TMR-dextran, lysosome), (**h**) calculate position properties, (**i**) calculate morphology properties, (**j**) select Population (live cells), (**k**) results

5. Add *Find Cytoplasm block* (Fig. 1d): This is the building block for detecting cytoplasm around nuclei (*see* **Note 9**). In Channel, choose "DAPI"; Nuclei, choose "All Nuclei"; Method, choose A. You can adjust "individual threshold" to get the best result of cytoplasm finding. Check "Restrictive Region" and choose "Whole Image Region".

6. Add *Select Region block* (Fig. 1e): This is a building block to use mask to remove nuclei from image region. In Population, choose "Whole Image"; Region, choose "Whole Image Region"; Method, choose "Restrict by Mask"; Population, choose "All Nuclei"; Mask Region, select "Nucleus"; check "use inverted mask"; named Output Region is "Non-Nucleus Region".

7. Add *Find Spots block* (Fig. 1f): This is a building block for detecting spots and quantifying their properties. We will first find endosomes. In Channel, choose "Alexa647"; ROI, choose "Whole Image"—"Cytoplasm"; Method, choose D (*see* **Note 10**). Fine-tune its second and third-level parameters—"-Detection Sensitivity," "Splitting Sensitivity" and "Background Correction" using visual feedback from the image with visible spots of your interest until you have found a satisfactory set of parameters. Check "Calculate spot properties". The Output Population is named "Endosome-647-dextran".
8. Add another *Find Spots block* (Fig. 1g): We then next find lysosomes. In Channel, choose "TMR-dextran"; ROI, choose "Whole Image" – "Cytoplasm"; Method, choose D. You can adjust "Detection Sensitivity", "Splitting Sensitivity" and "Background Correction" to get the best spots finding result. Check "Calculate Spot Properties". The Output Population is named "Lysosome-TMR-dextran".
9. Add *Calculate Position Properties block* (Fig. 1h): This is the building block we use to find overlapping between endosomes and lysosomes. In Population, choose "Endosome-647-dextran"; Region, choose "Spot"; Method, choose "Cross Population". In Population B, choose "Lysosome-TMR-dextran"; Region B, choose "Spot", check "Overlap". The Property Prefix is named "Late endosome vs Lysosome".
10. Add *Calculate Morphology Properties block* (Fig. 1i): This is a building block for calculating one or more morphology properties (area, roundness) of a cell region. We will use this block to help distinguish dead cells from live cells. In Population, choose "all nuclei"; Region, choose "Nucleus"; Method, choose "Standard," check "Area". The Property Prefix is named "Nucleus".
11. Add *Select Population block* (Fig. 1j): This is a building block for selecting a subpopulation of the input population by applying a condition or multiple conditions. In Population, choose "all Nuclei"; Method, choose "Filter by Property," select "Nucleus area (μm^2)" >40. Output Population is named "Live Cells".
12. The last building block *Define Results* is used for performing the statistical analysis of arbitrary populations. It also defines the assay readout values for each well of a microplate (results per well). All properties calculated by the preceding building blocks can be included in this readout, as well as any combination (by formula). In Method, choose "List of Outputs"; in "Population: Endosome-647-dextran," check "Number of Objectives"; in Apply to All, choose "individual selection"; in "Late Endosome vs Lysosome Overlap (%)," choose "mean"; In "Population: live cells", check "Number of Objectives".

13. Save the whole analysis sequence protocol. Do a test run in one selected well to examine the complete analysis sequence (*see* **Note 11**).

3.5 Run Analysis for the Whole Experiment

Go to the Evaluation tab, make sure that Single mode is active. In the Navigation – Measurement – Plate panel, mark the desired wells to be evaluated. In the Navigation – Measurement – Well panel, mark all the fields to be evaluated. In the Navigation – Measurement – Stack panel, mark all the planes. In the Navigation – Measurement – Timepoints panel, mark all the time points. Click "Start" to run the analysis. The readout values (well results) can be inspected in the Results pane (Fig. 1k).

3.6 Data Analysis and Graph

The value of the "Late endosome vs Lysosome Overlap [%]—Mean per Well" shows the quantification of the late endosome-lysosome fusion events, which is defined by endocytosed AlexaFluor-647-dextran puncta that colocalized with the lysosome marker TMR-dextran. Endosome-lysosome fusion % is expressed as % fold change to time point 0 for the same animal (Figs. 2 and 3). In our experiments, there was a significant interaction effect for time × treatment ($F_{(3,106)} = 21.87$, $p < 0.0001$, non-linear regression: one-phase exponential association). As expected, bafilomycin A1 treatment disrupts the trafficking of late endosomes to lysosomes [17] (impaired endosome-lysosome fusion in treated neurons).

4 Notes

1. Black CellCarrier-96 Ultra microplates should be pre-coated with 100 μg/mL poly-D-lysine at 37 °C for at least 1 h. Wash with filter-sterilized ddH_2O twice. Air dry the plate and then the plate is ready for seeding neurons.
2. Dextrans are large molecules. Conjugated dextrans are actively taken up by endocytosis, and once loaded, they are finally accumulated in lysosomal compartment for long periods [18].
3. For optimal results, cells should be maintained in a CO_2 chamber held at 37 °C during the 2 h imaging. Thus, the TCO settings of Harmony Software needs to be turned on at least 30 min before imaging.
4. Bafilomycin A1 (BafA1) is a macrolide antibiotic. BafA1 by binding to the membrane-spanning domain of the V-ATPase specifically inhibits the V-ATPase activity. Study showed that the degradative pathway was blocked by BafA1 at the step from late endosomes to lysosomes [17], which makes it a nice control for the experiments.

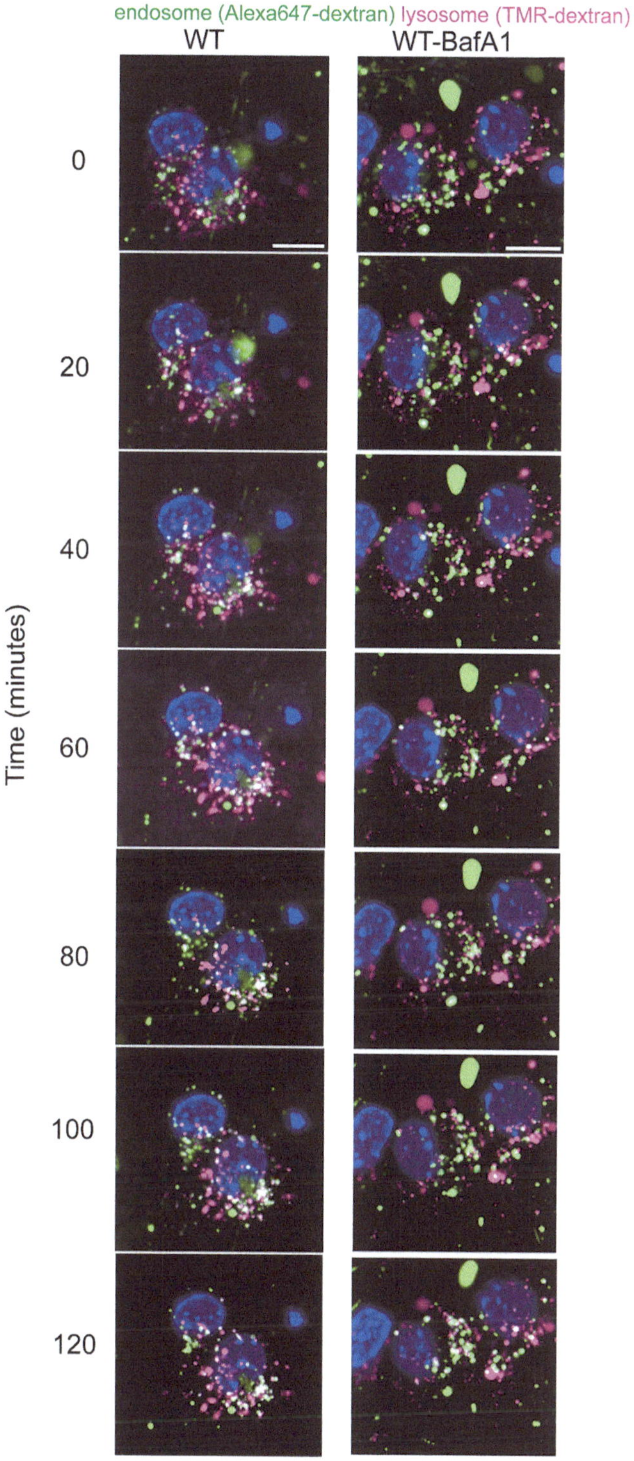

Fig. 2 Representative images showing endosome-lysosome fusion in wt neurons and bafilomycin A1-treated neurons in vitro. Live cell confocal microscopy imaging (Opera Phenix) of endosome-lysosome fusion in wild type mouse primary hippocampal neurons at 5 DIV with and without bafilomycin A treatment. The following time points were measured following incubation with AlexaFluor-647-dextran: 0, 20, 40, 60, 80, 100, and 120 min. Green: endosome labeled by Alexa647-dextran; Magenta: lysosome labeled by TMR-dextran; Blue: Hoechst. Fused endosome-lysosome will show white color (scale bar 10 μm)

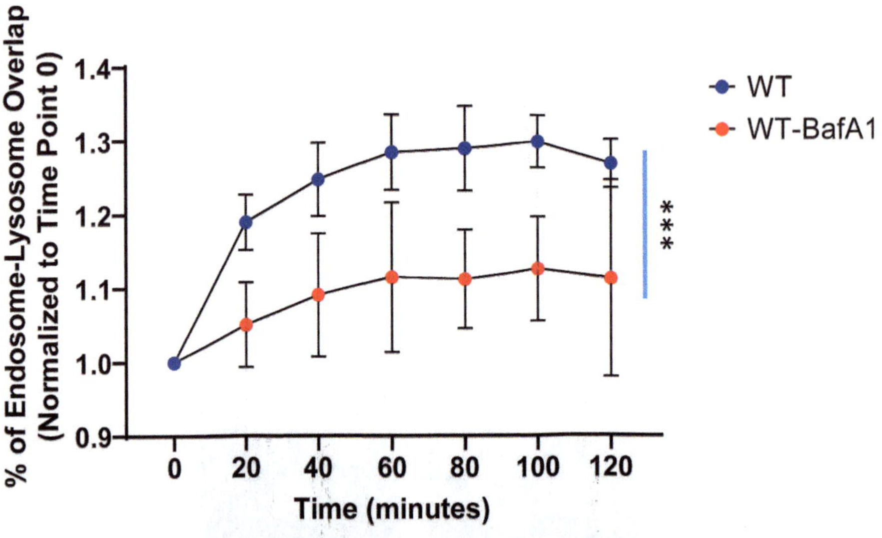

Fig. 3 Quantification of endosome-lysosome fusion (WT $n = 8$ animals, WT-BafA1 $n = 8$ animals. 3 litters). Endosome-lysosome fusion % is expressed as % fold change to time point 0 for the same animal. There was a significant interaction effect for time X treatment ($F_{(3,106)} = 21.87$, $p < 0.0001$). Data are mean ± SEM, non-linear regression

5. In the global experiment settings-Binning: With binning 1 the full resolution (2160 × 2160 px) of the camera is used. With binning 2 the signal of 2 × 2 camera pixels is integrated into 1 pixel in the final image. Although the effective resolution of the cameras is reduced accordingly (1080 × 1080 px), the signal to noise ratio gets a four-fold increase. We found this is more beneficial for collecting fluorescent signals in small vesicles.
6. Perform Snapshot to test channels' settings and adjust parameters accordingly. When using the Test function, make sure the Height for each channel is set to the same Height (Focus height).
7. When setting up Time Series, make sure the Fixed Interval should be at least as long as the calculated Shortest Interval (the maximum duration of imaging for a single time point). If you enter shorter times, the scheduled time will not be kept. The next time point will then be measured as fast as possible. In the case of the Shortest Interval is longer than the Fixed Interval (20 min), either reduce the number of fields per well or increase the vertical distance between the stack planes and reduce total number of planes to achieve a shorter Shortest Interval.
8. Method C in Find Nuclei block provides good results for images with low background or with size variations of nuclei. It also supports images with large variations in intensity or contrast of nuclei.

9. Most of the late endosome-lysosome fusion events happen in the soma of neurons. Find Cytoplasm block will help us to define neuron soma. To use Find Cytoplasm block, nuclei must have been detected beforehand. Nucleus-stained images can be used to detect cytoplasm. Method A is the most robust method of Find Cytoplasm. It is most applicable in cases when the intensity decreases with the distance from nucleus.
10. Spots are searched only in the Cytoplasm region.
11. In Image Analysis, apply building block parameter at each step to check the fitting result and fine-tune accordingly. Tune Parameter allows the researcher to find the best value for a parameter by testing and previewing different settings.

Acknowledgments

This work was supported by the following grants: NIH NINDS grants (R01NS113141, R01NS121618) to E.M.M. The authors wish to acknowledge the Leduc Bioimaging Facility of Brown for providing the Opera Phenix® High Content Screening System. The Opera Phenix in the Center for Animal Alternatives in Testing is supported by the National Institute of Health (U01 ES028184), the National Science Foundation (EPSCoR award 1655221), and a generous gift by Donna McGraw Weiss ('89) and Jason Weiss.

References

1. Goldstein JL, Brown MS, Anderson RGW et al (1985) Receptor mediated endocytosis: concepts emerging from the LDL receptor system. Annu Rev Cell Biol 1:1–39
2. Gruenberg JE, Howell KE (1989) Membrane traffic in endocytosis: insights from cell-free assays. Annu Rev Cell Biol 5:453–481
3. Kornfeld S, Mellman I (1989) The biogenesis of lysosomes. Annu Rev Cell Biol 5:483–525
4. Langemeyer L, Fröhlich F, Ungermann C (2018) Rab GTPase function in endosome and lysosome biogenesis. Trends Cell Biol 28(11):957–970
5. Huotari J, Helenius A (2011) Endosome maturation. EMBO J 30:3481–3500
6. Fraldi A, Klein AD, Medina DL, Settembre C (2016) Brain disorders due to lysosomal dysfunction. Annu Rev Neurosci 39:277–295
7. Sharma J, di Ronza A, Lotfi P et al (2018) Lysosomes and brain health. Annu Rev Neurosci 41:255–276
8. Winckler B, Faundez V, Maday S et al (2018) The endolysosomal system and proteostasis: from development to degeneration. J Neurosci 38:9364–9374
9. Otomo A, Pan L, Hadano S (2012) Dysregulation of the autophagy-endolysosomal system in amyotrophic lateral sclerosis and related motor neuron diseases. Neurol Res Int 2012:498428. https://doi.org/10.1155/2012/498428
10. Ouyang Q, Lizarraga SB, Schmidt M et al (2013) Christianson syndrome protein NHE6 modulates TrkB endosomal signaling required for neuronal circuit development. Neuron 80: 97–112
11. Pescosolido MF, Ouyang Q, Liu JS et al (2021) Loss of Christianson Syndrome Na^+/H^+ exchanger 6 (NHE6) causes abnormal endosome maturation and trafficking underlying lysosome dysfunction in neurons. J Neurosci 41(44):9235–9256
12. Lloyd-Evans E, Morgan AJ, He XX et al (2008) Niemann-Pick disease type C1 is a sphingosine storage disease that causes deregulation of lysosomal calcium. Nat Med 14(11):1247–1255
13. Urwin H, Authier A, Nielsen JE et al (2010) Disruption of endocytic trafficking in

frontotemporal dementia with CHMP2B mutations. Hum Mol Genet 19(11): 2228–2238

14. Emans N, Biwersi J, Verkman AS (1995) Imaging of endosome fusion in BHK fibroblasts based on a novel fluorimetric avidin-biotin binding assay. Biophys J 69:716–728
15. Biwersi J, Emans N, Verkman AS (1996) Cystic fibrosis transmembrane conductance regulator activation stimulates endosome fusion in vivo. Proc Natl Acad Sci 93:12484–12489
16. Ba Q, Raghavan G, Kiselyov K et al (2018) Whole-Cell scale dynamic organization of lysosomes revealed by spatial statistical analysis. Cell Rep 23(12):3591–3606
17. van Weert AW, Dunn KW, Geuze HJ et al (1995) Transport from late endosomes to lysosomes, but not sorting of integral membrane proteins in endosomes, depends on the vacuolar proton pump. J Cell Biol 130:821–834
18. Eriksson I, Öllinger K, Appelqvist H (2017) Analysis of lysosomal pH by flow cytometry using FITC-dextran loaded cells. Methods Mol Biol 1594:179–189

Chapter 17

Live-Imaging Detection of Multivesicular Body-Plasma Membrane Fusion and Exosome Release in Cultured Primary Neurons

Matthew F. Pescosolido, Qing Ouyang, Judy S. Liu, and Eric M. Morrow

Abstract

Exosomes represent a class of extracellular vesicles (EVs) derived from the endocytic pathway that is important for cell-cell communication and implicated in the spread of pathogenic protein aggregates associated with neurological diseases. Exosomes are released extracellularly when multivesicular bodies (also known as late endosomes) fuse with the plasma membrane (PM). An important breakthrough in exosome research is the ability to capture MVB-PM fusion and exosome release simultaneously in individual cells using live-imaging microscopy techniques. Specifically, researchers have created a construct fusing CD63, a tetraspanin enriched in exosomes, with the pH-sensitive reporter pHluorin whereby CD63-pHluorin fluorescence is quenched in the acidic MVB lumen and only fluoresces when released into the less acidic extracellular environment. Here, we describe a method using this CD63-pHluorin construct to visualize MVB-PM fusion/exosome secretion in primary neurons using total internal reflection fluorescence (TIRF) microscopy.

Key words Extracellular vesicles (EVs), Exosomes, Multivesicular bodies (MVBs), Endocytic pathway, Total internal reflection fluorescence (TIRF), Primary neuronal culture

1 Introduction

Extracellular vesicles (EVs) are a diverse group of membrane-containing compartments, that may contain proteins, nucleic acids, and lipids, and are released by cells. Although EVs were originally considered an alternative pathway for cells to discard their contents, it is now clear that EVs represent an important form of intercellular communication involved in a wide range of biological processes [1, 2]. Exosomes (30–150 nm) are an EV subgroup originating from the endocytic pathway wherein multi-vesicular bodies (MVBs, also known as late endosomes) fuse with the plasma membrane (PM). During endosome maturation, intra-

Yu-Wen Alvin Huang and ChangHui Pak (eds.), *Stem Cell-Based Neural Model Systems for Brain Disorders*, Methods in Molecular Biology, vol. 2683, https://doi.org/10.1007/978-1-0716-3287-1_17,

luminal vesicles (ILVs) are formed by the inward budding of endosomal membrane. MVB-PM fusion releases ILVs extracellularly, which are now referred to as exosomes.

EV research is particularly challenging given the heterogeneity in their biogenesis, their nano size, and contents [3]. The EV field has generally relied on separating large, heterogenous EV populations via ultracentrifugation and/or size-exclusion chromatography (SEC) from biological fluids, tissue, or *in vitro* cell culture media to characterize their function. Unfortunately, these methods are currently unable to fully isolate EV subtypes like exosomes nor provide the spatial and temporal resolution for single EV release in an individual cell [3]. Recent advances in EV/exosome labeling now allow greater insight into their fusion and release dynamics in single cells [4].

A major breakthrough for visualizing exosome release occurred when researchers fused CD63, a tetraspanin associated with exosomes, to the pH-sensitive fluorescent green protein pHluorin [5]. These researchers demonstrated this construct labels CD63-positive ILVs/secreted exosomes, which fluoresce when MVBs fuse with the plasma membrane to release their ILVs extracellularly. Specifically, the pHluorin fluorescence is quenched in the more acidic MVB lumen, and fluoresces during MVB-PM fusion as it is exposed to the higher pH extracellular environment. Researchers have visualized real-time MVB-PM fusion/exosome release with this CD63-pHluorin fusion protein using both total internal reflection fluorescence (TIRF) and spinning-disc confocal microscopy [5, 6]. TIRF microscopy excites fluorophores on a cell's surface close to a cover slip while limiting intracellular fluorescent signal [7]. These properties allow for greater signal-to-noise ratio and lower phototoxicity compared to other types of microscopes. Furthermore, spinning-disc confocal microscopy captures fewer CD63-positive fusion events compared to TIRF microscopy [6]. Since Verweij et al. [5] published their seminal technique, additional tools have been developed to visualize CD63 trafficking, release, and uptake [8] as well as CD81- and CD9-positive exosomes [6].

Extracellular vesicles play an important role in cellular communication in the nervous system [9]. For example, EVs (particularly exosomes) mediate a range of processes including synaptic function (e.g., neuronal firing/plasticity), myelin formation, and neurodegeneration (e.g., spreading pathogenic proteins such as amyloid β and tau) [10, 11]. Therefore, there is a need to harness EV/exosome techniques to study their activity in the nervous system. Here we have adapted CD63 methodology to visualize and measure CD63-positive MVB-PM fusion and exosome release in mammalian primary neurons using TIRF microscopy [12].

2 Materials

2.1 Primary Culture of Mouse Hippocampal Neurons

1. Dissecting solution: Hank's balanced salt solution (HBSS).
2. Primary neuronal culture media: Neurobasal A supplemented with B27 (2%) and GlutaMAX (1%).
3. Papain Dissociation Kit (Worthington).
4. 15 mL conical tubes.
5. 50 mL conical tubes.
6. 70 um nylon mesh.
7. Cell count materials: Trypan blue & hemocytometer chamber.
8. Coverslip: 22 mm poly-D-lysine-coated coverslips (Neuvitro, GG-22-1.5-PDL).
9. 6-well tissue culture plate (i.e., 1 well per coverslip).

2.2 Transfecting Primary Neurons

1. Lipofectamine 2000.
2. Opti-MEM.
3. Plasmids:
   (a) CD63-pHluorin (Addgene: Plasmid #130901).
   (b) pmCherry-N1 (Clontech: Cat. No. 632523).

2.3 Visualizing MVB-PM/Exosome Release

1. TIRF microscope such as the DeltaVision OMX SR imaging system (GE).
2. TIRF objective (Apo 60×/1.49 oil).
3. Tyrode's solution: NaCl (124 mM), KCl (3 mM), $CaCl_2$ (2 mM), $MgCl_2$ (1 mM), HEPES (10 mM, pH = 7.4), D-glucose (5 mM).
4. Treatments such as Bafilomycin A and U18666A.

3 Methods

3.1 Primary Culture of Mouse Hippocampal Neurons

1. Primary hippocampal neurons isolated from mouse strain (e.g., C57BL6) at postnatal day 0–1 (P0–P1) (*see* **Note 1**).
2. For each dissected mouse, prepare the following:
   (a) Papain solution (500 ul) in a 15 mL canonical tube. Keep on ice.
   (b) Inhibitor cushion (1 mL) in a 15 mL canonical tube. Store in dissection hood at room temperature (RT).
3. Prewarm supplemented Neurobasal A media in water bath (37 °C).

4. Dissect mouse pup brain in ice-cold HBSS.
5. Remove brain from skull.
6. Dissect both hippocampi.
7. Put hippocampi in tube with 500 ul papain solution.
8. Put on ice until finished dissecting all mice.
9. Put all tubes into water bath (37 °C) for 15 min.
10. Gently pipette tissue (*see* **Note 2**).
11. Put tubes back into water bath (37 °C) for 15 min.
12. Gently pipette tissue.
13. Let tissue settle to bottom of tube.
14. Transfer as much of supernatant as possible (i.e., not a lot of tissue) to a new 15 mL conical tube.
15. Centrifuge at 300 g for 6 min at RT.
16. Remove as much supernatant as possible.
17. Add 630 ul inhibitor solution (from papain dissociation kit) to each tissue pellet.
18. Gently resuspend pellet and collect all liquid.
19. Filter through 70 um nylon mesh into a 50 mL canonical tube.
20. Layer filtered solution on 1 mL inhibitor solution.
21. Centrifuge at 70 g for 10 min at RT.
22. Remove supernatant.
23. Resuspend in 500 ul of warm supplemented Neurobasal A media.
24. Count cells using Trypan blue (5:1 ratio of Trypan blue to neuron solution).
25. Plate neurons at 3.5×10^5 cells/mL density on poly-D-lysine coated coverslips (*see* **Notes 3 and 4**).
26. Next day: Change entire culture media volume with fresh primary culture media.
27. Change 1/2 of culture media volume every other day until transfection.

3.2 Transfecting Primary Hippocampal Neurons

1. Transfect cells with CD63-pHluorin and empty mCherry plasmids the day before imaging (i.e. if planning to image at DIV 14, transfect cell at DIV 13).
2. Combine 4 ul of Lipofectamine 2000 and 250 ul of Opti-MEM into a 1.5 mL tube for each transfected well.
3. Let Lipofectamine 2000 + Opti-MEM solution sit at RT for 5 min.

4. Mix 1 ug of CD63-pHluorin and 1 ug mCherry (e.g., 1 ul of 1 ug/ul plasmid stock) into 250 ul Opti-MEM (*see* **Note 5**).
5. Add 250 ul Lipofectamine 2000 + Opti-MEM solution to each Opti-MEM tube with CD63 and mCherry plasmids.
6. Put in 37 °C incubator for 20 min.
7. Collect most of the conditioned media from each well (~2 mL), put in tube labeled with sample ID, and put tubes in 37 °C incubator (*see* **Note 6**).
8. Add 500 ul of Lipofectamine 2000 + Opti-MEM + plasmids solution to each well.
9. Incubate for 1 h at 37 °C.
10. Aspirate all Lipofectamine 2000 + Opti-MEM + plasmids solution.
11. Replace with previously collected conditioned media.

3.3 Visualizing MVB-PM/Exosome Release Using Total Internal Reflection Fluorescence (TIRF) Microscopy

1. Prepare Tyrode's solution.
2. Warm Tyrode's solution to 37 °C in a warm bath.
3. Set environmental controls on TIRF microscope (*see* **Note 7**).
4. Approximately 15–30 min prior to imaging, move coverslip with primary neurons to metal/microscope holder, and add warm Tyrode's solution. Keep in incubator until ready to image.
5. Prior to loading sample, apply oil to lens.
6. Identify a transfected neuron.
7. Collect an image with the neuron's morphology as outlined by mCherry construct (*see* **Note 8**).
8. Locate flat region of plasma membrane suitable for TIRF imaging using TIRF objective (*see* **Note 9**).
9. Collect 5-min videos at 2 Hz (i.e. every 500 ms) (*see* **Note 10**).
10. Treatments:
    (a) Bafilomycin: Pretreat neurons for 2 h with 100 nm Bafilomycin in supplemented Neurobasal media.
    (b) U18666A: Pretreat neurons overnight (~16 h) with 1.5 ug/mL in supplemented Neurobasal media.
11. Continue imaging transfected neurons from multiple coverslips (e.g., 1–4 neurons per coverslip) (*see* **Note 11**).

3.4 Quantifying MVB-PM/Exosome Release

1. Create an Analysis spreadsheet for recording: Randomized file name, Start time, Stop time, # events, etc. (*see* **Note 12**).
2. Open an imaging file in FIJI (ImageJ) (*see* **Note 13**).

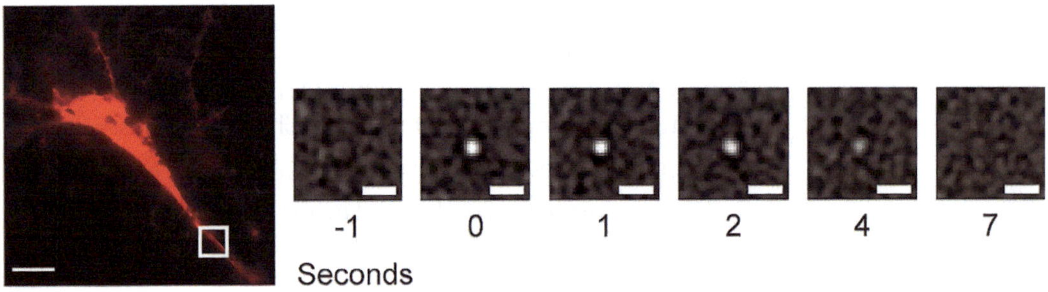

Fig. 1 Full MVB-PM fusion and exosome release. Widefield image of neuron cotransfected with mCherry and CD63-pHluorin plasmids. White inset indicates the location of fusion event shown in panel. Panel images depict before, during, and after full fusion event

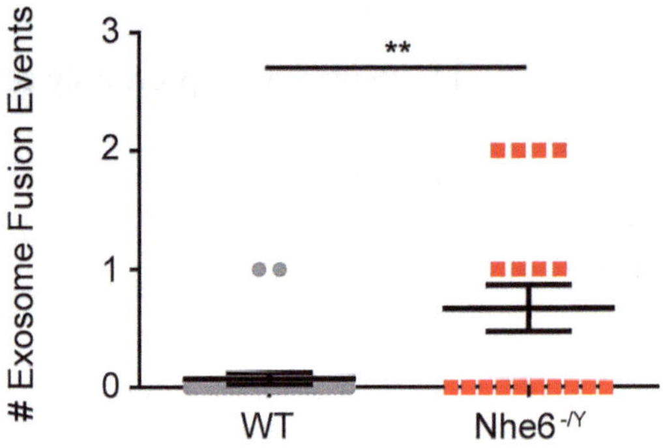

Fig. 2 MVB-PM fusion and exosome release quantification. Number of full fusion events over 5 min per neuron in WT and *Nhe6*$^{-/Y}$ male littermate primary hippocampal neurons at 14 DIV (WT $n = 28$ cells from 7 mice, *Nhe6*$^{-/Y}$ $n = 18$ cells from 7 mice)

3. Determine when video starts (i.e., earliest time point when recording is properly positioned to flat section of plasma membrane). Record this Start time in Analysis spreadsheet as well as the Stop time (i.e., 5 min after Start time).
4. Manually count full fusion events using the following criteria reported by Bebelman et al. [6] (Fig. 1, *see* **Note 14**):
   (a) Pronounced signal strongly above background,
   (b) Lasts more than 2 s at 2 Hz (i.e., >4 consecutive frames),
   (c) Peak intensity in maximum of 3 frames,
   (d) Minimal movement following initial burst,
5. Calculate the total number of full fusion events per 5-min video.
6. Report the total number of full fusion events per 5-min video per neuron (Fig. 2, *see* **Note 15**).

4 Notes

1. Can also use mouse cortical neurons, although protocol(s) may need to be adapted. For example, use 1 mL of papain solution per mouse neocortex.
2. Approximately 8 stroked with 1000 ul pipette tips. Do not push too hard/make too many bubbles.
3. If not enough neurons survive transfection, may need to plate more cells.
4. Coverslip size may need to be adapted to microscope. We strongly suggest using pre-coated poly-D-lysine coverslips to promote neuronal health and survival.
5. Make sure to sufficiently mix plasmids in Opti-MEM by pipetting up and down.
6. Make sure to leave ~1 mL of conditioned media in each well.
7. We use the following settings: O_2 (20%), CO_2 (5%), and humidity (50%).
8. For the DeltaVision OMX SR, we identified transfected neurons using widefield fluorescence. A single-plane, multi-tiled image of the neuron was collected using the 561 nm laser at 60× in order to capture mCherry fluorescence.
9. Select an imaging field where the neuron's plasma membrane is relatively flat. We typically select an imaging area encompassing the cell body at 1024 × 1024 resolution. Although, it is possible to investigate the subcellular localization of MVB-PM fusion/exosome release in other areas such as axons and dendrites.
10. Laser power must be identical across all experiments. Use lowest laser power possible to prevent phototoxicity and photobleaching. For example, we used 35% laser power using our DeltaVision OMX SR imaging system.
11. Image enough neurons from multiple coverslips and animals until sufficiently powered. For example, we imaged >15 cells from 7 mice per genotype. Bebelman et al. [6] recommend at least 10 cells per condition.
12. Experimenter should be blinded to genotype, treatment, etc. status of each video. Columns can be added such as Real file name, genotype, treatment, etc. after the analysis is complete.
13. Raw images may need to be processed (e.g., deconvolved). Refer to TIRF imaging instrument/software for instructions.

14. Be careful to clearly differentiate between bona fide full fusion events and other rapid increases in fluorescence intensity such as kiss-and-run fusion events or neutral vesicles entering TIRF field. Kiss-and-run fusion events may represent partial exosome secretion or full fusion followed by quick endocytosis of exosomes (for a more detailed description *see* Bebelman et al. [6]. Kiss-and-run fusion events are more likely to have (1) a strong signal with little to no decay in signal and (2) signal movement. Bebelman et al. [6] also developed a FIJI (ImageJ) macro Analyzer of MVB Exocytosis (AMvBE) for quantifying MVB-PM fusion.
15. We have observed an average of <1 full fusion event per 5 min in wildtype neuronal cultures. Alternatively, fusion events can be reported as # events per minute.

Acknowledgments

This work was supported by NIH NINDS grants (R01NS113141, R01NS121618) to EMM. MFP was supported by National Institutes of Health National Research Service Award (F31NS093880). The authors declare no competing financial interests.

References

1. Maas SLN, Breakefield XO, Weaver AM (2017) Extracellular vesicles: unique intercellular delivery vehicles. Trends Cell Biol 27: 172–188
2. Van Niel G, D'angelo G, Raposo G (2018) Shedding light on the cell biology of extracellular vesicles. Nat Rev Mol Cell Biol 19:213–228
3. Van Niel G, Carter DRF, Clayton A et al (2022) Challenges and directions in studying cell-cell communication by extracellular vesicles. Nat Rev Mol Cell Biol 23:369–382
4. Verweij FJ, Balaj L, Boulanger CM et al (2021) The power of imaging to understand extracellular vesicle biology in vivo. Nat Methods 18: 1013–1026
5. Verweij FJ, Bebelman MP, Jimenez CR et al (2018) Quantifying exosome secretion from single cells reveals a modulatory role for GPCR signaling. J Cell Biol 217:1129–1142
6. Bebelman MP, Bun P, Huveneers S et al (2020) Real-time imaging of multivesicular body-plasma membrane fusion to quantify exosome release from single cells. Nat Protoc 15:102–121
7. Mattheyses AL, Simon SM, Rappoport JZ (2010) Imaging with total internal reflection fluorescence microscopy for the cell biologist. J Cell Sci 123:3621–3628
8. Sung BH, Von Lersner A, Guerrero J et al (2020) A live cell reporter of exosome secretion and uptake reveals pathfinding behavior of migrating cells. Nat Commun 11:2092
9. Budnik V, Ruiz-Canada C, Wendler F (2016) Extracellular vesicles round off communication in the nervous system. Nat Rev Neurosci 17: 160–172
10. Holm MM, Kaiser J, Schwab ME (2018) Extracellular vesicles: multimodal envoys in neural maintenance and repair. Trends Neurosci 41:360–372
11. Song Z, Xu Y, Deng W et al (2020) Brain derived exosomes are a double-edged sword in Alzheimer's disease. Front Mol Neurosci 13:79
12. Pescosolido MF, Ouyang Q, Liu JS et al (2021) Loss of Christianson syndrome Na(+)/H(+) exchanger 6 (NHE6) causes abnormal endosome maturation and trafficking underlying lysosome dysfunction in neurons. J Neurosci 41:9235–9256

Chapter 18

Assays of Monitoring and Measuring Autophagic Flux for iPSC-Derived Human Neurons and Other Brain Cell Types

Ryan O'Rourke, Guzide Ayse Erdemir, and Yu-Wen Alvin Huang

Abstract

Autophagy is a highly conserved, cytoprotective, catabolic process induced in response to conditions of cellular stress and nutrient deprivation. It is responsible for the degradation of large intracellular substrates such as misfolded or aggregated proteins and organelles. This self-degradative mechanism is crucial for proteostasis in post-mitotic neurons, requiring its careful regulation. Due to its homeostatic role and the implications, it has for certain disease pathologies, autophagy has become a growing area of research. We describe here two assays that can be used as part of a tool kit for measuring autophagy-lysosomal flux in human iPSC-derived neurons.

One way to measure autophagic flux is through a western blotting assay, which can be used to analyze two important autophagy proteins: microtubule-associated protein 1 light chain 3 (LC3) and p62. In this chapter, we describe a western blotting assay for use in human iPSC neurons that can be used to quantify these two proteins of interest to measure autophagic flux.

In addition to conventional western blotting techniques, more sophisticated tools have come available to readout autophagic flux in a sensitive and high-throughput manner. In the latter portion of this chapter, we describe a flow cytometry assay which utilizes a pH-sensitive fluorescent reporter which can also be used to measure autophagic flux.

Key words Induced pluripotent stem cells (iPS cells), Neurons, Autophagy, Aitophagic flux, Microtubule associated protein 1 light chain 3 (LC3), p62, Immunoblotting flow cytometry

1 Introduction

Autophagy is a highly conserved and fundamental intracellular process that is crucial for the survival and proteostasis of neurons [1]. During this catabolic process, cytoplasmic cargo is engulfed within lipid phagophores and delivered to the lysosome for hydrolytic degradation via double-membraned vesicles termed autophagosomes [2]. This process by which autophagosomes fuse with lysosomes for cargo degradation is termed autophagic flux [3]. Measurement of autophagic flux has become increasingly prevalent over simple quantification of autophagic structures since such

Yu-Wen Alvin Huang and ChangHui Pak (eds.), *Stem Cell-Based Neural Model Systems for Brain Disorders*, Methods in Molecular Biology, vol. 2683, https://doi.org/10.1007/978-1-0716-3287-1_18,

a crude measurement cannot differentiate, for example, increased autophagosome biogenesis versus impaired lysosomal degradation. The increasing implications of autophagic impairment in disease pathologies such as neurodegeneration and cancer have drawn significant attention to its place in research. Hence, there is a need for validated and reliable assays to study autophagic flux in human iPSC-derived neurons. Here we describe two such assays.

The first assay we describe utilizes western blotting to measure the levels of two key autophagy proteins: microtubule-associated protein 1 light chain 3 (LC3) and p62. LC3B is a member of the autophagy-related ATG8 family of proteins and it is the most well-studied LC3 species to date [4]. Upon autophagy induction, soluble LC3B-I protein is incorporated into autophagosomes via conjugation to phosphatidylethanolamine (PE) to form the lipidated LC3B-II species [5]. Lipidated LC3B-II is a marker of mature autophagosomes and migrates at a lower molecular weight than LC3-I due to its hydrophobic nature (14 kDa and 16 kDa, respectively). An increase in the levels of LC3-II can be interpreted as an increase in autophagosome formation or a decrease in turnover where lysosomal degradation is impaired. Likewise, a decrease in LC3-II protein levels could indicate an inhibition of autophagosome biogenesis or an increased rate of autophagic clearance at lysosomes. The inclusion of p62 into this assay aids researchers in distinguishing between these scenarios. The p62 protein is an autophagy receptor that binds directly to LC3B through its LC3-interacting (LIR) domain, bringing with it cargo destined for degradation at the lysosome [1]. Increased p62 levels by western blot indicate impaired lysosomal degradation whereas a decrease represents increased lysosomal degradation. The combination of LC3B and p62 western blotting can therefore inform researchers about the degree of autophagic flux in vitro. The assay described includes the lysosomal alkalinizing agent, Chloroquine, as a necessary control to confirm that changes in protein levels are due to autophagy-lysosomal flux versus transcriptional changes.

The second assay described in this chapter details a pH-sensitive flow cytometry assay that can be used in a high-throughput manner to measure autophagic flux. This assay leverages the use of a tandem fluorescent (dual RFP and GFP) reporter tagged to the aforementioned LC3B. This assay has been widely documented as a tool that can be used to measure autophagic flux due to the different stabilities of each fluorescent reporter in the acidic environment of the lysosome [6]. Upon fusion of RFP-GFP-LC3B+ autophagosomes with lysosomes, the low pH of the lysosomal compartment quenches the GFP signal while the RFP signal remains stable [7]. Thus, the ratio of RFP signal to GFP signal can be used as a readout for autophagic flux. Such an assay has been widely documented in the literature for its use in cell imaging experiments [6]. Here we describe another use for this reporter system in a flow cytometry-based assay. The advantage of a flow cytometry assay is its ability to be used in a high throughput manner as well as the high sensitivity of the readout. The data shown at the end of this chapter demonstrates that the GFP signal is quenched upon autophagy induction, while the RFP signal remains stable. Therefore, the RFP/GFP ratio can be used as a measure of autophagic flux with a higher ratio indicating greater flux due to GFP quenching in the highly acidic lysosomal compartment.

2 Materials

2.1 Chemical Modulation of Autophagy

1. Vehicle control/DMSO: Invitrogen (D12345).
2. Chloroquine: Selleck Chemicals (S6999).

2.2 Cell Lysis and Western Blotting

1. 6 well TC-treated plate: Corning (3516).
2. 1× Laemmli SDS sample buffer: Boston BioProducts (BP-111R).
3. Cell scrapers: CytoOne (CC7600-0220).
4. Microcentrifuge tubes: USA Scientific (1615-5510).
5. Heating block: Eppendorf (ThermoMixer C).
6. Benchtop microcentrifuge.
7. Electrophoresis gel tank, Midi size: Bio-Rad (1656001).
8. 10× Tris/Glycine/SDS buffer: Bio-Rad (1610732).
9. 12% polyacrylamide protein gel, Midi size: Bio-Rad (5671044).
10. Protein ladder: Bio-Rad (1610376).
11. Semi-dry protein transfer system: Bio-Rad (Transblot Turbo).
12. 0.2 um nitrocellulose transfer stack: Bio-Rad (1704159).
13. 1% casein blocking buffer: Bio-Rad (1610782).

2.3 Antibodies and Detection Reagents

1. GAPDH housekeeping primary antibody: CST (5174S).
2. LC3B primary antibody: Abcam (ab192890).
3. P62 primary antibody: Novus Biologicals (NBP1-48320).
4. 10× TBS: Bio-Rad (1706435).
5. Tween 20: Sigma (P9416-50ML).
6. Anti-Rabbit HRP conjugated secondary antibody: CST (7074).
7. Streptactin-HRP conjugate: Bio-Rad (1610380).
8. ECL detection kit: Bio-Rad (1705062).
9. Membrane imaging instrument: Invitrogen (iBright 1500).

2.4 Human RFP-GFP-LC3B Reporter Plasmid Cloning

1. Human RFP-GFP-LC3B gene fragment sequence (5' to 3'):
GGCCGTTTTTGGCTTTTTTGTTAGACAG
GATCCCCGGGTACCGGTCCACCATGGCGTTCTCT
GAAGACGTGATCAAGGAGTTTATGCGCTTTAAGGTCA
GAATGGAGGGAAGTGTAAACGGTCATGAATTT
GAAATTGAAGGCGAAGGGGAGGGAAGACCCTAT
GAAGGTACACAGACCGCAAAACTGAAAGTGA
CAAAAGGCGGCCCCCTGCCTTTTGCATGGGA
TATCCTTTCTCCACAATTTCAGTATGGGTCCAAGGCC
TACGTTAAGCACCCGGCAGATATCCCTGATTACCT
CAAACTCAGCTTTCCAGAAGGGTTCAAGTGGGAAC
GAGTTATGAACTTTGAGGATGGAGGTGTAGTTACGGT
CACTCAAGACAGCAGTCTGCAAGATGGGGAGTTTATT
TATAAAGTAAAATTGCGAGGCACTAACTTCCCTTCC
GACGGCCCAGTTATGCAGAAGAAGACTATGGGTTGG
GAGGCATCTACCGAGCGAATGTACCCAGAG
GATGGGGCACTTAAGGGCGAAATCAAGATGCGCCT
CAAATTGAAGGATGGAGGACATTATGATGCTGAAGT
CAAGACCACATACATGGCCAAGAAACCAGTG
CAACTGCCGGGAGCCTACAAAACTGACATAAAACTG
GATATAACTAGCCATAATGAGGATTACACAATAGTC
GAACAGTATGAAAGGGCCGAAGGTCGCCATAGCA
CAGGCGCCGGTGGACCGGTCGCCACCATGGTTAG
TAAAGGCGAAGAGTTGTTTACTGGAGTAGTACC
CATTCTTGTCGAACTCGATGGCGATGTTAATGGGCA
TAAGTTTAGTGTCTCTGGGGAGGGCGAAGGCGACGC
TACATACGGGAAATTGACCTTGAAATTCATCTGTAC
GACTGGAAAATTGCCCGTGCCGTGGCCCACACTGG
TAACTACCCTGACGTACGGGGTCCAGTGCTTTTCTC
GATACCCTGACCATATGAAACAACACGACTTTTT
TAAAAGTGCGATGCCAGAGGGGTACGTGCAGGAAA
GAACAATATTTTTCAAGGATGACGGGAATTA
TAAAACCCGAGCGGAGGTTAAGTTTGAAGGTGATA
CACTGGTAAATCGCATAGAACTTAAAGGGATTGATTT
CAAAGAAGATGGCAACATACTTGGCCATAAGCTGGAA
TACAATTATAACAGTCACAATGTGTACATAATGGCAGA
TAAACAAAAGAATGGAATTAAGGTTAATTTCAAAA
TAAGGCACAATATAGAAGATGGCAGTGTA
CAACTGGCCGATCATTACCAACAAAATACTCCGATCG
GAGATGGTCCTGTCCTCTTGCCAGACAACCAC
TACCTCTCAACCCAAAGCGCACTCTCAAAAGATCC
GAATGAGAAGCGCGACCATATGGTATTGTTG
GAATTTGTTACCGCAGCCGGAATTACCCTTGGAATG
GATGAGCTTTATAAGTCCGGACTCAGATC
TATGCCGTCGGAGAAGACCTTCAAGCAGCGCCG
CACCTTCGAACAAAGAGTAGAAGATGTCCGACT
TATTCGAGAGCAGCATCCAACCAAAATCCCGGTGA

TAATAGAACGATACAAGGGTGAGAAG
CAGCTTCCTGTTCTGGATAAAACAAAGTTCCTTG
TACCTGACCATGTCAACATGAGTGAGCTCATCAAGA
TAATTAGAAGGCGCTTACAGCTCAATGCTAAT
CAGGCCTTCTTCCTGTTGGTGAACGGACACAG
CATGGTCAGCGTCTCCACACCAATCTCAGAGGTGTAT
GAGAGTGAGAAAGATGAAGATGGATTCCTGTA
CATGGTCTATGCCTCCCAGGAGACGTTCGGGAT
GAAATTGTCAGTGTAAGAATTCGATATCAAGCTTATC
GATAATCAACCTCTGGATTAC.

2. FUGW backbone plasmid: Addgene (14883).
3. EcoRI restriction enzyme: NEB (R0101S).
4. AgeI restriction enzyme: NEB (R0552S).
5. 10× Buffer 3.1: NEB (B7203S).
6. Heating block: Eppendorf (ThermoMixer C).
7. Agarose: Fisher Scientific (BP160-500).
8. 50× TAE buffer: Bio-Rad (1610773).
9. SYBR Safe DNA Gel Stain: Invitrogen (S33102).
10. 6× loading dye: Promega (G190A).
11. 1kB DNA ladder: Promega (G571A).
12. iBright DNA gel imager: Invitrogen (A33828).
13. DNA gel extraction kit: Qiagen (28704).
14. DNA spectrophotometer (i.e. Thermo Fisher NanoDrop instruments).
15. GeneArt Gibson Assembly HiFi cloning kit with TOP10 chemically competent cells: Thermo Fisher (A46624).
16. Ampicillin 100 ug/mL agar plates: Teknova (L5104).
17. LB agar powder: Beckton, Dickinson, & Company (244520).
18. DNA plasmid Mini prep kit: Qiagen (27104).
19. DNA plasmid Maxi prep kit: Qiagen (12162).

2.5 Stable Transduction of Cell Line Using Lentivirus

1. HEK293 cell line: ATCC (CRL-1573).
2. 150 mm × 25 mm TC-treated dish: Corning (430599).
3. DMEM culture medium: Gibco (11965-092).
4. Opti-MEM Reduced Serum media: Gibco (31985070).
5. Fetal bovine serum: Sigma (F4135).
6. Anti/Anti 100×: Gibco (15240062).
7. pRSV-REV: Addgene (12253).
8. pMDLg/pRRE: Addgene (12251).
9. VSV-G: Addgene (12259).

10. PEI: VWR (71002-812).
11. Lenti-X Concentrator: Takara Bio (631-231).
12. 1× PBS: StemCell Technologies (37350).

2.6 Chemical Modulation of Autophagy

1. mTOR inhibitor/AZD8055: Selleckchem (S1555).
2. Lysosomal vATPase inhibitor/Bafilomycin: CST (54645S).
3. Vehicle control/DMSO: Invitrogen (D12345).

2.7 Sample Preparation and Acquisition

1. 1× PBS: StemCell Technologies (37350).
2. Trypsin (0.25%): Gibco (25200056).
3. Bovine serum albumin: Sigma (A9418).
4. Cell strainers: Fisher Scientific (22-363-549).
5. Flow cytometer: We present data here acquired using the Beckman Coulter CytoFlex S complete with a blue (GFP) and yellow (RFP) laser.

3 Methods

3.1 Chemical Modulation of Autophagy in Human iPSC Neurons

1. As a control to block autophagy-lysosomal flux, treat the cells with Chloroquine at a final concentration of 40 uM.
2. Treat vehicle control condition with DMSO only.
3. Allow cells to incubate in 5% CO2 incubator for 20 h.

3.2 Cell Lysis and Western Blotting (Fig. 1)

1. After the 20 h incubation, aspirate media from cells and wash gently with 1× PBS one time.
2. Aspirate the PBS completely from the cells.
3. Add 1× Laemmli SDS buffer directly to the cells for lysis and rotate the plate to ensure the buffer completely covers the cells. The volume of buffer used will depend on the cell density. For a confluent 6-well we recommend a 400 uL buffer volume per well.
4. Immediately scrape the lysed cells into labeled microcentrifuge tubes.
5. Boil the lysates at 95 °C for 6 min.
6. Using a benchtop microcentrifuge, quickly spin down the contents of each tube.
7. Remove the plastic strip and comb from the protein gel and place into the gel tank.
8. Make up 250 mL of 1× Tris/Glycine/SDS running buffer by adding 25 mL of buffer to 225 mL milliQ water.

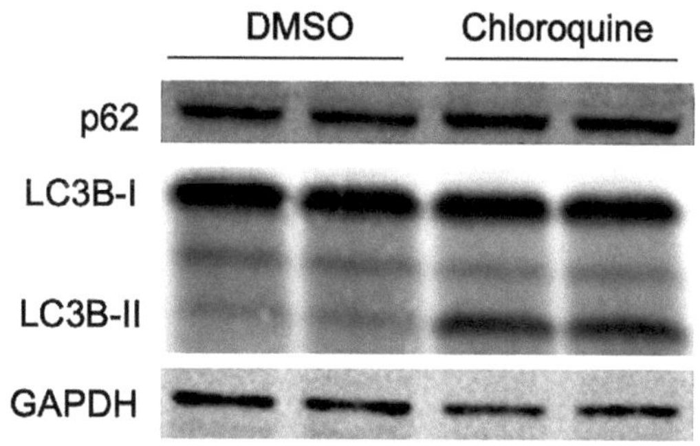

Fig. 1 Western blotting images shown for the autophagy-related proteins LC3B and p62 Changes in autophagic flux can be assessed by quantification of these proteins from cellular lysate. The lysosomal alkalinizing agent, Chloroquine, is included to confirm that observed protein changes are due to autophagic flux versus transcriptional changes

9. Add 1× running buffer to the gel reservoir until full, then pour the remaining buffer into the front of the gel tank.
10. Pre-clear the wells by pipetting a volume of running buffer into each lane well several times.
11. Add 5 uL protein ladder per lane as desired.
12. Add up to 28 uL protein lysate per lane.
13. Run at 140 V until the dye front reaches the bottom of the gel (approximately 1 h).
14. Crack open the gel, remove the combs from the top of the gel, and remove the raised foot at the bottom of the gel.
15. Open the transfer stack package and place the bottom portion of the transfer stack (membrane on top) into the transfer tray.
16. Carefully place the protein gel onto the membrane within the transfer tray.
17. Place the top portion of the transfer stack on top of the gel.
18. Using a roller, firmly but carefully roll the stack with light pressure to remove any air bubbles and ensure close contact of the gel with the membrane.
19. Transfer the proteins from the gel to the membrane at 25 V, 1.0 A, for 30 min. (There is a pre-set protocol on the TransBlot Turbo with these conditions.)
20. When the transfer is complete, carefully remove the transfer stack from the transfer tray.
21. Remove the stack components and peel away the gel from the membrane. Ensure the transfer worked by checking to see if the ladder transferred efficiently from the gel onto the membrane.
22. Using scissors, cut the membrane at the appropriate molecular weight markers to probe with primary antibodies for LC3

(~14-16 kDa), GAPDH (~37 kDa), and p62 (~62 kDa). Do this step without allowing the membrane to dry out.

23. Place the cut membrane sections into a membrane tray and block for 1 h using the 1% casein buffer in TBS. This step is done on a rocker at room temperature.

3.3 Antibodies and Imaging

1. After blocking, make up primary antibody solutions in the same blocking buffer at the desired dilutions.
   (a) LC3 1:500.
   (b) GAPDH 1:1000.
   (c) P62 1:1000.
2. Remove the blocking buffer from the membranes and add the primary antibody-containing buffer to the appropriate membranes. Let the membranes incubate on a rocker at 4 C overnight.
3. The following day, make up 1× TBS-Tween by diluting 10× TBS to 1× by adding 100 mL 10× TBS to 900 mL MQ water, plus 500 uL Tween 20. Swirl the buffer to ensure proper mixing of components.
4. Remove the primary antibody from the membranes and wash each membrane 3 times. Repeat this step 3 times with a 5 min incubation on the rocker at room temperature between each step.
5. During the third TBS-Tween incubation, make up the anti-Rabbit HRP secondary antibody solution (all primary antibodies are Rabbit).
   (a) Anti-Rabbit HRP 1:2000.
   (b) If desired, add 1 uL Streptactin-HRP per every 10 mL secondary antibody solution. This reagent ensures proper visualization of the ladder under chemiluminescence.
6. Add the secondary antibody/Steptactin-HRP solution to each membrane for 1 h on the rocker at room temperature.
7. After incubation, repeat washing steps from Subheading 3.3, **step 4**.
8. During the final TBS-Tween incubation, make up the ECL detection reagent by adding equal parts of each buffer.
9. Add ECL detection buffer to each membrane and incubate protected from light on the rocker at room temperature for 3–5 min.
10. Use membrane imaging instrument to capture images for each membrane. Prior to adding the membrane to the imager, use a KimWipe to gently allow excess ECL reagent to run off the membrane and onto the wipe.

11. Image bands can be quantified using ImageJ or other similar software.

3.4 Human RFP-GFP-LC3B Reporter Plasmid Cloning

1. Restriction enzyme digestion of FUGW plasmid using EcoRI and AgeI enzymes.
   (a) Follow NEB cloner recommendations for digestion reagent mixture volumes with the referenced enzymes.
2. Gel extraction of digested backbone.
   (a) Prepare 0.5% agarose gel using 1× TAE buffer containing SYBR Safe DNA Gel Stain.
   (b) Combine 25 uL digestion reaction with 5 uL 6× loading dye, load into gel along with 1kB DNA ladder.
   (c) Run gel at 80 V until dye front reaches 75% down the gel.
   (d) Place gel onto the imaging device and cut out the larger band which represents the digested FUGW backbone. There should be a large band (approximately 9kB) and a small band (approximately 1kB) visible.
   (e) Extract the digested backbone from the gel slice using a gel extraction kit.
3. Gibson Assembly ligation and plasmid preparation.
   (a) Follow GeneArt protocol for specific ligation instructions and bacterial transformation. Include negative control reaction which contains water in place of the RFP-GFP-LC3B insert gene fragment.
   (b) Streak the transformed bacteria onto Ampicillin (100ug/mL) coated agar plates and incubate overnight at 37 C for 16–18 h. Include the negative control to observe the rate of formation of break-through colonies.
   (c) The following morning, pick colonies and add them to starter cultures (5 mL LB media containing 100 ug/mL ampicillin). Allow them to shake at 225 RPM in a 37 °C incubator for 12–16 h. If not preparing starter cultures immediately then the plates can be stored at 4 °C temporarily.
   (d) Prep plasmid DNA using Mini prep kit and send the samples for Sanger sequencing to verify the presence of the RFP-GFP-LC3 insert.
   (e) Upon verification, the plasmid can be prepped using a Maxi prep kit to increase the amount of plasmid DNA for lentivirus production.

3.5 Stable Transduction of Cell Line Using Lentivirus

1. Lentivirus Production.
   (a) Day 0: Seed 17.5 million HEK293 cells into a 150 mm × 25 mm TC-treated dish.
   (b) Day 1: Transfect HEK293s with lentiviral plasmids.
      (i) To a tube add 1 mL Opti-MEM media, 20 ug pMDLg/pRRE, 11 ug pRSV-REV, 8 ug VSV-G, and 30 ug RFP-GFP-LC3 transfer plasmid.
      (ii) To another tube add 1 mL Opti-MEM media plus 172.5 uL PEI transfection reagent.
      (iii) Allow the tubes to incubate separately in the hood for 5 min, then combine the tubes together and allow for another 5–10 min incubation.
      (iv) During the latter incubation, replace the media on the HEK293 cells with 15 mL fresh DMEM media.
      (v) When 5–10 min incubation is complete, add the entire volume of the transfection reaction (approximately 2 mL) to the plate of cells dropwise so as not to physically disturb them.
      (vi) Replace the media with 15 mL fresh DMEM media 4 h post-transfection.
   (c) Day 2: Observe the transfected HEK293s under the microscope to confirm RFP and GFP fluorescence.
   (d) Day 3: Harvest lentivirus (first batch).
      (i) Collect the 15 mL media containing lentivirus, transfer to a 50 mL conical tube, and store at 4 °C. Replace the media with 15 mL fresh DMEM.
   (e) Day 4: Harvest lentivirus (second batch), and concentrate.
      (i) Collect the 15 mL media from the HEK293 cells and transfer to the same 50 mL conical tube for a final volume of 30 mL lentiviral supernatant.
      (ii) Spin the lentiviral supernatant at 500 × g 10 min at 25 °C. Transfer the supernatant to a new tube and leave behind the cell pellet.
      (iii) Add 1 part (10 mL) LentiX concentrator solution to 3 parts (30 mL) lentiviral supernatant, mix gently, and let the tube cool at 4 °C for at least 30 min according to Takara protocol.
      (iv) Spin the tube at 1500 × g, 45 min, 4 °C, aspirate the supernatant, and resuspend the viral pellet in 1/100th to 1/1000th original volume using 1× PBS.
2. Transduce the cell line of interest with RFP-GFP-LC3 lentivirus at desired MOI. The data shown below was acquired using the H4 neuroglioma cell line from ATCC.

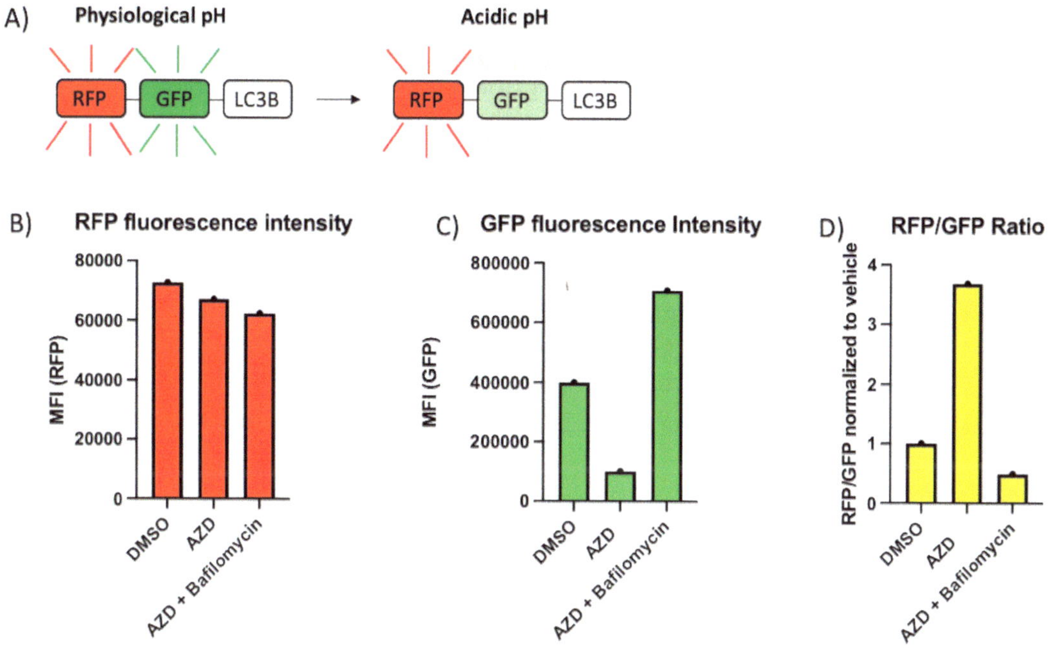

Fig. 2 Flow cytometry data representing the impact of chemical modulation of autophagy on RFP-GFP-LC3B reporter fluorescence. Panel (**a**) displays a schematic of the differential fluorescence activity of GFP and RFP, with the former being quenched at low pH and the latter being stable at low pH. Panel (**b**) shows a graph of the RFP intensity after treatment of cells with control compounds. The mean fluorescence intensity (MFI) of the RFP is only slightly variable between the different treatment groups. Panel (**b**) shows a graph of the GFP intensity after treatment of cells with control compounds. The mean fluorescence intensity (MFI) of the GFP is dramatically altered by the different treatment conditions. Treatment with AZD8055, which induces the autophagy pathway through mTOR inhibition, results in a nearly 75% reduction in the GFP signal. When cells are treated with a combination of AZD8055 and the lysosomal vATPase inhibitor Bafilomycin, the GFP signal increases. This is likely due to a combinatorial effect of the increased autophagosome biogenesis from AZD8055 and inhibited degradation at lysosomes from Bafilomycin. Panel (**c**) shows a graph of the ratio of RFP mean fluorescence intensity (MFI) to GFP MFI between the different treatment conditions. The height of the bars is normalized to the DMSO vehicle control. The increased ratio with AZD8055 treatment is the predicted result of the GFP quenching of the LC3B reporter in lysosomes coupled with the stability of RFP signal in this acidic compartment. The decreased ratio with combinatorial AZD8055 and Bafilomycin treatment is the predicted result of increased autophagosome biogenesis in tandem with lysosomal degradation blockade

3.6 Chemical Modulation of Autophagy (Fig. 2)

1. To adequately modulate the autophagy pathway to confirm appropriate observation of reporter characteristics, treat cells with control compounds. The experiment described here utilizes the mTOR inhibitor AZD8055 at a final concentration of 2 uM and lysosomal alkalinizing agent Bafilomycin at a final concentration of 200 nM.
2. Treat the vehicle control condition with DMSO only.
3. Allow cells to incubate in 5% CO2 incubator for at least 18 h. This is the minimum duration required to observe GFP quenching.

4. After treatment is complete, wash cells once with 1× PBS, dettach the cells using a reagent such as trypsin, pellet the cells at 300 × g for 5 min at room temperature, wash cells one more time with 1× PBS, and pellet once again.
5. Resuspend the final cell pellet in 1× PBS supplemented with 0.1% BSA.
6. Filter this cell mixture through a cell strainer to further prevent clumping.
7. Acquire the samples on a flow cytometer using the blue laser for GFP acquisition and the yellow laser for RFP acquisition. Record a minimum of 10,000 events.
8. Data can be analyzed using FlowJo or similar software.

4 Notes

1. Direct lysis in 1× Laemmli SDS buffer has proven advantageous over other cell lysis reagents for detecting LC3-II protein in neurons.
2. One drawback of using 1× Laemmli SDS buffer for direct cell lysis is that protein quantification of protein lysate prior to gel loading is not feasible.
3. Therefore, careful estimation of cell density should be taken into account when adding the appropriate volume of Laemmli buffer to each well.
4. Using a 12% gel allows for greater resolution of small molecular weight proteins. Since LC3-I and LC3-II are small (14 and 16 kDa, respectively), and similar in size, these gels help to separate the proteins for easier quantification.
5. Flow cytometry controls have ensured there is no cross-contaminating fluorescence between the GFP and RFP emission signals.
6. The RFP/GFP ratio is decreased in the combination treatment with AZD8055 and Bafilomycin compared to vehicle alone, potentially due to a couple of reasons. The increased autophagosome biogenesis with AZD8055 treatment, combined with the increased fluorescence intensity of GFP compared to RFP, results in a lower ratio due to greater GFP signal. Additionally, there is some basal rate of autophagic flux in the vehicle treated cells, whereas the Bafilomycin inhibits degradation resulting in a lower ratio due to greater GFP signal.
7. The slight variability in RFP signal between the different treatment conditions may be due to the cytostatic effects of AZD8055.

Acknowledgments

This work was supported by grants from National Institutes of Health (R21AG077697 and P20GM119943 [Subaward 7137296] to Y.A.H), and the Brown UTRA SPRINT Summer Award (to G.A.E.). Y.A.H. is a GFL Translational Professor from Center for Translational Neuroscience at the Brown Institute for Translational Sciences.

References

1. Dikic I, Elazar Z (2018) Mechanism and medical implications of mammalian autophagy. Nat Rev Mol Cell Biol 6:349–364
2. Glick D, Barth S, Macleod KF (2010) Autophagy: cellular and molecular mechanisms. J Pathol 1:3–12
3. Loos B, du Toit A, Hofmeyr JH (2014) Defining and measuring autophagosome flux – concept and reality. Autophagy 11:2087–2096
4. Wesch N, Kirkin V, Rogov V (2020) Atg8-family proteins – structural features and molecular interactions in autophagy and beyond. Cell 9: 2008
5. Nakatogawa H (2020) Mechanisms governing autophagosome biogenesis. Nat Rev Mol Cell Biol 8:439–458
6. Lee JH, Rao MV, Yang DS et al (2019) Transgenic expression of a ratiometric autophagy probe specifically in neurons enables the interrogation of brain autophagy in vivo. Autophagy 3:543–557
7. Kimura S, Noda T, Yoshimori T (2007) Dissection of the autophagosome maturation process by a novel reporter protein, tandem fluorescent-tagged LC3. Autophagy 5:452–460

Chapter 19

Measuring Neuronal Network Activity Using Human Induced Neuronal Cells

Danny McSweeney, Jay English, Ethan Howell, Fumiko Ribbe, and ChangHui Pak

Abstract

Synchronous firing of neurons, often referred to as "network activity" or "network bursting," is an indication of a mature and synaptically connected network of neurons. We previously reported this phenomenon in 2D human neuronal in vitro models (McSweeney et al. iScience 25:105187, 2022). Using induced neurons (iNs) differentiated from human pluripotent stem cells (hPSCs) coupled with high-density microelectrodes arrays (HD-MEAs), we probed the underlying patterns of neuronal activity and found irregularities in network signaling across mutant states (McSweeney et al. iScience 25:105187, 2022). Here, we describe methods for plating cortical excitatory iNs differentiated from hPSCs on top of HD-MEAs and culturing iNs to maturity, examples of representative human wild-type Ngn2-iN data, and troubleshooting tips and tricks for the experimenter interested in integrating HD-MEAs into one's research approach.

Key words High-density microelectrode arrays, Ngn2-iNs, Neuronal network assay, Activity assay, Complementary metaloxide-semiconductor (CMOS) chip technology

1 Introduction

In the developing human brain, maturing neurons begin to form synaptic connections with neighboring cells, which allow for direct cell-to-cell communication, network burst activity, and higher-order brain circuitry. Network bursting activity represents synchronous action potential firing by neurons, which leads to a single clear and coordinated message traveling through and between brain regions [1]. Disruptions in synaptic signaling and neural connectivity are common in neurodevelopmental disorders, such as autism spectrum disorders, intellectual disabilities, and schizophrenia [2, 3]. Determining the differences in the underlying patterns of synchrony may inform our understanding of the physiological drivers and clinical outcomes of these neurodevelopmental disorders.

Yu-Wen Alvin Huang and ChangHui Pak (eds.), *Stem Cell-Based Neural Model Systems for Brain Disorders*, Methods in Molecular Biology, vol. 2683, https://doi.org/10.1007/978-1-0716-3287-1_19,

In a recent study, we have optimized approaches to the plating and recording of our 2D iN co-culture model to high-density microelectrode arrays (HD-MEAs) [4]. This approach allows quantification of electrophysiological changes in firing activity at the single-spike level as well as network burst synchrony in neuronal cells [5, 6]. While we describe our protocol optimization using the MaxWell Biosystems MaxOne HD-MEA platform, the principles of plating and recording are applicable to other recording platforms.

For our culture model, we utilized transcription factor-based differentiation methods to generate homogeneous populations of layer-2/3-like cortical excitatory neurons from human embryonic stem cells in a dish [7] (*see* Chapter 3, Subheading 3.4 for Ngn2-iN differentiation protocol). Three days following differentiation (d3), Ngn2-iNs were co-cultured with mouse glia (to aid in neuronal maturity) directly on top of HD-MEA arrays, and recordings were performed once a week, beginning at d21, when robust synaptic activity has initiated, to d42, well into neuronal maturity. Tracking spike- and burst-level network activity across time allows for establishment and refinement of neural connections and tighter regulation of bursting synchrony over time. For our approach, we performed all functional assays without stimulation conditions, thereby recording spontaneous neuronal firing and bursting patterns.

2 Materials

2.1 Cells

H1 human embryonic stem cells (WA-01).

Primary mouse glia (*see* Chapter 3 Subheading 3.3 for mouse glial dissection protocol).

2.2 Plating and Culturing Reagents

Chip preparation and maintenance	
Reagent	**Final concentration**
Mouse laminin	0.02 mg/mL diluted in plating media
DPBS	As prepared by manufacturer
Terg-a-zyme	1% in diH_2O
Trypan Blue	As prepared by manufacturer
20× Borate buffer	1× in sterile diH_2O
Poly(Ethyleneimine) (PEI)	0.07% in 1× borate buffer
Trypsin-EDTA	As prepared by manufacturer

B27 Co-culture and plating medium (d3–d9)	
Reagent	**Final concentration**
Neurobasal medium	–
B27 Supplement	2%
Glutamax	1%
BDNF	10 ug/L
NT-3	10 ug/L
Laminin	0.2 mg/L
Doxycycline	2 mg/L
A Cytosine b-D-arabinofuranoside hydrochloride (AraC)	2 uM

2AraC neuronal maturation media (d10–end)	
Reagent	**Final concentration**
Minimal Essential Media (MEM)	–
Glucose	0.5%
Sodium bicarbonate ($NaHCO_3$)	0.02%
Transferrin	0.1 mg/mL
Hyclone FBS	5%
B27 supplement	2%
L-glutamine	0.5 mM
A Cytosine b-D-arabinofuranoside hydrochloride (AraC)	2 uM

3 Methods

3.1 Chip Preparation (Fig. 1)

1. Perform background noise assessment to estimate quality of electrode array before using new chips and after each round of culturing.
2. One day before cell plating, add 1% Terg-a-zyme to chips.
3. Incubate at room temperature for at least 16 h.
4. Aspirate Terg-a-zyme.
5. Wash three times with diH_2O.
6. Submerge in 70% EtOH to sterilize chips, and incubate in BSC hood for 30 min.

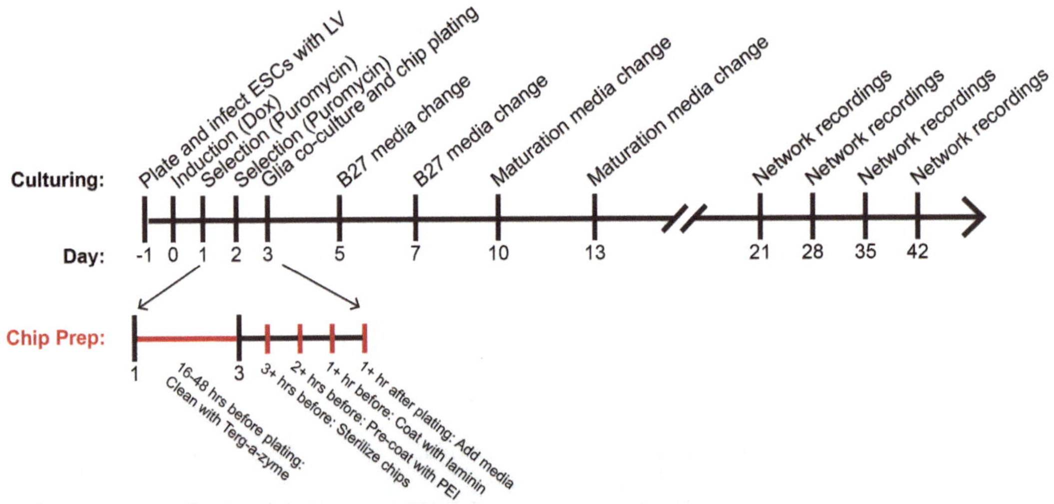

Fig. 1 Chip preparation and culture timeline: prepping and plating HD-MEA chips occurs over a 24-h period starting on d1. On d3, co-culturing and plating of Ngn2-iNs and glia should be performed using B27-supplemented plating medium, and 50% of the medium should be replaced every 2–3 days. On d10, 50% of medium should be replaced by 2AraC neuronal maturation medium. Media changes should occur weekly until maturity is reached. Activity scans and networks assays should be performed once a week, starting on day 21

7. In BSC hood, rinse 3 times with sterile diH_2O.
8. Carefully and thoroughly dry chips with vacuum.
9. Place chip in sterile 10 cm dish. Add an autoclaved lid to the chip and diH_2O reservoir next to chip.
10. Prepare 1× borate buffer.
11. Prepare ~7% PEI solution by scooping ~1 mL 50% PEI into a 15 mL conical, allowing it to settle, and resuspending in 6 mL 1× borate buffer (*see* **Note 1**).
12. Prepare ~0.07% PEI using ~7% PEI and 1× borate buffer, then 0.22-um sterile filter.
13. Add 50 uL ~0.07% PEI to each array, and gently tap the side of the chip to spread PEI across the full plating area.
14. Incubate chips for 1 h at 37 °C.
15. Aspirate PEI.
16. Rinse 3 times with sterile diH_2O.
17. Prepare 50 uL (per chip) of 0.02 mg/mL laminin in plating medium.
18. Add 50 uL laminin to each array.
19. Incubate 1 h at 37 °C.

3.2 Cell Plating on Chips

1. For each chip, mix 150,000 Ngn2-iNs with 150,000 glia in 50 uL of plating medium, including room for error (*see* **Notes 2** and **3**).
2. Aspirate laminin coat from each chip.
3. Plate 50 uL of cell mixture, gently tapping side of chip to spread volume evenly across full array.
4. Incubate chips for 1 h in a 37 °C incubator to allow cells to adhere.
5. Carefully add 700 uL plating medium to the side of each chip well.
6. Follow the Ngn2-iN culture protocol to grow cells to day 21, and perform activity scans and network assays (*see* **Notes 4** and **5**).

3.3 Chip Recording: Activity Assay

1. Starting on d21, perform activity assays once a week for 4–6 weeks to track cell culture maturity and signaling patterns over time (*see* **Note 6**).
2. Perform a sparse activity scan as a first-pass test to gauge successful and even plating across the array (*see* **Notes** 7 and **8**).
3. If cultures are more than 40% active and activity is evenly spread across the array, perform a full configuration activity scan to identify electrodes covered by active, signaling neurons. This subset of electrodes should be used for recording the network assay (*see* **Note 9**).

3.4 Chip Recording: Network Scan

1. Using only active electrodes identified during the activity scan, perform a 10-min network scan to measure changes in spike-level activity and network synchrony (*see* **Notes** 7 and **10**).

3.5 Cleaning Chips After Use

1. Incubate chip in warmed Trypsin-EDTA for 10+ min at 37 °C to dissociate cells from the chip.
2. Rinse chips and lids 3 times with diH_2O.
3. Use a 70% EtOH squeeze bottle to dislodge remaining residue from chips.
4. Rinse chips and lids 3 times with diH_2O.
5. Store chips in diH_2O at 4 °C.
6. Autoclave lids only and store in sterile container.
7. Re-use chips within 6 months.

3.6 Analysis Parameters

1. For activity scan analysis, call signals as "events" using the following thresholds: active area > 40%; 0.10 Hz (firing rate); 20 mV (amplitude); 200 ms (inter-spike interval) (*see* **Note 11** for definitions of each metric).

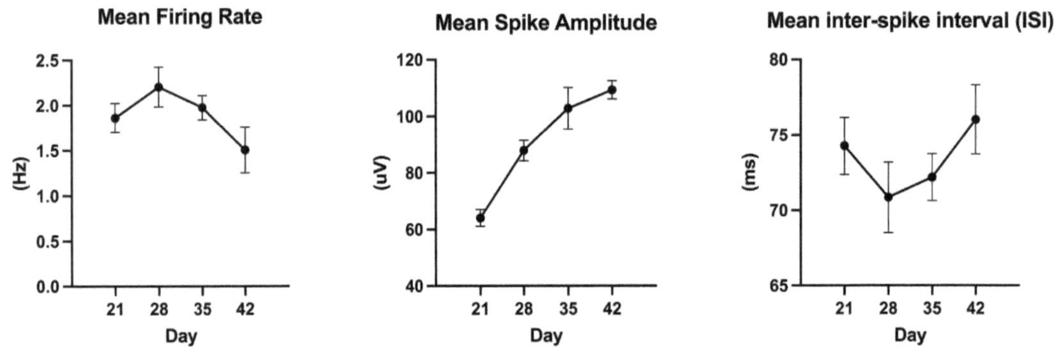

Fig. 2 Anticipated results for Ngn2-iN cortical spike-level activity during synapse maturation: weekly measurements of spontaneous synchronized neuronal activity were performed for maturing cultures of Ngn2-iNs from day 21 to 42. Data was collected using an HD-MEA platform (MaxWell Biosystems). First, a full-scan configuration (20 min) for an activity assay was performed to identify only active electrodes. Then, spike-level activity data was extracted from a 10-min network scan of the active electrodes. Data is represented as averages across 4–5 independent Ngn2-iN cell differentiations (culture batches) +/− SEM. All data shown here had active areas >40%. For details on experimental set-up, *see* McSweeney et al. [4]

2. For the network assay, use the following thresholding parameters for identifying action potentials and network bursts: 0.3 s (smoothing window size); 1.2 (burst detection threshold), 1 s (minimum peak distance); 0.3 (start-stop threshold). (*see* **Notes 12** and **13**).

3. Through our analysis, we found the following metrics to be most helpful in representative characterization of Ngn2-iN electrophysiology: Spike-level—mean firing rate, mean spike amplitude, and mean inter-event interval; Network-level—burst frequency, percent of spikes within bursts, mean spikes per burst, mean burst duration, mean burst peak firing rate, and mean inter-burst interval (*see* **Note 14**; *see* Figs. 2 and 3 for representative graphs).

4 Notes

1. 50% PEI is very viscous. Use a flat, rectangular measuring spatula to scoop 50% PEI and scrape PEI onto the side of a 15 mL conical. Allow it to settle to the 1 mL mark and remove extra PEI before adding borate buffer. Mix thoroughly with a pipette or use a vortex. Aliquot extra ~7% PEI into 1.5 mL microcentrifuge tubes, and store at −20 °C for 1 month.

2. A 1:1 iN:glia ratio may need to be optimized for each cell type, especially if no synchrony is observed or if culture peeling occurs. Do not exceed more than 300,000 cells per chip. In addition to the 150 k iN: 150 k glia ratio, we also explored 125 k iNs: 125 k glia and 200 k iNs: 200 k glia. We found plating too many cells led to premature sloughing off the array.

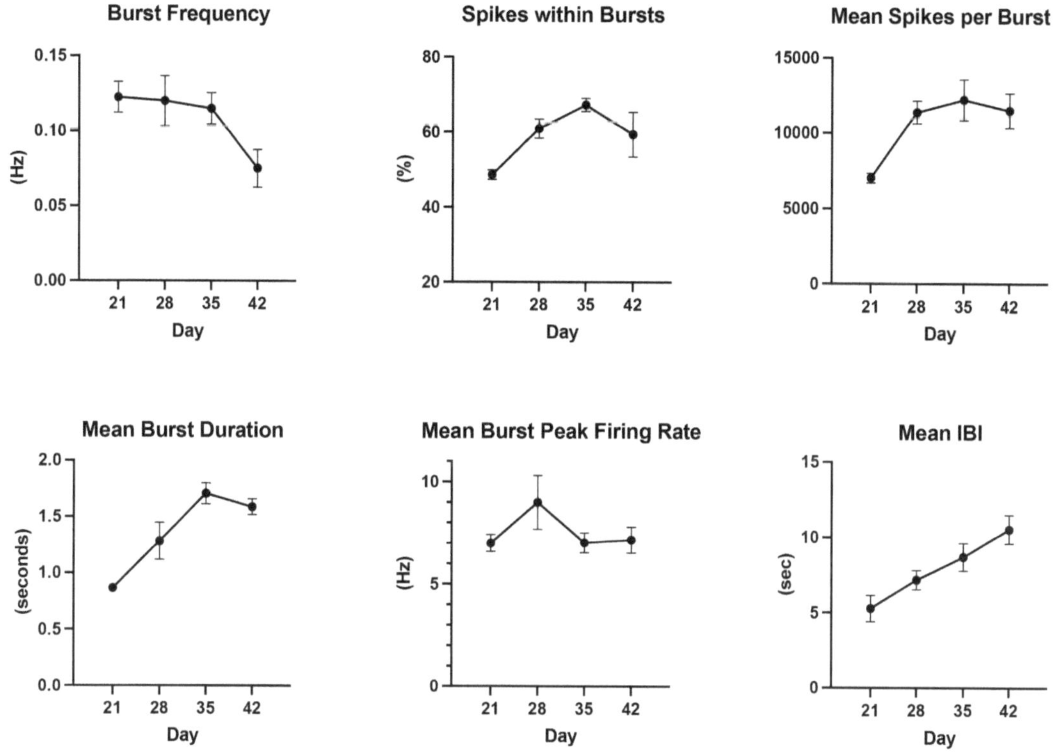

Fig. 3 Anticipated results for Ngn2-iN cortical network activity during synapse maturation: weekly measurements of spontaneous synchronized neuronal activity were performed for maturing cultures of Ngn2-iNs from day 21 to 42. Data was collected using an HD-MEA platform (MaxWell Biosystems). First, a full-scan configuration (20 min) for an activity assay was performed to identify only active electrodes. Then, bursting activity was recorded using a 10-min network scan. Data is represented as averages across 4–5 independent iN cell differentiations (culture batches) +/− SEM. For details on experimental set-up, *see* McSweeney et al. [4]

However, plating too little will prevent neuronal network connectivity from forming. Optimization of plating density may be necessary.

3. To reach 300,000 cells in 50 uL total volume, mix desired 1.5× volumes of iNs + glia into a 1.5 mL microcentrifuge tube, and spin for 8 min at 1200 rpm. If 150,000 iN + 150,000 glia combined volume is more than 1.5 mL, re-spin down cells before aliquoting and resuspend each pellet in a smaller volume of media to concentrate cells. After re-counting, combining iNs with glia, and pelleting, carefully remove supernatant; the cell pellet will be very small. Resuspend in 75 uL of plating medium to account for the 1.5× error, and plate 50 uL of cells per chip.

4. Common culturing and plating issues can include: (1) Cultures peel over time. This may be due to PEI or laminin not being ideal as a pre-coat or final coat for your culture type. If you suspect PEI, try one of the following: Option 1: Poly-L-

Ornithine (PLO), 0.005% final concentration diluted in diH_2O, incubated for 2 h at 37 °C. Option 2: Poly-D-Lysine (PDL), 0.1 mg/mL final concentration (ThermoFisher Sci sells PDL at final concentration ready to use), incubated for 1 h at 37 °C. If you suspect laminin to be the issue, try one of the following alternatives: Option 1: Matrigel, 0.4 mg/mL diluted in cold plating media, incubated for 30 min at 37 °C. Option 2: Geltrex, 1× (ThermoFisher sells Geltrex at final concentration, ready to use), incubated for 1 h at 37 °C.

5. Liquid build-up between chip lid and upper edge of well or along outer chip surface can lead to contamination. To avoid this, move slowly when moving chips from secondary containment into the recording chamber, or consider putting the recording chamber itself in secondary containment. Additionally, thoroughly vacuum the chip well's upper edge and any other moisture, and replace with a new sterile lid if any moisture build-up between chip and lid is suspected.
6. While activity assays and network scans were performed over 4 weeks starting with d21 (*see* Figs. 2 and 3 for representative H1 Ngn2-iN time-course data), d28 was found to be the fully mature timepoint and was further evaluated using both immunocytochemistry and whole-cell patch-clamp electrophysiology for changes in synaptic connectivity [4]. The fully mature timepoint may differ with each cell type, and a time-course study would aid in identification.
7. As part of the MaxWell Biosystems MaxOne platform, we performed recordings and analysis for activity scans and network assays using the built-in MaxLive software package. However, the fundamental approach and recording details are translatable to other MEA platforms and analysis software.
8. Use the following thresholding parameters for the activity assay:

Parameter	Setting	Description
Firing rate threshold [Hz]	0.1	Minimum threshold of firing rate for a single electrode, to consider electrode active. Increasing the firing rate threshold will exclude more electrodes
Amplitude threshold [uV]	20	Minimum threshold of amplitude for a single electrode, to consider electrode active. Increasing the amplitude threshold will exclude more electrodes
ISI threshold [ms]	200	Maximum threshold of ISI for a single electrode, to consider electrode active

9. If low or no activity is detected, this may be due to the need for cell density optimization. Alternatively, cells may have been lost during the final spin-down and pellet resuspension. If no cell pellet was detected after spinning down iNs + glia, aliquot more than 1.5× cells until you visually see a pellet. Be careful to only remove the supernatant before resuspending the pellet. It is not recommended to perform the co-culture step in a 15 mL conical, as a cell pellet is less likely to form or be detectable.
10. Use the following thresholding parameters for the network scan:

Parameter	Setting	Description
Smoothing window size [s]	0.3	Standard deviation of the Gaussian convolution kernel used for smoothing the network firing rate
Burst detection threshold	1.2	The minimum value that the network firing rate must reach to be considered as a network burst
Minimum peak distance [s]	1.0	The minimum time by which the detected peaks must be separated in order to be considered bursts
Start-stop threshold	0.3	The upper fraction of the total peak height at which the burst start and stop times are defined

11. Below are key representative metrics for characterizing spike-level activity:

Metric	Unit	Description
Active area	%	Electrodes called as "active" by thresholding parameter settings
Firing rate	Hz	Average firing rate across active electrodes only
Spike amplitude	uV	Average spike amplitude across active electrodes only
Inter-spike interval (ISI)	ms	Average amount of time between spikes, measured across active electrodes only

12. Below are key representative metrics for characterizing burst-level activity:

Metric	Unit	Description
Burst frequency	Hz	Number of detected network bursts divided by the total recording time (in seconds)
Spikes within bursts	%	Percentage of detected spikes in network bursts with respect to the total number of detected spikes across the whole recording
Mean spikes per burst	–	Number of spikes within individual bursts, for all bursts detected
Mean burst duration	s	The time between the start and stop times of a given burst at a certain percentage of the peak amplitude, for all bursts detected
Mean burst peak firing rate	Hz	Peak amplitude of a given burst, derived from the Network Activity plot for all bursts detected
Mean inter-burst interval (IBI)	s	The time interval between the peak times of a given burst and the following one (IBI = tpeak(n + 1)—tpeak(n)), for all bursts detected. The IBI of the last burst is assigned to 0

13. Some cell types may go through waves of synchrony throughout the day or may be more active during certain times of the day. If no bursting is detectable over a 10-min recording, perform a 10-min network scan every 2 h over a 12-h period to gauge whether synchrony occurs during specific times of the day or whether there are long periods of low activity between bursts.
14. If using other cell types, average activity or network values may need to be determined across several batches and multiple recordings. Results may also be verified through pairing with other techniques, such as structural confirmation using immunocytochemistry or whole-cell patch-clamp electrophysiology for functional agreement [4].

Acknowledgments

This work was supported by grants from the NIH. Pak lab is supported by NIMH R01MH122519. D.M. was supported by NIGMS T32 GM135096. We would like to thank Giulio Zorzi, Marie Obien, Elena Gronskaya, Martina Elena De Gennaro, and David Jackel at MaxWell Biosystems for their support and assistance.

References

1. Gansel KS (2022) Neural synchrony in cortical networks: mechanisms and implications for neural information processing and coding. Front Integr Neurosci 16:900715. https://doi.org/10.3389/fnint.2022.900715
2. Zoghbi HY, Bear MF (2012) Synaptic dysfunction in neurodevelopmental disorders associated with autism and intellectual disabilities. Cold Spring Harb Perspect Biol 4:a009886. https://doi.org/10.1101/cshperspect.a009886
3. Kern JK, Geier DA, King PG, Sykes LK, Mehta JA, Geier MR (2015) Shared brain connectivity issues, symptoms, and comorbidities in autism spectrum disorder, attention deficit/hyperactivity disorder, and Tourette syndrome. Brain Connect 5:321–335. https://doi.org/10.1089/brain.2014.0324
4. McSweeney D, Gabriel R, Jin K, Pang ZP, Aronow B, Pak C (2022) CASK loss of function differentially regulates neuronal maturation and synaptic function in human induced cortical excitatory neurons. iScience 25:105187. https://doi.org/10.1016/j.isci.2022.105187
5. Muller J, Ballini M, Livi P, Chen Y, Radivojevic M, Shadmani A, Viswam V, Jones IL, Fiscella M, Diggelmann R et al (2015) High-resolution CMOS MEA platform to study neurons at subcellular, cellular, and network levels. Lab Chip 15:2767–2780. https://doi.org/10.1039/c5lc00133a
6. Emmenegger V, Obien ME, Franke F, Hierlemann A (2019) Technologies to study action potential propagation with a focus on HD-MEAs. Front Cell Neurosci. https://doi.org/10.3389/fncel.2019.00159
7. Zhang Y, Pak C, Han Y, Ahlenius H, Zhang Z, Chanda S, Marro S, Patzke C, Acuna C, Covy J et al (2013) Rapid single-step induction of functional neurons from human pluripotent stem cells. Neuron 78:785–798. https://doi.org/10.1016/j.neuron.2013.05.029

Chapter 20

A Simple Ca^{2+}-Imaging Approach of Network-Activity Analyses for Human Neurons

Zijun Sun

Abstract

Rapid advances in light microscopy and development of all-optical electrophysiological imaging tools have greatly leveraged the speed and the depth of neurobiology studies. Calcium imaging is a common method that is useful for measuring calcium signals in cells and has been used as a functional proxy for neuronal activity. Here I describe a simple, stimulation-free approach that measures neuronal network activity and single-neuron dynamics in human neurons. This protocol provides the experimental workflow that includes step-wise illustrations of sample preparations, data processing, and analyses that can be used for quick phenotypical assessment and serves as a quick functional readout for mutagenesis or screen effort for neurodegenerative studies.

Key words Ca^{2+}-imaging, Network activity, Automated image analysis, Human neuron, Matlab

1 Introduction

Calcium imaging is a well-established tool for neurobiology studies. It is greatly improved by the development of calcium indicators, the advancement in fluorescence microscopy, and automated analysis software [1, 2]. Because of its optical nature, calcium imaging offers a wide range of applications in neuroscience: from monitoring activity of large neural ensembles [3–5] to visualization of synaptic dynamics at spines or boutons [6, 7].

While calcium imaging is a powerful method, one challenge in measuring activity of in vitro neuronal culture is that there is a big variety of spontaneous neural activities which depend on culture conditions such as neuron density [8], neuron age, and neuron types [9]. Moreover, the network activity, which is typically represented by the synchronous firing event in culture, occurs stochastically and can be rare in human embryonic stem cell (hES) derived human neurons, rendering it an unreliable assay. These issues demand simple approaches to improve Ca^{2+}-imaging in cultured

Yu-Wen Alvin Huang and ChangHui Pak (eds.), *Stem Cell-Based Neural Model Systems for Brain Disorders*, Methods in Molecular Biology, vol. 2683, https://doi.org/10.1007/978-1-0716-3287-1_20,

neurons and to establish reliable analyses of neuronal activity in induced neurons (iN). One approach to induce neuronal activity is to apply external stimuli such as optogenetic stimulation or pharmacological agonist [10, 11]. However, the disadvantage of using external stimuli is that the variations in the efficacy of stimuli may add complexity to the system.

In order to overcome the abovementioned limitation, we develop a stimulation-free Ca^{2+}-imaging protocol that potentiates spontaneous network activity in human neurons and provide analysis algorithms that enable Ca^{2+}-imaging a readily used, functional assay that is amenable to be scaled up. We observe that the network activity of cultured human neurons can be robustly induced in the optimized Ca^{2+}-imaging conditions and have successfully tested the sensitivity of this approach using mutants that are shown to exhibit defects in synaptic connectivity or transmission [12, 13]. We also validated our analysis algorithms using established Ca^{2+}-imaging software programs [14]. Here, I describe the experimental procedures of sample preparation, calcium imaging, and step-wise illustrations of data analyses.

2 Materials

2.1 Equipment

1. Laminar flow hood.
2. Tissue culture incubator (37 °C, 100% humidity, 5% CO_2).
3. Multi-well, #1.5 glass-bottom plate, or #1.5 glass-bottom petri dish.
4. Dissection microscope.
5. Water bath (37 °C).
6. Automated cell counter or hemocytometer.
7. Inverted epifluorescence microscope equipped with a digital or CCD camera and live-cell imaging environment.

2.2 Plasmids

1. pCMV-VSV-G (Addgene, #8454).
2. pRRE (Addgene, #12251).
3. pRSV-REV (Addgene, #12253).
4. FUW-rtTA (Addgene, #20342).
5. pTetO-Ngn2 (Addgene, #52047).
6. pSyn-GCaMP6m (constructed in the lab, original GCaMP6m plasmid: Addgene #100838).

2.3 Reagents Setup

Carry out all procedures for preparing and handling lentiviruses in Biological Safety Level 2 (BSL-2) environments.

1. Lentivirus production: we use lentiviral infection method to express Ngn2 and rtTA factors for iN generation and to express GCaMP. The lentiviruses are made in HEK293T cells, as previously described [15].
2. hES plating medium: add Thiazovivin (BioVision) into mTeSR Plus supplemented medium (Stem Cell Technologies) and store at 4 °C for up to 2 weeks, 1 month in −20 °C.
3. hES growth medium: add mTeSR Plus Supplement into mTeSR, aliquot and store at 4 °C for up to 2 weeks, 1 month in −20 °C.
4. iN induction medium: add 5 mL N2 supplement, 5 mL NEAA, 100 uL BDNF/Laminin/NT-3 each, and 500 uL Doxycycline (Dox) into DMEM/F12 medium to make a final volume of 500 mL medium. Filter with a 0.22 μm filter and store at 4 °C for up to 1 month.
5. iN growth medium with 2-AraC: add 10 mL B27 Supplement, 10 mL Gluatmax, 25 mL FBS (final concentration, 5%), 0.25 mL 4 mM Arabionside (AraC, final concentration, 2 μM) into Neurobasal-A Medium to make a final volume of 500 mL medium. Filter with a 0.22 μm filter and store at 4 °C for up to 1 month.

2.4 Imaging Microscope

We use an inverted Nikon EclipseTS2R epifluorescence microscope equipped with a DS-Qi2 digital camera (*see* **Note 1**). For imaging, 20×, 40× air objectives, or 60× oil immersion objective with N.A. (numerical aperture) above 1.2 can be used.

3 Methods

Timeline for a typical calcium-imaging workflow is illustrated in Fig. 1. Ca^{2+}-imaging can be done at room temperature if the recording duration for each culture takes less than half an hour, or else it should be done at a microscope equipped with live-imaging environment (i.e., with a 37 °C, 5% CO_2 chamber).

3.1 Mouse Glia Culture

Confluent mouse glia culture (passage 1, P1) should be prepared 1–2 weeks ahead of iN culture generation. The same dissection method for glia culture is used as what is used for generating primary cortical or hippocampal neuronal culture, e.g., one is described in ref. [15]. Brief steps are described below.

1. Dissect the cortices region from 2 pups that are born postnatal day 0–2 for significant amount of glia tissue.
2. After dissociation of the tissue in papain-containing Hank's Balanced Salt Solution (HBSS), remove the dissociation buffer and wash the tissue twice with 10% FBS-supplemented DMEM (*see* **Note 2**).

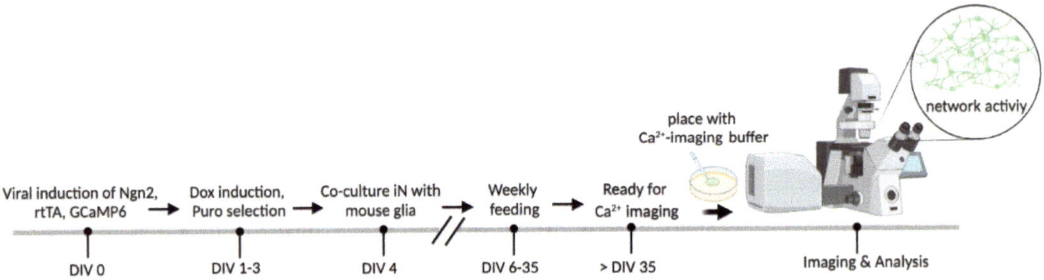

Fig. 1 iN generation and calcium-imaging timeline. The timeline starts from the induction of iN from human embryonic stem cells (hES) cells. GCaMP is induced by viral infection together with Ngn2 and rtTA on DIV 0. DIV35 or above iN cells are used for Ca^{2+}-imaging. Imaging is done in Ca^{2+}-imaging buffer on an inverted microscope

3. Add 1 mL DMEM/10% FBS medium after washing and use pipette to triturate the tissue, then add 11 mL DMEM/10% FBS and move the tissue into T-75 flask.
4. Put the culture into 37 °C incubator.
5. Change media to DMEM/10% FBS the next day. This glia culture is at passage 0 (P0).
6. Check confluency of P0 glia in 6–7 days and passage the confluent P0 glia at splitting ratio of 1:3 into T-75 flasks—one T-75 confluent P0 glia culture can be passaged to three P1 glia T-75 flasks, which typically reach confluency in 1 week.
7. Change media once a week with DMEM/10% FBS and avoid contamination.
8. P1 glia culture can be kept for up to 2 months for optimal condition. Confluent P0 or P1 can be frozen down depending on experiment needs.

3.2 iN Generation from Human ES Cells

iN cells for Ca^{2+}-imaging are generated using established techniques [14, 16]. Here I outline the workflow of iN preparations.

1. (DIV -1, optional): coat 6-well plate with Matrigel 1 day before hES disassociation. Alternatively, Matrigel can also be coated 1 h ahead of hES disassociation and plating.
2. (DIV 0, lentiviral infection) Dissociate hES cells with Accutase, centrifuge to pellet down cells in 15 mL falcon tube.
3. Resuspend hES cells in 1 mL plating medium containing Thiazovivin, mix well with lentiviruses needed (e.g., Ngn2, rtTA, and GCaMP6) (*see* **Note 3**).
4. (DIV 1, Dox induction) After 16–24 h, change the infection media to add 2 mL/well of iN induction medium.

5. (DIV 2–3, Puro selection) After 24 h, change the media to 2 mL/well of iN induction medium with 1× puromycin (Puro) for selection.
6. Repeat this selection step for 2 days.
7. In the meantime, prepare glia culture as follows: dissociate P1 glia culture from one T-75 flask with Trypsin/EDTA, and one confluent T-75 flask glia is typically used for 2 24-well plates.
8. Re-plate glia onto the overnight Matrigel-coated coverslips or the glass-bottom plate (*see* **Note 4**).
9. (DIV 4, co-culture with glia) Dissociate iN cells with Accutase.
10. Count the fully suspended cells using an automated cell counter or a hemocytometer.
11. Collect the amount of 1.8–2 × 10^5 cells per one 24-well coverslip (*see* **Note 5**).
12. Resuspend the amount of iN cells needed in iN growth medium, and add 1 mL of the iN culture into one well of 24-well plate, which is plated with glia at least 1 day before.
13. Let the iN-glia co-culture stay for one another day in 37 °C incubator.
14. (DIV 6) Change half of the media with iN growth medium.
15. (DIV 8) Change half of the media with 2 μM AraC-containing iN growth medium (*see* **Note 6**).
16. (DIV 10) Change half of the media with 2 μM AraC-containing iN growth medium.

    Change half of the media weekly from DIV10 onwards until iN cells mature to DIV 35 or above (*see* **Note 7**).

3.3 Calcium Imaging

1. Prepare the Tyrode's solution and Ca^{2+}-imaging buffer according to Table 1 (*see* **Note 8**).
2. Warm up the Tyrode's and imaging solution in 37 °C water bath before imaging (*see* **Note 9**).
3. Gently aspire the iN growth media and wash the coverslip twice with pre-warmed 0.5 mL Tyrode's solution.
4. If the iN culture is grown on the coverslip, take out the coverslip by tweezer and put into a glass-bottom imaging plate or petri dish that contains the pre-warmed imaging solution.
5. Let the coverslip equilibrate at 37 °C for 2 min. If the cells have grown on the glass-bottom plate, replace growth medium, wash twice, add imaging buffer, and equilibrate.
6. Mount the imaging plate/petri dish onto microscope (*see* **Note 10**).

Table 1
Tyrode's and Ca^{2+}-imaging solution

	normal Tyrode's solution		Ca^{2+}-imaging solution	
Compound (*MW*)	**Required quantity**	**Final concentration (*mM*)**	**Required quantity**	**Final concentration (*mM*)**
NaCl (58.44)	7.54 *g*	129	7.54 *g*	129
KCl (74.55)	5 mL of 1 *M* stock solution	5	8 mL of 1 *M* stock solution	8
Glycine (75.07)	1 mL of 10 *mM* stock solution	10 *μM*	1 mL of 10 *mM* stock solution	10 *μM*
$MgCl_2 \cdot 6H_2O$ (203.3)	1 mL of 1 *M* stock solution	1	1 mL of 1 *M* stock solution	1
$CaCl_2 \cdot 2H_2O$ (147.01)	2 mL of 1 *M* stock solution	2	4 mL of 1 *M* stock solution	4
HEPES (238.3)	5.96 *g*	25	5.96 *g*	25
D-Glucose (180.16)	2.7028 *g*	15	2.7028 *g*	15
Ultra-pure water	Add Ultura-pure water or double distilled water to final volume: 1000 *mL*			

Note: adjust pH from ~5.31 to 7.2–7.4 with NaOH, filter the solution with 0.22 μm filter, store at 4 °C

7. Adjust the light intensity, detection gain, frame rate, and recording duration using the control culture (*see* **Note 11**).
8. Perform time-lapse imaging with a 10× or 20× air objective (*see* **Note 12**).

3.4 Analysis

A Macintosh or Windows computer running ImageJ and Matlab is used for data analyses. We use Matlab to process image and to perform quantifications of network activity and single-neuron dynamics (Fig. 2a).

To quantify the network activity, follow Subheading 3.4, **step 1**, and follow Subheading 3.4, **step 2** for quantifications of single-neuron dynamics. All described Matlab programs in this protocol can be downloaded in the supplementary materials from refs. [14, 17] (*see* **Note 13**).

3.4.1 Quantification of Network Activity

Amplitude and rate of synchronous Ca^{2+} peaks is quantified as readout of network activity [14]. Essential steps of quantification are illustrated as follows:

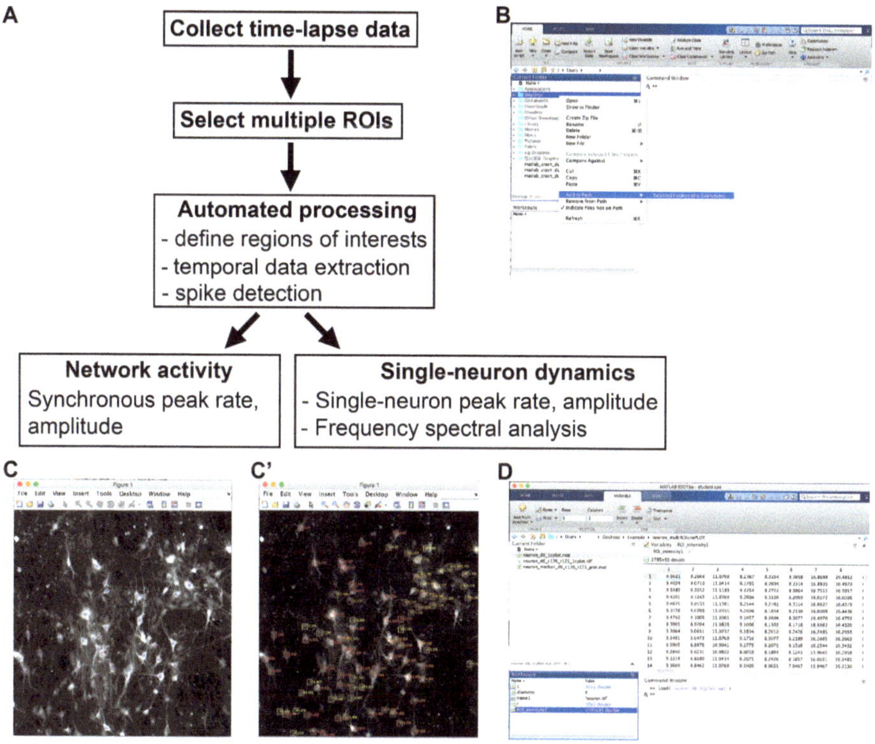

Fig. 2 Analyses pipeline of calcium-imaging data: (**a**) pipeline of calcium-imaging analyses, (**b**) window showing Matlab working environment for analysis, (**c**)–(**c'**) detection of ROIs corresponding to selected neurons, (**c**) example image of a maximum projection of a Ca^{2+}-imaging file, (**c'**) example image of image in (**c**) overlayed with selected ROIs and their X-Y coordinations (numbers besides each ROI box), (**d**) window showing data file of raw Ca^{2+} intensity computed from selected ROIs

1. Covert the time-lapse data files to .tiff/.tif format using ImageJ (https://imagej.nih.gov/ij/).
2. Install Matlab and the Image Processing Toolbox (MathWorks) for following quantifications.
3. Start Matlab.
4. Place data files to be analyzed under the Matlab working path as shown in Fig. 2b.
5. Choose a data file and run Matlab function *GCAMP_multiROI.m*. An image will appear in 10–20 s (*see* **Note 14**). This image is the maximum intensity-projected figure of the selected *.tiff* data (see an example in Fig. 2c).
6. On this image, navigate and click at the center of each neuron soma.
7. Finish all selections by pressing the Enter key, and a second image will appear showing ROIs-overlayed image (Fig. 2c'). Running this program computes the raw calcium intensity as a function of time of selected ROIs and compiles the data into a .

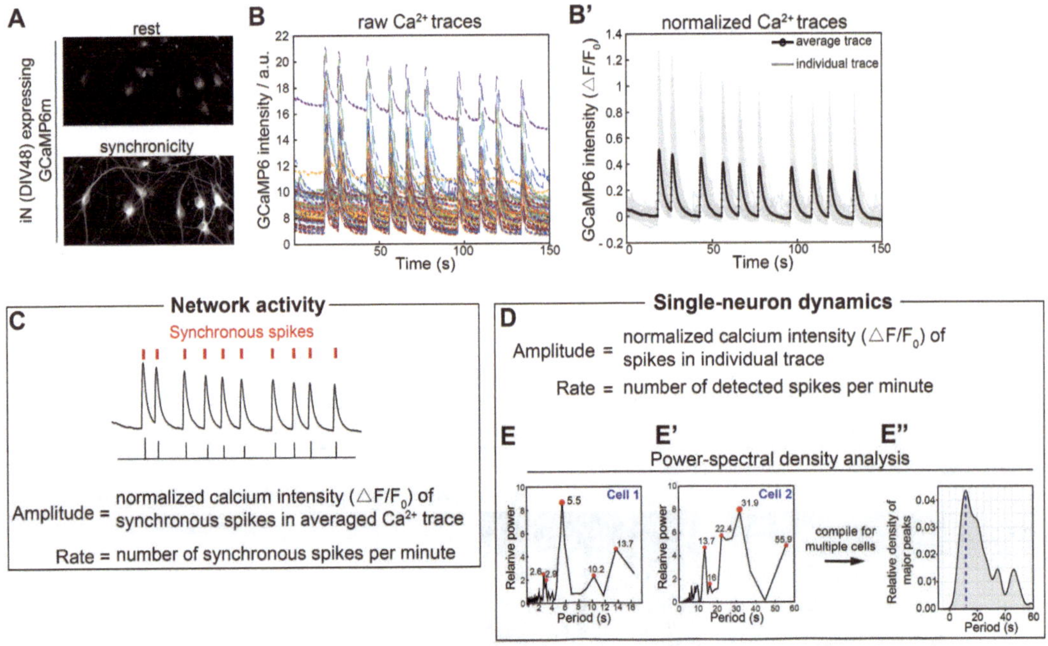

Fig. 3 Illustrations of network activity and single-neuron dynamics analyses. (**a**) Representative images of iN culture expressing GCaMP6m in the rest state (upper panel) and the synchronous firing state (lower panel). (**b–b'**) Representative plots of raw Ca^{2+} traces (**b**) and normalized Ca^{2+} traces (**b'**). (**c**) Schematic illustration of network activity analysis and the equations that are used to compute spike amplitude and rate. The illustrative Ca^{2+} trace represents average Ca^{2+} intensity calculated from all Ca^{2+} traces in one FOV. The spikes detected in this trace are synchronous firing spikes. (**d–e''**) Quantifications of single-neuron dynamics. (**d**) Equations that are used to compute amplitude and rate. (**e–e''**) Example graphs of the power-spectral density analysis. (**e–e'**) Example graphs of the power-spectrum analysis extracting top five peaks by fast Fourier transform. In each cell, the five peaks at the highest power are highlighted in red dots. (**e**) Cell 1 and (**e'**) Cell 2 in one FOV exhibit different oscillation periods in their calcium transients but have a common oscillation period at 13.7 s. (**e''**) Example of a density plot from frequencies of top 5 peaks compiled from a population of neurons. The period that presents in most cells (i.e., at high density) in this example is around 10 s, e–e'' are modified from Ref. [17]

mat file, which is named after the data name. Running this program will also automatically create a folder and save all data into it (*see* **Note 15**).

8. Double-click the *.mat* file in the command window and this will load several data matrixes. The matrix *ROI_intensity1* stores raw data sheet in which each column is an ROI and the row is the time-course (Fig. 2d).
9. Run the script *Raw_Intensity_plot.m,* and this will return a figure showing raw Ca^{2+} traces (colored lines in Fig. 3b).
10. To compute the amplitude and rate of synchronous Ca^{2+} spikes, load the *ROI_intensity1* matrix generated in the above step.

11. Run the script *Synchronous_rate_amplitude.m* and this returns two plots: a plot of detected spikes above baseline (middle panel in Fig. 3c, *see* **Note 16**) and the other is the plot of normalized calcium traces (Fig. 3b'). The amplitude value is saved in data matrix *Sync_amplitude_dFF* and the rate value is saved in data matrix *Sync_rate* (*see* **Note 17**).

3.4.2 Analysis of Single-Neuron Dynamics

Three parameters of single-neuron dynamics are quantified: the amplitude and rate of single-neuron Ca^{2+} activity, and the power-spectral density analysis [18, 19].

1. Load the *ROI_intensity1* matrix and run the script *SingleNeuron_Amplitude_ddF0_Frequency.m*. The amplitude of Ca^{2+} spikes of each neuron is computed in the *dFF0* matrix, and the firing rate of each neuron is in the *neuron_freq* matrix (Fig. 3d).
2. To do power-spectral density analysis, first load the *ROI_intensity1* matrix.
3. Specify variables in the function *FFT_TOP5peaks.m* and run the function for one cell at a time [17]. This function plots the five most significant (i.e., at the highest power) frequency wavelengths that are exhibited in the analyzed neuron (see examples in Fig. 3e-e'). Running this function displays the value in the command window and saves the values to the matrix *freq_period_pow_perc.mat*.
4. Analyze as many cells as needed.
5. Compile the frequency data from single-neuron for a density plot (Fig. 3e") (*see* **Note 18**).

4 Notes

1. Acquisition frame rate can be 10 100 ms/frame, depending on experiment need. But the camera should have the filming rate at 100 ms/frame or above in order to have sufficient time resolution for Ca^{2+} spike acquisition using this protocol.
2. The tissue mixture is sticky, so remove the solution carefully to avoid losing glia.
3. We find one well of confluent 6-well of hES can be used to generate iN for one 6-well plate efficiently, or else calculate the amount needed at the splitting ratio 1:18.
4. Overnight Matrigel coating is essential to ensure the long-term culture stays attached because it typically takes around 5–6 weeks for iN culture to reach maturity for functional assays such as calcium imaging.

5. The density and age of iN culture is important for their network activity and should be constantly monitored throughout the culture course. If the density is too low, there is little network activity in the culture even when using the Ca^{2+}-imaging buffer. We find the plating density of 1.8×10^5/24-well or above could generate frequent synchronicity event in our hands. However, the culture will tend to cluster if the density goes too high.
6. The timing of addition of AraC should be carefully adjusted depending on the glia density, such that it should be added after the glia is fully confluent but is added once the glia is confluent to prevent it overtaking iN cells.
7. Avoid complete medium exchange, which can result in neuronal cell death. It is also recommended that functional assays should be done within 2–3 days after a media exchange, allowing for culture to reach homeostasis.
8. We observe that iN culture show low spontaneous network activity in growth medium or in normal Tyrode's solution. Therefore, we use ambient Ca^{2+} and K^+ levels at 4 mM and 8 mM, respectively, in the calcium-imaging solution to potentiate network activity. Under this condition, frequent synchronous firing was observed in cultured neurons.
9. Stock solution of Tyrode's and Ca^{2+}-imaging buffer should be aliquoted to small volumes and kept at 4 °C, warm up aliquot each time.
10. Locate to a field of view (FOV) where iN cells are evenly distributed and where the soma of individual cell is clear—avoid regions with clustered cells or is uncovered by glia.
11. Use the same recording conditions (e.g., light power, detection gain, frame rate, camera exposure time, etc.) of control and experiment groups for the accuracy of acquisition and quantitative analysis.
12. The choice of objective depends on resolution needs. 20× or 40× air objective provides low-magnification view but can yield sufficient resolution for image presentation and analysis. We use 20× objective to obtain large FOV that contains large amount of neurons.
13. Make sure a Matlab plugin named "spike_detection" is installed before running analysis.
14. The time delay depends on the size of the data file.
15. For example, running the algorithm for a data named *neuron.tiff* will create a folder named "neuron_multiROIplot," which contains an ROI-overlayed image and its intensity matrix *neuron.mat*.

16. Adjust the threshold for spike detection according to the data quality. For example, if the culture exhibits very low activity, the threshold should be considerably adjusted towards the lower end. Typically, we set the threshold which is within the range of mean ± 1–3 × s.d. It is important to note that once the threshold is set, the same threshold should be applied consistently for control and experiment groups.
17. The synchronous firing peaks are quantified as the detected peaks in the average intensity trace. The amplitude is defined as the ratio of fluorescence change (delta F, ΔF) to the baseline fluorescence (F_0) (i.e., $\Delta F/F_0$ of calcium peaks, Fig. 3c). The synchronization rate is defined as the number of synchronous peaks within 1 min (i.e., synchronous Ca^{2+} spikes/min, Fig. 3c).
18. It is important to collect large enough sample size to approximate a normal distribution in the density plot. For further statistical analysis, the density plot can be binned to compare specific frequency periods [20]. This method converts periodic signal from its time domain to its frequency domain. It is useful when little information can be directly deduced from temporal data.

References

1. Grienberger C, Konnerth A (2012) Imaging calcium in neurons. Neuron 73(5):862–885
2. Stringer C, Pachitariu M (2019) Computational processing of neural recordings from calcium imaging data. Curr Opin Neurobiol 55: 22–31
3. Ahrens MB, Orger MB, Robson DN, Li JM, Keller PJ (2013) Whole-brain functional imaging at cellular resolution using light-sheet microscopy. Nat Methods 10(5):413–420
4. Ziv Y, Burns LD, Cocker ED, Hamel EO, Ghosh KK, Kitch LJ, El Gamal A, Schnitzer MJ (2013) Long-term dynamics of CA1 hippocampal place codes. Nat Neurosci 16(3): 264–266
5. Dombeck DA, Harvey CD, Tian L, Looger LL, Tank DW (2010) Functional imaging of hippocampal place cells at cellular resolution during virtual navigation. Nat Neurosci 13(11):1433–1440
6. Sheffield ME, Dombeck DA (2015) Calcium transient prevalence across the dendritic arbour predicts place field properties. Nature 517(7533):200–204
7. Chen TW, Wardill TJ, Sun Y, Pulver SR, Renninger SL, Baohan A, Schreiter ER, Kerr RA, Orger MB, Jayaraman V, Looger LL, Svoboda K, Kim DS (2013) Ultrasensitive fluorescent proteins for imaging neuronal activity. Nature 499(7458):295–300
8. Brewer GJ, Boehler MD, Pearson RA, DeMaris AA, Ide AN, Wheeler BC (2009) Neuron network activity scales exponentially with synapse density. J Neural Eng 6(1):014001
9. Verstraelen P, Pintelon I, Nuydens R, Cornelissen F, Meert T, Timmermans JP (2014) Pharmacological characterization of cultivated neuronal networks: relevance to synaptogenesis and synaptic connectivity. Cell Mol Neurobiol 34(5):757–776
10. Fan LZ, Nehme R, Adam Y, Jung ES, Wu H, Eggan K, Arnold DB, Cohen AE (2018) All-optical synaptic electrophysiology probes mechanism of ketamine-induced disinhibition. Nat Methods 15(10):823–831
11. Williams LA, Joshi V, Murphy M, Ferrante J, Werley CA, Brookings T, McManus O, Grosse J, Davies CH, Dempsey GT (2019) Scalable measurements of intrinsic excitability in human iPS cell-derived excitatory neurons using all-optical electrophysiology. Neurochem Res 44(3):714–725
12. Wang J, Miao Y, Wicklein R, Sun Z, Wang J, Jude KM, Fernandes RA, Merrill SA, Wernig M, Garcia KC, Sudhof TC (2022)

RTN4/NoGo-receptor binding to BAI adhesion-GPCRs regulates neuronal development. Cell 185(1):218

13. Patzke C, Dai J, Brockmann MM, Sun Z, Fenske P, Rosenmund C, Sudhof TC (2021) Cannabinoid receptor activation acutely increases synaptic vesicle numbers by activating synapsins in human synapses. Mol Psychiatry 26(11):6253–6268
14. Sun Z, Sudhof TC (2021) A simple Ca(2+)-imaging approach to neural network analyses in cultured neurons. J Neurosci Methods 349: 109041
15. Tsetsenis T (2017) Monitoring synapses via trans-synaptic GFP complementation. Methods Mol Biol 1538:45–52
16. Zhang Y, Pak C, Han Y, Ahlenius H, Zhang Z, Chanda S, Marro S, Patzke C, Acuna C, Covy J, Xu W, Yang N, Danko T, Chen L, Wernig M, Sudhof TC (2013) Rapid single-step induction of functional neurons from human pluripotent stem cells. Neuron 78(5): 785–798
17. Sun Z, Südhof TC (2020) A simple Ca2+-imaging approach to neural network analysis in cultured neurons. bioRxiv 2020.08.09.243576; https://doi.org/10.1101/2020.08.09.243576
18. Uhlen P (2004) Spectral analysis of calcium oscillations. Sci STKE 2004(258):pl15
19. Wu M, Wu X, De Camilli P (2013) Calcium oscillations-coupled conversion of actin travelling waves to standing oscillations. Proc Natl Acad Sci U S A 110(4):1339–1344
20. Bortone D, Polleux F (2009) KCC2 expression promotes the termination of cortical interneuron migration in a voltage-sensitive calcium-dependent manner. Neuron 62(1):53–71

Chapter 21

Whole Cell Patch Clamp Electrophysiology in Human Neuronal Cells

Rafael Gabriel III, Andrew J. Boreland, and Zhiping P. Pang

Abstract

Whole cell patch clamp recording techniques are commonly used to assay membrane excitability, ion channel function, and synaptic activity in neurons. However, assaying these functional properties of human neurons remains difficult because of the difficulty in obtaining human neuronal cells. Recent advents in stem cell biology, especially the development of the induced pluripotent stem cells, made it possible to generate human neuronal cells in both 2-dimensional (2D) monolayer cultures and 3D brain-organoid cultures. Here, we describe the whole cell patch clamp methods of recording neuronal physiology from human neuronal cells.

Key words Induced pluripotent stem cells, Human induced neurons, Neuronal physiology, Brain organoids, Brain slice, Synaptic transmission, Whole cell patch clamp electrophysiology, iPSC, NPC

1 Introduction

Elucidating the functional properties of human neurons is crucial for understanding aberrant neuron physiology, including membrane excitability, ion channel, and synaptic activities underlying neurodevelopmental and neurodegenerative disorders. Patch clamp techniques [1–3] revolutionized our ability to investigate these functional properties by allowing researchers to dissect the ionic components of synaptic transmission and neuronal excitation [4, 5]. This is achieved by placing a micropipette in extremely close contact with a neuronal cell membrane forming a high-resistance seal (Gigaseal) to allow the recording of tiny current-voltage changes caused by ionic fluxes across cell membranes via "single" ion channels. Taken a step further, by applying brief negative pressure through the micropipette, one can rupture the membrane and create a continuous flow between the internal solution of the pipette and the cellular cytosol. This so-called "whole cell" configuration allows for stable, intracellular recording of ion

Yu-Wen Alvin Huang and ChangHui Pak (eds.), *Stem Cell-Based Neural Model Systems for Brain Disorders*, Methods in Molecular Biology, vol. 2683, https://doi.org/10.1007/978-1-0716-3287-1_21,

channel activities across the whole cell membranes. Whole-cell recordings also allow diffusional exchange of material from the micropipette into the cell cytosol for pharmacological manipulation and labeling of the cell morphology.

The whole cell patch clamp recordings have two different configurations: voltage clamp and current clamp, to record current and voltage changes across the cell membranes, respectively. The voltage clamp mode holds the cell's membrane voltage constant, allowing one to measure changes in currents generated by ion fluxes across the membrane when channels open or close. Under voltage clamp recording mode, whole cell current responses mediated by ion channels, including voltage-dependent sodium, potassium, and calcium channels could be recorded by holding the cell membrane at different voltages. Synaptic currents including both excitatory postsynaptic currents (EPSCs) and inhibitory postsynaptic currents (IPSCs) mediated by neurotransmitters such as glutamate and γ-amino butyric acid (GABA) could also be recorded at certain holding potentials depends on different internal solutions compositions.

The current clamp mode holds the current constant, enabling one to measure membrane voltage changes. Resting membrane potentials, action potentials (either spontaneous or elicited by depolarization of cell membrane), and spontaneous or evoked synaptic responses could be recorded. This powerful technique has greatly contributed to the understanding of neuronal electrophysiology and how it relates to neuronal health and disease [6–12].

The high prevalence of neurodevelopmental, neuropsychiatric, and neurodegenerative disorders necessitates investigation of neuronal physiology in human-based systems. Human neuronal models also allow representation of heterogeneous genetic backgrounds potentially relevant to a given disease. The advent of induced pluripotent stem cell (iPSC) technology [13] and then further reprogramming to neurons and other brain cells now allows this investigation. One method of making monolayer 2D human neuron cultures is by ectopic expression of specific transcription factors, such as *Ngn2* and *Ascl1*, that drive a stem cell to become a neuron [12, 14]. Alternatively, neurons can also be generated using growth factors and small molecules through a process with greater mimicry to normal development [15–19]. Together, these two methods for generating human stem cell derived neurons hold great promise for understanding human-specific physiology and pathology. We have used these genetically-defined iPSC-derived neurons and cerebral organoids to model opioid-receptor variants [20], nicotine addiction [21], ethanol's effect on synaptic activity [22], neurodevelopmental disorders [23–25], and integrated with microelectromechanical systems [26, 27].

In comparison to 2D cell cultures, cerebral organoids offer increased complexity and physiological relevance because of their

3D structure. Brain/cerebral organoids are derived from human iPSCs that aggregate in suspension, undergo neural induction, and finally differentiate and mature into functional neural tissue. Furthermore, these organoids can be "patterned" to certain brain areas such as cortical, thalamic, or even fused organoids of two separate regions [28–30]. These cortical organoids develop cytoarchitecture and anatomical structures reminiscent of the developing human brain's ventricular zones [31]. Thus, the ability of 3D organoids to recapitulate human neural development during its embryonic stage shows itself to be a very promising and powerful model. Recent work has integrated microglia into these brain organoids to facilitate study of neuroimmune interactions and how they relate to neural function [32]. One major challenge with long term organoid culture is hypoxia in the organoid interior. Due to the lack of vascularization, cells within the interior of the organoid suffer from lack of oxygen, nutrient delivery, and waste disposal. Cellular hypoxia can be observed as early as two months into culture leading to cell death in the interior of the developing organoid resulting in limited corticogenesis. To solve this issue a culture system called "sliced neocortical organoids" (SNO) was developed [33], bypassing the diffusion limit to prevent cell death over long-term cultures by slicing organoids and culturing in a manner similar to organotypic brain slice culture. This culture method effectively reduces cell hypoxia and fosters enhanced neural development and cortical lamination. Patch clamp recording techniques can be applied to organoid slices to assess neuron physiology [32, 33]. However, because of high cell density, cell type heterogeneity, and a functional maturity gradient throughout the cortical layers, patching organoids can be labor-intensive and low throughput.

In this chapter, we detail the setup and basic patch clamp technique to achieve whole cell configuration in human neuronal cells in both 2D (Fig. 1) and 3D (i.e., organoids) cultures (Fig. 2, Adapted from Xu et al. [32]). We will describe four basic protocols to reveal the membrane properties and synaptic activities in human neuronal cells.

2 Materials

2.1 Solutions

Prepare all solutions with double deionized water (ddH_20) (Resistivity of 18 MΩ at 25 °C) and laboratory grade reagents.

1. External solution—Artificial Cerebral Spinal Fluid (ACSF) contains (in mM):

   NaCl 125, KCl 2.5, $NaH_2PO_4 \cdot H_2O$ 1.25, $NaHCO_3$ 25, $MgCl_2$ 1.2, $CaCl_2$ 2.5, glucose ($C_6H_{12}O_6$) 2.5, sucrose ($C_{12}H_{22}O_{11}$) 22.5. Bubbled with 95% oxygen 5%

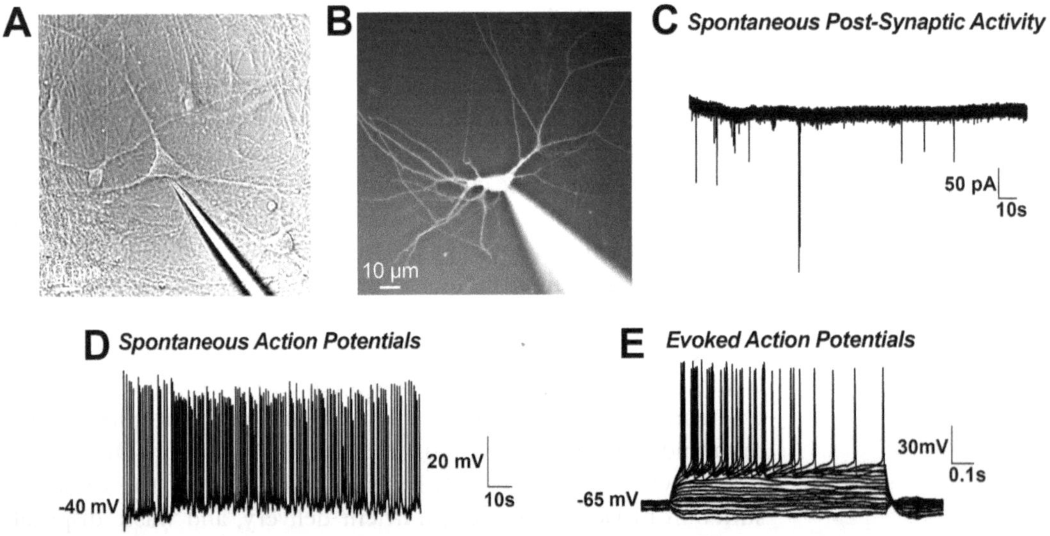

Fig. 1 Patch-clamp recording in iPSC-derived neurons. (**a**) A representative image showing a patched neuron in an NGN2 induced neuron (iN) 2D culture. (**b**) A representative image of a Dextran-AlexaFluor488 filled iN. (**c**) A representative trace of spontaneous post-synaptic currents (PSCs) recorded from an iN. (**d**) A representative trace of spontaneous action potentials recorded from an iN. (**e**) Example of repetitive action potentials evoked by stepwise currents injections in an iN

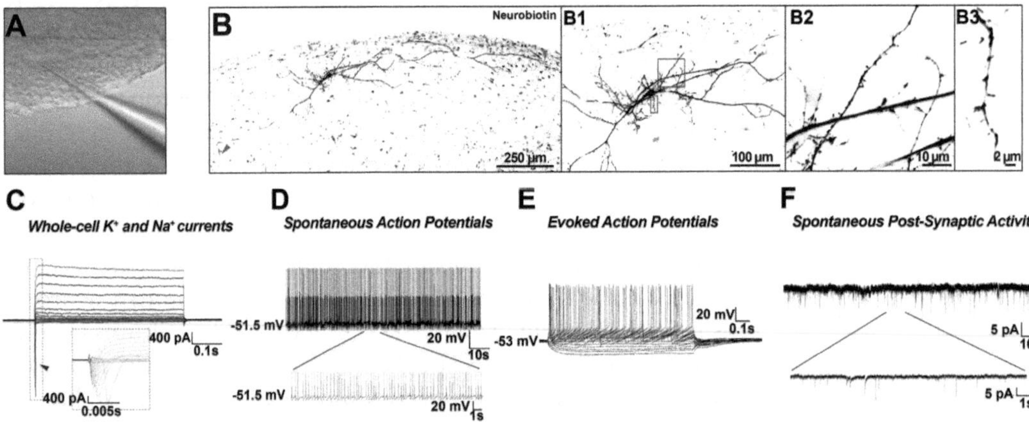

Fig. 2 Patch-clamp recording in iPSC-derived brain organoid slice. (**a**) A representative image showing a patched neuron in acute organoid slice. (**b**) Sample images of a neuron filled with biotin ethylenediamine HBR from a day 92 microglia-containing organoid after whole-cell recording. Complex arborization (B1) and spine-like structures (B2, B3) were identified. The images are inverted to a white background to help visualize neuron morphology. (**c**) Representative traces of whole-cell currents recorded from neurons in acutely sectioned organoid (90 day old in culture) slices. Insert: fast activation/inactivation voltage-dependent sodium currents. (**d**) Representative traces of spontaneous action potentials recorded from neurons in organoid (90 day old in culture) slices. (**e**) Example of repetitive action potentials evoked by stepwise currents injections. (**f**) Representative traces of spontaneous post-synaptic currents (PSCs) recorded from neurons in sliced collected from a day 92 organoid (Adapted from Xu et al.[32] with permission)

CO_2 and perfused at 1–2 mL per min during recordings. Osmolarity ~300 mOsmol. External solution is prepared before each experiment. A 10× stock solution could be made and stored (up to 1 week) to speed up the preparation of the external solution (*see* **Note 1**).

2. Potassium-Gluconate internal solution contains (in mM):

K-gluconate 126, KCl 4, HEPES 10, ATP-Mg 4, GTP-Na_2 0.3, Phosphocreatine 10, adjust pH to 7.2 with KOH. Measure and ensure final osmolarity is between 270–290 mOsmol. Aliquot solution and store at −20 °C for up to 6 months.

2.2 Equipment

The electrophysiology rig is modular and versatile, allowing multipatch setups, extracellular recordings, and fluorescent visualization with the simple addition of different equipment. The following list of equipment is the minimum necessary components to build a rig for basic patch-clamp electrophysiology recordings. Adjust your setup accordingly depending on your experimental needs.

1. Air table.
2. Faraday cage.
3. Microscope (upright scope with a 40× water immersion lens, an inverted microscope can be used for 2D neuronal cells).
4. Micromanipulator.
5. Patch clamp amplifier with headstage and micropipette holder.
6. Digitizer.
7. Computer and acquisition software.
8. Inline solution heater.
9. Recording chamber.
10. Perfusion system.
    - (a) Luer lock couplings.
    - (b) Perfusion reservoir gas bubbler.
    - (c) On/Off valves.
    - (d) Syringe filters.
    - (e) I.V. perfusion accessories.
    - (f) Flexible Tygon tubing.
    - (g) Peristaltic pump.
11. Pipette puller.

2.3 Additional Materials

1. Thin-walled borosilicate glass capillaries.
2. MicroFil flexible needle.
3. Eppendorf tubes.

4. 10 mL and 1 mL syringes.
5. P10, P20, P200, P1000 pipettes and pipette tips.
6. Forceps.
7. Kimwipes.
8. Bleach.

3 Methods

3.1 Recording Protocols (Samples Provided Using Molecular Devices Multiclamp 700B and pClamp 10 Software)

1. Amplifier settings
   (a) Bessel: 2 kHz.
   (b) Zap: 25 μs, increase if needed.
2. Spontaneous post-synaptic currents (sPSCs) recording protocol:
   (a) Acquisition mode: Gap free.
   (b) Sampling rate per signal (Hz): 10000.
   (c) 3–5 min duration.
3. V-clamp step-depolarization protocol (whole-cell current responses):
   (a) Acquisition mode: Episodic stimulation.
   (b) Holding potential −70 mV (should be adjusted accordingly).
   (c) Runs / trials: 1.
   (d) Sweeps / run: 20.
   (e) Sweep duration (s): 1.
   (f) Start-to-Start Intervals.
   (i) Sweep (s): 5.
   (g) First level (mV): −100.
   (h) Delta level (mV): 10 (or 5 depending on the input resistances of the neuronal cell types).
   (i) First duration (ms): 500.
4. I-clamp spontaneous action potential and postsynaptic potential recording protocol (I=0):
   (a) Acquisition mode: Gap free.
   (b) 3–5 min duration.
   (c) Sampling rate per signal (Hz): 10000.
5. I-clamp step current injection protocol (to elicit action potentials):
   (a) Acquisition mode: Episodic stimulation.
   (b) Runs / trials: 1.

(c) Sweeps / run: 16.
(d) Sweep duration (s): 2.
(e) Start-to-Start Intervals.
(f) Sweep (s): Minimum.
(g) First level (mV): −50.
(h) Delta level (mV): 10.

3.2 Setup

Carry out all procedures at room temperature unless specified otherwise. *See* **Notes 2–6** for instruction on reducing background electrical noise during recordings.

1. Bleach the recording and ground electrode (silver wires) overnight to provide a chloride coating on the silver wires. Thoroughly rinse the electrodes with ddH_20 and dry with a Kimwipe prior to reassembling.
2. Prepare ACSF solution for recording by diluting the 10× stock solution to 1× with ddH20 in a volumetric flask obtaining a final volume of 500 mL. Add and dissolve 0.225 g glucose and 3.847 g sucrose. Bubble the patch-solution for 10 min with carbogen (95% O2 / 5% CO2). Add bivalent cations: 600 μL 1 M $MgCl_2$ and 1250 μL 1 M $CaCl_2$. Patch solution must be kept bubbling for the duration of the recording. Without proper bubbling, the calcium may precipitate out of solution.
3. Prepare K-Glu internal solution. Thaw an aliquot on ice and filter the internal solution through a 0.22 μm filter system. Transfer the filtered internal solution to a 100 μL stainless steel hypodermic needle. Keep the syringe on ice for the duration of the experiments.
4. Perfuse oxygenated patch ACSF solution at 2 mL/min and a temperature of 34 °C (*see* **Note 7**).
5. Pull glass patch pipettes. Aim for a resistance of about 5–7 MΩ (*see* **Note 8**).
6. Using a sterile transfer pipet, cut the tip off and gently transfer an organoid slice or coverslip to the rig chamber.
   (a) Procedures for organoid slicing can be found in this publication [33].
   (b) Ensure the perfusion system is turned off upon transferring to avoid inadvertent aspiration of the slice.
   (c) Using a pair of forceps, gently move the slice or coverslip to the center of the perfusion bath.
   (d) Using the forceps, set a harp on top of the slice, positioning it so at least two harp strings are anchoring the slice (*see* **Note 9**).

3.3 Obtaining the Whole Cell Patch Configuration

1. Using the 4× objective, visualize and capture an image of the whole slice for documentation.
2. Switch to the 40× water immersion objective, ensuring that the objective is immersed in the ACSF solution.
3. Identify a cell to patch and center the cell on your screen (*see* **Note 10**).
4. Slightly raise the objective to provide room for the glass pipette to be inserted into the chamber and recording solution. Take care to ensure the objective remains submerged in the ACSF solution.
5. Use a clean MicroFil/microloader tip on a 100 μL or 1 mL syringe and fill a glass pipette with approximately 5 μL of internal solution.
   (a) The microloader tip must be inserted as far down to the end of the glass pipette as possible.
   (b) Flick your wrist to ensure the solution travels to the tip of the pipette, then gently flick the pipette several times to eliminate any bubbles in the solution (*see* **Note 11**).
   (c) Internal solution must contact the inner electrode.
6. Attach the glass pipette to the pipette holder on the headstage (*see* **Note 12**).
7. Submerge the tip of the pipette into the bath solution and adjust the focus of the objective to visualize the pipette. Apply positive pressure (30–70 mbar) to eliminate any remaining air bubbles within the tip. The pipette resistance without positive pressure should measure to about 6–7 MΩ. Apply constant positive pressure to ensure no debris attaches onto the tip of the pipette prior to patching. A dirty tip will not form a high-quality seal.
   (a) Applying positive pressure will push any debris in the internal solution out over time. Check for any debris clogging the internal tip of the pipette. Any clogs will be detrimental to the patch, and the pipette should be discarded and replaced.
8. Ensure that the Multiclamp is set to *voltage clamp mode* at this point.
9. Slowly lower the pipette in tandem with the objective. Take great care to keep the focus of the objective below the tip of the pipette. (This avoids crashing the pipette tip). However, if the focus goes significantly lower than the tip, you risk crashing the objective to the base of the pipette tip and breaking it off.
10. Lower the pipette to just above the slice surface. Use the "Auto Pipette Offset" function to adjust the pipette offset current back to zero in voltage clamp mode.

11. Still applying positive pressure, approach the cell, taking care to not make any contact with other cells or debris. The pressure from the internal solution should clear the way (*see* **Note 13**).
12. In one smooth motion, bring the pipette tip in contact with the cell membrane. The membrane will form a small "dimple" from the pressure.
    (a) Check for a 0.1–0.2 MΩ access resistance increase. Immediately release the positive pressure.
    (b) The dimple should relax and form a seal around the tip of the pipette. The access resistance should quickly rise. If not, apply a small amount of negative pressure.
13. Once the access resistance is over 100 MΩ, apply a holding voltage of −45 mV.
    (a) If the access resistance does not increase to over 100 MΩ within a few seconds, abort the patch and try again with a different cell and a fresh pipette.
    (b) Continue to apply small amounts of negative pressure until the access resistance reaches 1 GΩ or above, forming a gigaseal.
    (c) Gently pull the tip of the pipette slightly backwards and up from the slice until the membrane looks relaxed and in the state it was prior to patching.
14. Once a gigaseal has been achieved, apply a holding voltage of −70 mV.
15. Use Auto Cp Fast on Multiclamp to compensate for pipette capacitance.
16. Apply small amounts of negative pressure in tandem with the zap function (25–50 μs) to break into the cell (*see* **Note 14**).
    (a) Upon breaking into the cell, the membrane test should show a capacitive transient resulting from the discharge of the membrane capacitance. Let the cell relax for about 5 min. If the current amplitude decreases, the patch is starting to seal back up. Apply a very small amount of negative pressure to open it back up.
    (b) The access resistance should be less than 20 MΩ and remain unchanged or change less than 15%. Applying more negative pressure may help decrease the access resistance. Consider compensating for access resistance for whole cell current recordings.
    (c) If the access resistance remains too high or if the capacitive current looks unstable, abort the recording and try again with a different cell and fresh pipette.

3.4 Patch-Clamp Recordings

1. After successfully establishing whole cell configuration, run the *sPSCs protocol* for 3–5 min to record any spontaneous currents while on *voltage clamp mode* (at a holding potential of −70 mV, for example).
   (a) Record the membrane properties (membrane capacitance, membrane resistance, access resistance, tau, and the holding current) in a lab notebook.
   (b) Abort the protocol if the baseline does not remain stable for the duration of the recording.
2. Check the membrane test again to ensure the patch is still open and the access resistance is still less than 20 MΩ.
   (a) Apply negative pressure to reopen the patch if the access resistance is too high. If the access resistance remains too high or the capacitive current becomes unstable, abort the patch and try again with a different cell and fresh pipette.
3. Apply whole cell compensation to adjust for cell membrane capacitance. Consider compensate access resistance for voltage-dependent whole-cell current recordings.
4. Run the *V-clamp step-depolarization* protocol.
5. Disable whole cell compensation and switch to the *I-clamp spontaneous protocol.*
6. Switch the configuration to *current clamp (I = 0) mode.*
   (a) The amplifier's internal circuitry must briefly reset switching between current and voltage clamp mode. It is crucial to always go through I = 0 mode when switching between current and voltage clamp.
7. Run the *I-clamp spontaneous protocol* to record spontaneous action potentials as well as postsynaptic potentials.
   (a) After switching to I = 0, let the cell's resting membrane stabilize before recording.
   (b) If the resting membrane continues to increase for the duration of the recording, the cell is unhealthy or dying and the recording must be aborted.
8. Run the *I-clamp step current injection protocol.* Inject currents to keep basal membrane potential closer to −65 mV if desired.
9. After all acquisition protocols have finished recording, release the negative pressure and carefully retract the pipette backwards, and replace the pipette to record a new cell. The data collected are to be analyzed using appropriate analysis software.

4 Notes

1. For your convenience, prepare a 10× stock ACSF solution. The stock ACSF solution must be stored at 4 °C and used within 1 week. The molecular weights of the reagents are as listed (g/mol): NaCl 58.44, KCl 74.55, $NaH_2PO_4{\cdot}H_2O$ 137.99, $NaHCO_3$ 84.01, $MgCl_2$ 95.21, $CaCl_2$ 110.98, glucose 180.16, and sucrose 342.30. For 1 L 10 × stock, NaCl 73 g, KCl 1.86 g, $NaH_2PO_4{\cdot}H_2O$ 1.73 g, $NaHCO_3$ 21 g, dissolved in ddH_2O to final volume of 1000 mL. Note that $CaCl_2$, $MgCl_2$, sucrose, and glucose are not added to the stock solution. It is imperative that the stock solution is diluted to 1× ACSF prior to the addition of the sucrose and glucose, and the 1× solution must be bubbled with carbogen prior to the addition of $CaCl_2$ and $MgCl_2$, otherwise precipitate will form, and the solution must be discarded. For 500 mL ACSF, add 0.225 g glucose and 3.847 g sucrose, 0.6 mL 1 M $MgCl_2$ and 1.25 mL 1 M $CaCl_2$.
2. Prior to the start of electrophysiology recordings, ensure that the rig has minimal recording noise to ensure high-quality recordings for the duration of the experiment. Most of the electrical noise will be eliminated by grounding the antivibration air table and the protective Faraday cage but take great care that nearby electrical devices such as cell phones, computers, and nearby power sources are not contributing to noise in your system. Ensure all electrical components are grounded and the grounding wires have not inadvertently formed grounding loops.
3. As the first course of action, hook up a "model-cell" to the amplifier headstage. A model-cell replicates the Resistor-Capacitor (RC) circuitry of the microelectrode in bath, the electrode in contact with a cell membrane, and the electrode after successfully breaking into the cell. The traces from all three configurations should be steady, if not, the noise issues originate from the software/rig hardware and all settings should be reset.
4. Switch off or disconnect all peripheral devices on the rig including the perfusion system, pipette holder, camera, and the manipulator. Test each device and watch if they are contributing to noise in the recordings.
5. Ensure the rig is entirely dry aside from the flow into the chamber from the perfusion system. Any solution inside the pipette holder or contacting electrical or metal components on the rig may be transmitting small amounts of current and causing noise.

6. A loose seal within the micropipette electrode holder can also contribute to noise. Check the O-rings to ensure they are not damaged and the seal is tight. With a micropipette inserted into the holder, apply positive pressure and visually check to see a constant flow of solution from the pipette into the bath. The stream of pressure should remain stable and not weaken over the course of at least 1 min.
7. Take care not to let the bath solution rise too high or fall too low in the recording chamber. Allowing the bath solution level to fluctuate may alter flow, leakage, and/or noise or vibrations while recording. Furthermore, make sure that the input/output of the perfusion system are directly opposite each other to ensure adequate flow over the center of the bath where the slice or coverslip is positioned.
8. The optimal resistance of the glass pipette electrodes depends on the size of the cells in the culture. Organoid cells are generally small and delicate, so a higher resistance of 8–9 MΩ allows for good patch formation. Larger pipettes tend to suck the whole cell into the pipette upon breaking into the membrane. In contrast, mouse brain slices and induced neuronal cultures on coverslips have cells that are larger and well anchored. Larger pipettes with lower resistance (4–7 MΩ) allow for lower access resistance and easier breaking into the cell.
9. Smaller organoid slices may not be large enough to pin down with 2 harp strings. Instead, place the harp string directly through the center of the slice. Rotate the harp to ensure that the harp strings are perpendicular to the micropipette when patching. This is the least obtrusive configuration of the harp, allowing you to patch cells close to the harp without the micropipette contacting the strings and disrupting movement. Remove the harp and rotate the slice to patch other regions, rather than moving the pipette to the other side of the slice.
10. Choosing a healthy neuronal cell is critical for successful patching. Under brightfield 40× magnification or DIC imaging, healthy cells will have a well-defined soma with a 3D cell shape. Avoid cells with a rougher or darker edge or shriveled “raisin” like appearances. These cells are unhealthy or dying and are unlikely to form good seals and produce quality recordings. Cells that look swollen or too circular should also be avoided. If your capture software has the feature, mark the cell so you know where to look for it after you refocus the objective on the slice.
11. It can be difficult to see the bubbles in the solution inside the patch pipette. Raise the pipette above your head and towards a light source to see the bubbles better.

12. Ensure the tip of the silver electrode attached to the headstage is submerged in the internal solution but not contacting the tip of the patch pipette. Tightening the glass seal compression caps and applying positive pressure may move the electrode slightly forward, breaking the tip of the pipette.
13. Neuronal cells in 3D organoids are quite fragile. Patching cells deep within the slice requires positive pressure to push away other cells and debris, but at the risk of agitating the cell of interest. Approaching the cell at an angle rather than directly on top of the cell will help keep the cell anchored and prevent rough first contact, greatly increasing the chance of a successful seal and patch. If the cell is deep within the slice, do not bring the pipette directly down near the top of where the cell is. Clearing the tissue on top of the cell for too long will stress the cell and displace synapses on the soma. Clear a path on the slice approximately 50 μm away from the cell of interest, moving the pipette diagonally towards the cell applying positive pressure. Quickly move the pipette back and forth in a prodding motion, clearing cells in the way. The first contact is critical, do not touch the cell of interest until no cells are obstructing the way and the angle is perfect.
14. It may be easier to use mouth suction to break into the cell. Using a p1000 pipette tip inserted into Tygon tubing leading to the side port of the pipette holder, apply a small amount of suction with your mouth and quickly pull out the pipette tip. This action will produce a "kissing" sound. One or two "kisses" should be enough to break into the cell. The zap function can be used in tandem with brief suction to help rupture the membrane. Experiment with the force and duration of suction depending on the cell's health as observed by the state of the capacitive current.

Acknowledgments

Pang lab is supported by NIAAA R01AA023797 and NIMH R21MH126420. A.J.B. was supported by NIGMS T32GM008339 and by NCATS TL1TR003019. The Child Health Institute of New Jersey is supported in part by the Robert Wood Johnson Foundation (grant #74260).

References

1. Fenwick EM, Marty A, Neher E (1982) A patch-clamp study of bovine chromaffin cells and of their sensitivity to acetylcholine. J Physiol 331:577–597. Epub 1982/10/01. https://doi.org/10.1113/jphysiol.1982.sp014393. PubMed PMID: 6296371; PMCID: PMC1197770
2. Hamill OP, Marty A, Neher E, Sakmann B, Sigworth FJ (1981) Improved patch-clamp techniques for high-resolution current

recording from cells and cell-free membrane patches. Pflugers Arch 391(2):85–100. Epub 1981/08/01. https://doi.org/10.1007/BF00656997

3. Sakmann B, Neher E (1984) Patch clamp techniques for studying ionic channels in excitable membranes. Annu Rev Physiol 46:455–472. Epub 1984/01/01. https://doi.org/10.1146/annurev.ph.46.030184.002323
4. Vyklický L Sr (1992) Ervin Neher and Bert Sakmann, 1991 Nobel prize laureates for physiology and medicine. Ion channels and the patch clamp technic. Cas Lek Cesk 131(2):33–41. Epub 1992/02/17
5. Todman D (2008) John Eccles (1903–97) and the experiment that proved chemical synaptic transmission in the central nervous system. J Clin Neurosci 15(9):972–977. https://doi.org/10.1016/j.jocn.2008.01.001
6. Chanda S, Ang CE, Davila J, Pak C, Mall M, Lee QY, Ahlenius H, Jung SW, Sudhof TC, Wernig M (2014) Generation of induced neuronal cells by the single reprogramming factor ASCL1. Stem Cell Reports 3(2):282–296. Epub 2014/09/26. https://doi.org/10.1016/j.stemcr.2014.05.020. PubMed PMID: 25254342; PMCID: PMC4176533
7. Halikere A, Popova D, Scarnati MS, Hamod A, Swerdel MR, Moore JC, Tischfield JA, Hart RP, Pang ZP (2020) Addiction associated N40D mu-opioid receptor variant modulates synaptic function in human neurons. Mol Psychiatry 25(7):1406–1419. Epub 2019/09/05. https://doi.org/10.1038/s41380-019-0507-0. PubMed PMID: 31481756; PMCID: PMC7051890
8. Pak C, Danko T, Mirabella VR, Wang J, Liu Y, Vangipuram M, Grieder S, Zhang X, Ward T, Huang YA, Jin K, Dexheimer P, Bardes E, Mitelpunkt A, Ma J, McLachlan M, Moore JC, Qu P, Purmann C, Dage JL, Swanson BJ, Urban AE, Aronow BJ, Pang ZP, Levinson DF, Wernig M, Sudhof TC (2021) Cross-platform validation of neurotransmitter release impairments in schizophrenia patient-derived NRXN1-mutant neurons. Proc Natl Acad Sci U S A 118(22). Epub 2021/05/27. https://doi.org/10.1073/pnas.2025598118. PubMed PMID: 34035170; PMCID: PMC8179243
9. Pak C, Danko T, Zhang Y, Aoto J, Anderson G, Maxeiner S, Yi F, Wernig M, Sudhof TC (2015) Human neuropsychiatric disease modeling using conditional deletion reveals synaptic transmission defects caused by heterozygous mutations in NRXN1. Cell Stem Cell 17(3):316–328. Epub 2015/08/19. https://doi.org/10.1016/j.stem.2015.07.017. PubMed PMID: 26279266; PMCID: PMC4560990
10. Wang L, Mirabella VR, Dai R, Su X, Xu R, Jadali A, Bernabucci M, Singh I, Chen Y, Tian J, Jiang P, Kwan KY, Pak C, Liu C, Comoletti D, Hart RP, Chen C, Sudhof TC, Pang ZP (2022) Analyses of the autism-associated neuroligin-3 R451C mutation in human neurons reveal a gain-of-function synaptic mechanism. Mol Psychiatry. Epub 2022/10/26. https://doi.org/10.1038/s41380-022-01834-x. PubMed PMID: 36280753
11. Yi F, Danko T, Botelho SC, Patzke C, Pak C, Wernig M, Sudhof TC (2016) Autism-associated SHANK3 haploinsufficiency causes Ih channelopathy in human neurons. Science 352(6286):aaf2669. Epub 2016/03/12. https://doi.org/10.1126/science.aaf2669. PubMed PMID: 26966193; PMCID: PMC4901875
12. Zhang Y, Pak C, Han Y, Ahlenius H, Zhang Z, Chanda S, Marro S, Patzke C, Acuna C, Covy J, Xu W, Yang N, Danko T, Chen L, Wernig M, Sudhof TC (2013) Rapid single-step induction of functional neurons from human pluripotent stem cells. Neuron 78(5):785–798. Epub 2013/06/15. https://doi.org/10.1016/j.neuron.2013.05.029. PubMed PMID: 23764284; PMCID: PMC3751803
13. Takahashi K, Tanabe K, Ohnuki M, Narita M, Ichisaka T, Tomoda K, Yamanaka S (2007) Induction of pluripotent stem cells from adult human fibroblasts by defined factors. Cell 131(5):861–872. https://doi.org/10.1016/j.cell.2007.11.019
14. Pang ZP, Yang N, Vierbuchen T, Ostermeier A, Fuentes DR, Yang TQ, Citri A, Sebastiano V, Marro S, Südhof TC, Wernig M (2011) Induction of human neuronal cells by defined transcription factors. Nature 476(7359):220–223. https://doi.org/10.1038/nature10202
15. Hu B-Y, Weick JP, Yu J, Ma L-X, Zhang X-Q, Thomson JA, Zhang S-C (2010) Neural differentiation of human induced pluripotent stem cells follows developmental principles but with variable potency. Proc Natl Acad Sci 107(9):4335–4340. https://doi.org/10.1073/pnas.0910012107
16. Chandrasekaran A, Avci HX, Ochalek A, Rösingh LN, Molnár K, László L, Bellák T, Téglási A, Pesti K, Mike A, Phanthong P, Bíró O, Hall V, Kitiyanant N, Krause K-H, Kobolák J, Dinnyés A (2017) Comparison of 2D and 3D neural induction methods for the generation of neural progenitor cells from human induced pluripotent stem cells. Stem

Cell Res 25:139–151. https://doi.org/10.1016/j.scr.2017.10.010
17. Muñoz-Sanjuán I, Brivanlou AH (2002) Neural induction, the default model and embryonic stem cells. Nat Rev Neurosci 3(4):271–280. https://doi.org/10.1038/nrn786
18. Chambers SM, Fasano CA, Papapetrou EP, Tomishima M, Sadelain M, Studer L (2009) Highly efficient neural conversion of human ES and iPS cells by dual inhibition of SMAD signaling. Nat Biotechnol 27(3):275–280. https://doi.org/10.1038/nbt.1529
19. Shi Y, Kirwan P, Livesey FJ (2012) Directed differentiation of human pluripotent stem cells to cerebral cortex neurons and neural networks. Nat Protoc 7(10):1836–1846. https://doi.org/10.1038/nprot.2012.116
20. Halikere A, Popova D, Scarnati MS, Hamod A, Swerdel MR, Moore JC, Tischfield JA, Hart RP, Pang ZP (2019) Addiction associated N40D mu-opioid receptor variant modulates synaptic function in human neurons. Mol Psychiatry 25:1406. https://doi.org/10.1038/s41380-019-0507-0
21. Oni EN, Halikere A, Li G, Toro-Ramos AJ, Swerdel MR, Verpeut JL, Moore JC, Bello NT, Bierut LJ, Goate A, Tischfield JA, Pang ZP, Hart RP (2016) Increased nicotine response in iPSC-derived human neurons carrying the CHRNA5 N398 allele. Sci Rep 6: 34341. https://doi.org/10.1038/srep34341
22. Scarnati MS, Boreland AJ, Joel M, Hart RP, Pang ZP (2020) Differential sensitivity of human neurons carrying μ opioid receptor (MOR) N40D variants in response to ethanol. Alcohol 87:97. https://doi.org/10.1016/j.alcohol.2020.05.004
23. Xu R, Brawner AT, Li S, Liu J-J, Kim H, Xue H, Pang ZP, Kim W-Y, Hart RP, Liu Y, Jiang P (2019) OLIG2 drives abnormal neurodevelopmental phenotypes in human iPSC-based organoid and chimeric mouse models of down syndrome. Cell Stem Cell 24(6):908–26.e8. https://doi.org/10.1016/j.stem.2019.04.014
24. Kim H, Xu R, Padmashri R, Dunaevsky A, Liu Y, Dreyfus CF, Jiang P (2019) Pluripotent stem cell-derived cerebral organoids reveal human oligodendrogenesis with dorsal and ventral origins. Stem Cell Reports 12(5): 890–905. https://doi.org/10.1016/j.stemcr.2019.04.011
25. Chen C, Jiang P, Xue H, Peterson SE, Tran HT, McCann AE, Parast MM, Li S, Pleasure DE, Laurent LC, Loring JF, Liu Y, Deng W (2014) Role of astroglia in Down's syndrome revealed by patient-derived human-induced pluripotent stem cells. Nat Commun 5(1): 4430. https://doi.org/10.1038/ncomms5430
26. Robles DA, Boreland AJ, Pang ZP, Zahn JD (2021) A cerebral organoid connectivity apparatus to model neuronal tract circuitry. Micromachines 12(12). https://doi.org/10.3390/mi12121574
27. Liu X, Bibineyshvili Y, Robles DA, Boreland AJ, Margolis DJ, Shreiber DI, Zahn JD (2021) Fabrication of a multilayer implantable cortical microelectrode probe to improve recording potential. J Microelectromech Syst 30(4):569–581. https://doi.org/10.1109/JMEMS.2021.3092230
28. Bagley JA, Reumann D, Bian S, Lévi-Strauss J, Knoblich JA (2017) Fused cerebral organoids model interactions between brain regions. Nat Methods 14:743. https://doi.org/10.1038/nmeth.4304. https://www.nature.com/articles/nmeth.4304#supplementary-information
29. Lancaster MA, Knoblich JA (2014) Organogenesis in a dish: modeling development and disease using organoid technologies. Science 345(6194):1247125
30. Lee C-T, Bendriem RM, Wu WW, Shen R-F (2017) 3D brain organoids derived from pluripotent stem cells: promising experimental models for brain development and neurodegenerative disorders. J Biomed Sci 24:59. https://doi.org/10.1186/s12929-017-0362-8
31. Lancaster MA, Renner M, Martin C-A, Wenzel D, Bicknell LS, Hurles ME, Homfray T, Penninger JM, Jackson AP, Knoblich JA (2013) Cerebral organoids model human brain development and microcephaly. Nature 501(7467):373–379. https://doi.org/10.1038/nature12517
32. Xu R, Boreland AJ, Li X, Erickson C, Jin M, Atkins C, Pang ZP, Daniels BP, Jiang P (2021) Developing human pluripotent stem cell-based cerebral organoids with a controllable microglia ratio for modeling brain development and pathology. Stem Cell Reports 16(8): 1923–1937. https://doi.org/10.1016/j.stemcr.2021.06.011
33. Qian X, Su Y, Adam CD, Deutschmann AU, Pather SR, Goldberg EM, Su K, Li S, Lu L, Jacob F, Nguyen PTT, Huh S, Hoke A, Swinford-Jackson SE, Wen Z, Gu X, Pierce RC, Wu H, Briand LA, Chen HI, Wolf JA, Song H, Ming G-l (2020) Sliced human cortical organoids for modeling distinct cortical layer formation. Cell Stem Cell 26(5): 766–81.e9. https://doi.org/10.1016/j.stem.2020.02.002

Chapter 22

Assaying Chemical Long-Term Potentiation in Human iPSC-Derived Neuronal Networks

Deborah Pré, Alexander T. Wooten, Haowen Zhou, Ashley Neil, and Anne G. Bang

Abstract

Impairment of long-term potentiation (LTP) is a common feature of many preclinical models of neurological disorders. Modeling LTP on human induced pluripotent stem cells (hiPSC) enables the investigation of this critical plasticity process in disease-specific genetic backgrounds. Here, we describe a method to chemically induce LTP across entire networks of hiPSC-derived neurons on multi-electrode arrays (MEAs) and investigate effects on neuronal network activity and associated molecular changes.

Key words Long-term potentiation, Synaptic plasticity, Multi-electrode array, Human-induced pluripotent stem cell-derived neurons

1 Introduction

The ability of neurons to modify the strength and efficacy of synaptic transmission through activity-dependent mechanisms is collectively referred to as synaptic plasticity. One of the most widely studied forms of synaptic plasticity is long-term potentiation (LTP), which leads to a long-lasting strengthening of synapses in response to specific types of stimulation [1, 2]. LTP is considered the biological substrate for learning and memory, and its impairment is a prominent feature of many pre-clinical models of neurological disorders [3–8]. While most of the advances in the understanding of LTP have come from murine models, the advent of human induced pluripotent stem cells (hiPSCs) [9, 10] provides a unique opportunity to investigate plasticity-related processes in the context of human neurological disorders.

LTP is typically induced through high-frequency tetanic stimulation of the presynaptic axon of a synapse [2]; however, chemical stimulation protocols have been developed to induce LTP across

Yu-Wen Alvin Huang and ChangHui Pak (eds.), *Stem Cell-Based Neural Model Systems for Brain Disorders*, Methods in Molecular Biology, vol. 2683, https://doi.org/10.1007/978-1-0716-3287-1_22,

many synapses of a neuronal network, thus increasing the feasibility of detecting associated molecular changes [11]. One approach to monitoring neuronal network activity is the use of multi-electrode array (MEA) technology. Like primary rodent neurons [12], hiPSC and human embryonic stem cell-derived neurons self-organize into networks that generate spontaneous, synchronous bursts of action potentials, termed network bursts (NBs), that can be detected across electrodes of a MEA and provide a read-out of synaptic activity [13–15]. MEAs have been broadly used to study synaptic plasticity on neuronal networks cultured in vitro, including stimulation by chemical LTP (cLTP) paradigms on rodent neurons [16–18].

Here we describe a protocol we developed that takes advantage of MEAs to study LTP in networks of hiPSC-derived neurons, where we use a cLTP paradigm based on the adenylate cyclase activator forskolin (FSK) and the phosphodiesterase inhibitor rolipram (ROL) to increase cyclic adenosine monophosphate (cAMP) levels, thus activating protein kinase A (PKA) and stimulating the biochemical machinery that underlies LTP [19]. Importantly, this label-free, non-destructive approach enables tracking of the late phase of LTP for days after induction and the 48-well format we used is amenable to higher throughput analyses. Using this approach, we showed that FSK/ROL treatment of either hiPSC-derived cortical or midbrain-dopaminergic neuronal networks, in co-culture with primary human astrocytes or hiPSC-derived astrocytes, respectively, results in increased firing and NB frequencies that last for up to 72 h after drug washout. In a demonstration of how the system can be manipulated using pharmacological and other blocking reagents to probe underlying molecular mechanisms, we also showed that FSK/ROL effects on midbrain-dopaminergic neuronal networks are largely independent of the N-methyl-D-aspartate receptor (NMDAR) subclass of glutamate receptors and partially dependent on brain-derived neurotrophic factor (BDNF), and other unidentified factors released into the medium [20]. Finally, we observed molecular hallmarks of LTP, including a rapid increase in phosphorylated cAMP-response element-binding protein (CREB) and induction of activity-regulated gene expression [20]. This protocol is highly reproducible: FSK/ROL-mediated potentiation was confirmed on seven different lots of commercially available hiPSC-derived midbrain dopaminergic neurons (iDopa, FujiFilm Cellular Dynamics, Inc.), as well as hiPSC-derived cortical neurons produced in-house from four independent lines (reference [20], and Pré, Zhou, and Bang, unpublished). Taken together, these results support the utility of this platform for investigation of LTP in hiPSC-derived neurons in different brain regions.

2 Materials

2.1 Co-culture of iDopa Neurons and iCell Astrocytes on MEAs

1. iCell Dopa neurons (iDopa, FujiFilm Cellular Dynamics, Inc.).
2. iCell Astrocytes (iAstro, FujiFilm Cellular Dynamics, Inc.).
3. NBA-K medium: Neurobasal A medium with 10% Knock-out serum. Replacement (KOSR), 1% Penicillin streptomycin (P/S), and 1 μg/mL of laminin from Engelbreth-Holm-Swarm murine sarcoma basement membrane.
4. BrainPhys Complete [21]: BrainPhys (Stemcell Technologies), 2% B27, 1% N2, 1% penicillin/streptomycin (P/S) (5000 U/mL), BrainPhys Supplements [BDNF (recombinant human/murine/rat BDNF, 20 ng/mL), Glia-Derived Neurotrophic Factor (recombinant human GDNF, 20 ng/mL), ascorbic acid (AA, 200 nM), dibutyryl cyclic AMP (cAMP, 1 mM) and laminin from Engelbreth-Holm-Swarm murine sarcoma basement membrane (1 μg/mL)].
5. Dulbecco's Phosphate buffered saline solution (DPBS).
6. Dulbecco's phosphate-buffered saline, without magnesium and calcium (DPBS−/−).
7. 50% Poly (ethyleneimine) (PEI) solution.
8. Boric acid.
9. Sodium tetraborate.
10. Maestro MEA system (Axion BioSystems).
11. 48-well Cytoview MEA plates (Axion BioSystems).

2.2 Co-culture hiPSC-Derived Cortical Neurons and Primary Human Astrocytes on MEAs

1. hiPSC lines.
2. Cortical neurons generated from hiPSC lines using a method developed by Shi et al. (2012) [22], with modifications (*see* Subheading 3.2.1).
3. Primary human astrocytes (ScienCell Research Laboratories).
4. ES medium: DMEM/Ham's F12 50/50 Mix, KnockOut serum replacer, 1% Penicillin streptomycin (P/S), 1% MEM non-essential amino acids, 0.2% 2-Mercaptoethanol (55 mM stock), 10 ng/mL basic fibroblast growth factor (bFGF).
5. Differentiation medium: 50/50 mix DMEM/Ham's F12 and Neurobasal (+) Phenol Red, with 1% N2 supplement, 2% B27 supplement, 1% Insulin-Transferrin-Selenium-A supplement, 1% MEM non-essential amino acids, 100 μM 2-Mercaptoethanol, and 1% P/S.
6. Differentiation supplements: 10 ng/mL BDNF, 10 ng/mL GDNF, 200 ng/mL ascorbic acid, and 1 μM cAMP.
7. Human astrocyte medium complete (ScienCell Research Laboratories).

8. Versene (Gibco|Thermo-Fisher Scientific).
9. Accutase.
10. 0.25% trypsin.
11. Cytosine arabinoside (AraC).
12. ROCK inhibitor Y-27632 2HCl.
13. 70 μm sterile cell strainers.
14. 35 and 60 mm TC-treated Culture Dishes.
15. 75 cm^2 Rectangular straight neck cell culture flask with vented cap.

2.3 cLTP Induction and Pharmacological and Other Blocking Reagents

1. Forskolin (FSK), 25 mM in DMSO.
2. Rolipram (ROL), 50 mM in DMSO.
3. (2R)-amino-5-phosphonovaleric acid (AP5), 50 mM in ddH_2O.
4. Human TrkB-Fc Chimera (R&D systems).
5. Human IgG.
6. LEGEND MAX Human BDNF ELISA kit (BioLegend).

3 Methods

3.1 Co-culture of iCell Dopaminergic Neurons (iDopa) and iCell Astrocytes (iAstro) on MEA Plates

Most of the thawing and plating procedures follow the manufacturer's recommendation, with some modifications.

3.1.1 Prepare 0.1% PEI Solution

1. Prepare a 0.1% PEI solution by diluting 50% PEI solution in borate buffer (*see* **Note 1**).
2. Stir for 20 min until the PEI is completely in solution.
3. Filter the 0.1% PEI solution through a 0.22 μm filter.
4. Use it fresh or make aliquots at appropriate volume and store at −20 °C.

3.1.2 Coating MEA Plates

1. One day before plating, coat the MEA wells with 80 μL 0.1% PEI solution. Move the plate to the incubator for 1 h, then wash it four times and let it dry overnight in the hood with the lid open (*see* **Note 2**).
2. The following day, dilute 80 μL of laminin (1 mg/mL) in 1 mL of NBA-K medium to obtain an 80 μg/mL laminin solution (*see* **Note 3**).

3. Spot 10 μL of 80 μg/mL laminin solution on the center of each well (*see* **Note 4**). Move the plate to the incubator for 45 min to 1 h.

3.1.3 Thawing and Plating iDopa and iAstro on MEA

1. Prepare 40 mL of NBA-K.
2. Follow the manufacturer's instructions to thaw the two vials of iDopa and iAstro in a 37 °C water bath.
3. Transfer the cells to 15 mL conical tubes (*see* **Note 5**).
4. Rinse vials with 1 mL of NBA-K and then slowly add to the 15 mL tubes with medium drop-by-drop, not faster than one drop per second to avoid osmotic shock.
5. Add 8 mL of NBA-K drop-by-drop to the 15 mL conical tubes (final volume 10 mL) and gently mix.
6. Take 20 μL of iDopa and iAstro cell suspension to count cells on a hemocytometer. Mix 1:1 with 0.2% trypan blue, count iDopa and iAstro in duplicate, and average the two measurements.
7. Centrifuge the thawed cells at 300 g for 5 min in a tabletop centrifuge.
8. Aspirate the medium but leave ~50–100 μL to avoid disturbing the pellet. Determine the leftover volume with a P200 pipette (*see* **Note 6**).
9. Add NBA-K for a final concentration of 1.6×10^7 cells/mL for iDopa and 2×10^6 cells/mL for iAstro.
10. Transfer iDopa:iAstro at a 1:1 ratio to a 1.5 mL tube.
11. Gently invert the tube 3–4 times to mix.
12. Add 10 μL of 1:1 iDopa:iAstro on top of the 10 μL of laminin spotted in the center of each MEA well such that 80,000 iDopa and 10,000 iAstro are plated per well.
13. Move the plate to the incubator for 45 min.
14. After incubation slowly add 150 μL of NBA-K to each well from the side. Repeat for a final volume of 300 μL of medium per well (*see* **Note 7**).
15. The day after plating, exchange 200 μL medium with fresh NBA-K per well.

3.1.4 Maintaining iDopa and iAstro Co-cultures on MEAs

1. Exchange 200 μL medium with fresh NBA-K every 2–3 days until day 7 after plating.
2. At day 7, remove 200 μL of NBA-K and add 200 μL of BrainPhys Complete.
3. Exchange 200 μL medium with fresh BrainPhys Complete every 2–3 days.

3.2 Co-culture of hiPSC-Derived Cortical Neurons and Primary Human Astrocytes on MEAs

Cortical neurons were produced using the protocol of Shi et al. (2012) [22] with modifications described below. Primary human astrocytes (ScienCell) were cultured following the manufacturer's instructions on a T75 mm flask.

3.2.1 Differentiation of hiPSC-Derived Cortical Neurons

1. Maintain hiPSC on irradiated mouse embryonic fibroblast feeder layers in ES medium. Passage culture weekly as small clusters of cells at a 1:4–1:8 split ratio using Versene as described [23].
2. Differentiate cortical neural progenitor cells (NPCs) from hiPSC as described [22], except use Versene, rather than Dispase, to passage cells as clusters at the initiation of neural induction and at day 7 after neural induction.
3. At ~15–16 days after neural induction, when neural rosettes appear, passage the NPCs. Two hours before passaging, coat 60 mm tissue culture treated dishes with 2 mL of 20 μg/mL laminin solution in DPBS for 2 h in the incubator. Aspirate laminin prior to plating cells.
4. To passage the NPCs, aspirate the medium, wash the cells gently with DPBS−/−, add 2 mL of Accutase, pre-warmed to 37 °C, and place in incubator for 5–10 min until cells begin to visibly detach.
5. Add 6 mL of Neural Maintenance Medium, transfer cells to a 15 mL conical tube, and triturate to a single cell suspension.
6. Centrifuge cells at 300 g for 5 min.
7. Aspirate the medium without disturbing the pellet. Resuspend the cells in 1 mL of Neural Maintenance Medium plus differentiation supplements.
8. Count the cells as described in Subheading 3.1.3, **step 6**, and resuspend in Neural Maintenance Medium plus differentiation supplements at 2×10^6 cells per 60 mm tissue culture dish coated with 20 μg/mL laminin in **step 3**.
9. Culture NPC for 2 weeks in Neural Maintenance Medium plus differentiation supplements. Exchange 66% of the medium for fresh medium every other day.
10. After two weeks, switch Neural Maintenance Medium to NBA-K with differentiation supplements for an additional 2 weeks, exchanging 66% of the medium every other day.
11. After 4 weeks, cells are ready to be replated on MEA plates.

3.2.2 Replating and Maintaining hiPSC-Derived Cortical Neurons on MEAs

1. Two hours before replating, coat 60 mm tissue culture treated dishes with 2 mL of 20 μg/mL laminin solution in DPBS for 2 h in the incubator (*see* **Note 8**). Aspirate laminin prior to plating cells.

2. To dissociate the cells, aspirate the medium, gently wash with DPBS−/−, aspirate, add 2 mL Accutase, pre-warmed to 37 °C, and place in incubator for 30–45 min (*see* **Note 9**).
3. Add 2 mL of NBA-K without laminin and pipette up and down to break up the clumps (*see* **Note 10**).
4. Pass the cells through a 70 μm cell strainer mounted on a 50 mL conical tube.
5. Wash the dish four times with 2 mL of warm NBA-K without laminin and pass the volume through the same strainer each time, for a final volume of 12 mL in the 50 mL conical tube.
6. Centrifuge the cells at 300 g for 5 min.
7. Aspirate the medium without disturbing the pellet. Resuspend the cells in 1 mL of NBA-K plus differentiation supplements.
8. Count the cells as described in Subheading 3.1.3, **step 6**, and replate at 150,000 cells/cm^2 in NBA-K-medium plus differentiation supplements and 10 μM ROCK inhibitor on 60 mm tissue culture plates coated with 20 μg/mL laminin in **step 1**.
9. The day after plating, feed the cells by exchanging 66% of the medium with fresh NBA-K plus differentiation supplements plus 5 μM cytosine arabinoside (Ara-C). Do not include ROCK inhibitor.
10. The following day, coat the MEA plate with PEI as described in Subheadings 3.1.1 and 3.1.2.
11. The day after **step 10**, coat MEA plates with laminin as described in Subheading 3.1.2.
12. At this point, the cells have been in Ara-C for 2 days. To dissociate the cells, aspirate the medium, gently wash with DPBS−/−, then aspirate the DPBS−/− and add 2 mL of Accutase, warmed to 37 °C, and incubate for 5–10 min until cells begin to visibly detach.
13. Add 2 mL of NBA-K without laminin, pipette up and down to break up the clumps (*see* **Notes 10** and **11**).
14. Pass the cells through a 70 μm cell strainer mounted on a 50 mL conical tube.
15. Wash the dish 4 times with 2 mL of warm NBA-K without laminin and each time pass the volume through the same strainer used for the cells, for a final volume of 12 mL in the 50 mL conical tube.
16. Before centrifuging the cells, count them as described in Subheading 3.1.3, **step 6**.
17. Calculate the volume to resuspend the cells for a concentration of 8000 cells/μL.
18. Centrifuge the cells at 300 g for 5 min.

19. Aspirate the medium leaving ~50–100 μL to avoid disturbing the pellet (determine the leftover volume with a P200 pipette) (*see* **Note 6**).
20. Add NBA-K plus differentiation supplements and 10 μM ROCK inhibitor for a final concentration of 8000 cells/μL.
21. Spot the cells on the MEA plate following the same procedure described in Subheading 3.1.2, with the exception that astrocytes are not included at this point.
22. The following day, change the medium with NBA-K plus differentiation supplements (without ROCK inhibitor).
23. Change 200 μL of medium every 2–3 days with fresh NBA-K plus differentiation supplements.
24. After 1 week of culture according to the manufacturer's instructions, detach primary human astrocytes from the T75 flask using 0.05% trypsin for 5 min, add human astrocyte complete medium at a 1:1 ratio, and collect cells in a 50 mL tube. Centrifuge at 300 g for 5 min and resuspend cells in 1 mL NBA-K plus differentiation supplements.
25. Count the astrocytes, as described in Subheading 3.1.3, **step 6**, and adjust the volume to obtain a concentration of 400,000 astrocytes/mL by adding NBA-K plus differentiation supplements.
26. On the MEA plate, remove 200 μL of medium per well and add 200 μL of medium with primary human astrocytes, so 80,000 astrocytes are added to each well. This will achieve an approximate 8:1 neuron to astrocyte co-culture ratio, since 80,000 neurons were spotted in an area that is approximately 8 times smaller than the whole well area where the astrocytes are plated.
27. The following day, exchange 200 μL medium with BrainPhys complete.
28. Exchange 200 μL medium with fresh BrainPhys complete every 2–3 days.

3.3 cLTP Induction and Manipulations to Investigate Underlying Molecular Mechanisms Using Pharmacological and Other Blocking Reagents

3.3.1 cLTP Induction

1. To perform cLTP experiments, BrainPhys Supplements are removed from the BrainPhys Complete medium, based on our observations that cAMP in the medium blocks potentiation induced by FSK/ROL, and BDNF in the medium reduces potentiation [20].
2. cLTP experiments are performed ~4 to 5 weeks after plating of cells on MEAs.
3. On the day of the FSK/ROL treatment, record activity. Place the MEA plate in the recording system and wait 5 min before recording to allow the plate to equilibrate after movement as described in Subheading 3.4.1.

4. Perform a full media exchange (300 μL/well) with BrainPhys Complete without BrainPhys supplements and move the place to the incubator for 2 h.
5. During this 2-h period, use the initial 5-min recording to assign wells to treatment groups to achieve a comparable distribution of baseline activity across all groups as described in Subheading 3.4.3.
6. During this 2-h period, also prepare the diluted drugs. The final concentrations of FSK and ROL are 50 μM and 0.1 μM, respectively. Prepare a 10× solution of FSK and ROL (500 μM and 1 μM, respectively) and 10× solution of vehicle only (DMSO) in BrainPhys Complete without BrainPhys supplements.
7. Two h after the last medium change, take the initial baseline 5-min recording (*see* **Note 12**).
8. Following the plate map prepared after assigning wells to treatment and control groups (*see* Subheading 3.4.3), add 50 μL of FSK/ROL and vehicle solutions to a 48-well "drug plate".
9. With a multichannel pipettor, remove 30 μL of medium per well (of a total of 300 μL), and quickly add 30 μL of medium from the FSK/ROL and vehicle drug plate, changing the tips after each column.
10. Record at 5 min and at 30 min after FSK/ROL and vehicle addition (let the plate equilibrate in the recording system for 5 min each time before recording).
11. After the 30-min recording, wash the drugs out as follows: remove 100 μL of medium/well, then 190 μL of medium/well and replace it with 190 μL of fresh BrainPhys medium without BrainPhys Supplements, then repeat the preceding step and finally remove 100 μL of medium/well replaced it with 200 μL of fresh BrainPhys Complete without BrainPhys Supplements.
12. Move the plate to the incubator.
13. Record the plate at 1 h and 4 h after washout (let the plate equilibrate in the recording system for 5 min each time before recording).
14. The following day, change 200 μL of medium with fresh BrainPhys Complete without BrainPhys Supplements and move the plate to the incubator.
15. After 2 h, take a recording (24-h recording). Repeat the same process at 48 h and 72 h.
16. Pharmacological blockers can be used prior to, along with, or after addition of FSK/ROL, using the drug-addition protocol described above. For instance, we tested the contribution of

NMDAR to FSK/ROL-mediated potentiation by taking a baseline recording, pre-treating with the NMDAR antagonist AP5 (50 μM) for 30 min prior to FSK/ROL addition, taking a second baseline recording, adding FSK/ROL, and then washing out the FSK/ROL/AP5 [20].

17. Parallel experiments to assess the effects of cLTP on activity regulated gene expression using RNAseq, or on phosphorylation of cAMP-response element-binding protein (CREB) and expression of CREB-dependent immediate-early genes using immuno-cytochemistry (ICC) can be performed after applying the same drug treatment protocol described for the MEA plates [20].

3.3.2 Conditioned Medium Experiments

1. We observed that removal of conditioned medium (CM) 1 h after cLTP induction and replacement with fresh BrainPhys Complete without supplements, strongly reduced potentiation; moreover, the cLTP-CM itself elicited potentiation in untreated cultures, showing that during the first hour after washout, excitatory factors necessary for cLTP maintenance are released into the medium [20]. To test whether BDNF is necessary for FSK/ROL induced potentiation, we used a TrkB-Fc receptor body, which acts as a BDNF scavenger. These studies showed that FSK/ROL-mediated cLTP is partially dependent on BDNF, as potentiation is decreased, but not blocked by TrkB-Fc [20].
2. Treat the cells with FSK/ROL as described in Subheading 3.3.1.
3. One hour after drug washout, take a recording as described in Subheading 3.4.1.
4. Remove 100% of the medium from cLTP-treated wells and vehicle-treated wells and move it to two 15 mL tubes, labeled "cLTP-CM" and "Control-CM", respectively. While preparing the media to treat the wells, add 300 μL of fresh BrainPhys Complete without BrainPhys supplements to each well.
5. To assess BNDF release with an ELISA assay, store the medium collected at 4 °C for up to 2 days (or at −80 °C for longer term storage) before proceeding with the immunoassay following the manufacturer's instruction.
6. To observe the potentiating effect of cLTP-CM on control wells, aspirate the medium and add 300 μL of cLTP-CM to vehicle-treated wells. Record the plate as described in Subheading 3.4.1 at 5 min and 30 min after medium addition, and additional timepoints over the following h and days, as appropriate, as described in Subheading 3.3.1.

7. To observe the effect of removal of the cLTP-CM from cLTP-treated wells, aspirate the medium and add 300 μL Control-CM to the cLTP-treated wells. Record the plate as described in Subheading 3.4.1 at 5 min and 30 min after medium addition, and additional timepoints over the following hours and days, as appropriate, as described in Subheading 3.3.1.
8. To test the contribution of BDNF to FSK/ROL-induced potentiation, move half of the cLTP-CM into a new 15 mL conical tube labeled "cLTP-CM + TrkB-Fc" and the other half to a 15 mL tube labeled "cLTP-CM + IgG". Set up analogous tubes for the Control-CM.
9. Add 2 μg/mL human TrkB-Fc or 2 μg/mL human IgG as a control to the appropriate tubes.
10. Aspirate the medium from FSK/ROL-treated MEA wells and add 300 μL of cLTP-CM + TrkB-Fc or cLTP-CM + IgG. In parallel, aspirate the medium from control vehicle treated MEA wells and add 300 μL of Control-CM + TrkB-Fc or Control-CM + IgG.
11. Record the plate as described in Subheading 3.4.1 at 5 min and 30 min after medium addition, and additional timepoints over the following hours and days, as appropriate, as described in Subheading 3.3.1.

3.4 MEA Recording and Data Analysis

Recordings are acquired with the Maestro recording system and Axion Integrated Studio (Axion Biosystems).

3.4.1 MEA Plate Recording

1. Set the temperature on the MEA plate recording system at 37 °C for the duration of the experiment.
2. Apply a Butterworth band-pass filter (10–2500 Hz) and adaptive threshold spike detector set to 5.5× standard deviations.
3. Enter a plate map using the Axion Integrated Studio tool and save the file for use in later analysis steps.
4. Place the MEA plate in the recording system and wait 5 min before recording to allow the plate to equilibrate after movement.

3.4.2 MEA Data Analysis

1. Open the spike (spk) files produced by the Axion Integrated Studio using the Axion Neural Metrics Tool.
2. Select the following parameters: active electrode criterion = 5 spikes/min; single electrode bursts, Poisson surprise, min surprise = 5; network bursts, envelope, threshold factor = 2, min IBI (ms) = 1000, min electrodes = 25%, burst inclusion = 75%; synchrony window (ms) =20. Parameters can be adjusted depending on the preferences of the user.

3. Export the "Supplemental Metrics" NeuralMetric file and open using the AxIS Metric Plotting Tool.
4. Select the appropriate plate map file, and the NeuralMetric files for baseline, and subsequent timepoints for comparison.
5. Export the "Supplemental Metrics to CSV file" and open it with Excel.
6. Select the metric of interest and copy into Prism or other graphing programs for statistical analysis and to generate graphs.

3.4.3 Assign Wells to Treatment and Control Groups to Achieve a Comparable Distribution of Baseline Activity

1. Open the spike (spk) files produced by the Axion Integrated Studio using the Axion Neural Metrics Tool.
2. Select the following parameters: active electrode criterion = 5 spikes/min; single electrode bursts, Poisson surprise, min surprise = 5; network bursts, envelope, threshold factor = 2, min IBI (ms) = 1000, min electrodes = 25%, burst inclusion = 75%; synchrony window (ms) = 20. Parameters can be adjusted depending on the preferences of the user.
3. Export the "Supplemental Metrics" NeuralMetric file and open in Excel.
4. Select the well name, Weighted Mean Firing Rate (WMFR), and the number of Network Bursts (NB) metrics and copy them in a separate Excel tab.
5. Sort the data based on number of NB (from highest to lowest) first and WMFR next.
6. Distribute wells across treatment groups to achieve a comparable distribution of baseline activity. For instance, for six treatment groups, the first six most active wells are each assigned to a different treatment group, followed by assignment of the next six most active wells each to a different treatment group, and so-forth.

4 Notes

1. PEI at high concentration is viscous. To help prepare an accurate solution, weigh the desired amount of PEI in 50 mL conical tube and then add the borate buffer to reach the final 0.1% PEI concentration.
2. PEI is highly toxic to cells. Carefully wash the plate after coating by aspirating one column at a time and adding DPBS immediately after.

3. If the cells appear to spread over the well, the concentration of laminin in the drop that is spotted over the electrodes can be increased up to 240 μg/mL.
4. It is important to place the drop of laminin in the center of the well to increase the likelihood of neurons adhering to the electrode area. For a better placement, use 20 μL tips, position the tip loaded with laminin solution as close as possible to the surface of the well without touching it to avoid damaging the electrodes and then quickly shoot the solution on the electrode surface. Changing tips after each well reduces the chances that the laminin solution sticks to the tip wall and allows easier placement.
5. Before transferring the cells from the cryovial to the conical tube, wash the 1 mL tip with NBA-K to avoid cells sticking to the internal wall of the tip.
6. Since the cells are in such a high concentration at this step, it is very important to calculate the exact volume in which the cells are resuspended. Using a 200 μL tip, by trial and error, calculate how much volume is left in the tube and then add the rest of the medium to reach the final volume of resuspension.
7. Timing is critical to avoid dislodging the cells or letting them dry. Add medium very gently and only along the wall of the well. It should take about 5 min to fill one plate with medium.
8. After 4 weeks of differentiation, a large number of cells are harvested from a 60 mm culture dish. Coat the appropriate number of 60 mm dishes for replating them accordingly.
9. After 4 weeks on the dish, cells will come off as a sheet. Leave them in the incubator for 30 min, and then check them every 5 min until the cells are floating in the medium.
10. This is a crucial step for replating. Clumps need to be broken down before passing the cells through the strainer but without over triturating. Use a 1 mL tip and gently pipette up and down not more than 25–30 times. At this point, if the clumps are still not all broken, stop triturating and proceed with filtering the cells through the strainer.
11. At the second replating, if no clumps are observed, the cells do not need to be filtered through the 70 μm strainer.
12. The full media change and removal of BrainPhys supplement can temporarily alter the activity. If after 2 h the activity is still affected, wait one more hour and then check the activity frequently to see when it is back to the initial recording before starting the experiment.

Acknowledgments

We thank Drs. Sandy Hinkley, Steven Biesmans, Sean Sherman, Priyanka Kakad, and Isabel Onofre for contributions they made to the development of the protocols described herein. This work was supported by grants to A.G.B from the United States Department of Defense, US Air Force (FA8650-18-2-6971), National Institutes of Health (U19MH106434), Janssen Pharmaceuticals, and the Viterbi Family Foundation of the Jewish Community Foundation San Diego.

References

1. Bliss TV, Lomo T (1973) Long-lasting potentiation of synaptic transmission in the dentate area of the anaesthetized rabbit following stimulation of the perforant path. J Physiol 232(2): 331–356
2. Nicoll RA (2017) A brief history of long-term potentiation. Neuron 93(2):281–290
3. Picconi B, Piccoli G, Calabresi P (2012) Synaptic dysfunction in Parkinson's disease. Adv Exp Med Biol 970:553–572
4. Jung NH, Janzarik WG, Delvendahl I, Munchau A, Biscaldi M, Mainberger F et al (2013) Impaired induction of long-term potentiation-like plasticity in patients with high-functioning autism and Asperger syndrome. Dev Med Child Neurol 55(1):83–89
5. Mould AW, Hall NA, Milosevic I, Tunbridge EM (2021) Targeting synaptic plasticity in schizophrenia: insights from genomic studies. Trends Mol Med 27(11):1022–1032
6. Zak N, Moberget T, Boen E, Boye B, Waage TR, Dietrichs E et al (2018) Longitudinal and cross-sectional investigations of long-term potentiation-like cortical plasticity in bipolar disorder type II and healthy individuals. Transl Psychiatry 8(1):103
7. Mameli M, Luscher C (2011) Synaptic plasticity and addiction: learning mechanisms gone awry. Neuropharmacology 61(7):1052–1059
8. Selkoe DJ (2002) Alzheimer's disease is a synaptic failure. Science 298(5594):789–791
9. Nakagawa M, Koyanagi M, Tanabe K, Takahashi K, Ichisaka T, Aoi T et al (2008) Generation of induced pluripotent stem cells without Myc from mouse and human fibroblasts. Nat Biotechnol 26(1):101–106
10. Takahashi K, Tanabe K, Ohnuki M, Narita M, Ichisaka T, Tomoda K et al (2007) Induction of pluripotent stem cells from adult human fibroblasts by defined factors. Cell 131(5): 861–872
11. Molnar E (2011) Long-term potentiation in cultured hippocampal neurons. Semin Cell Dev Biol 22(5):506–513
12. Opitz T, De Lima AD, Voigt T (2002) Spontaneous development of synchronous oscillatory activity during maturation of cortical networks in vitro. J Neurophysiol 88(5):2196–2206
13. Brofiga M, Pisano M, Raiteri R, Massobrio P (2021) On the road to the brain-on-a-chip: a review on strategies, methods, and applications. J Neural Eng 18(4):10.1088/1741-2552/ac15e4
14. Odawara A, Saitoh Y, Alhebshi AH, Gotoh M, Suzuki I (2014) Long-term electrophysiological activity and pharmacological response of a human induced pluripotent stem cell-derived neuron and astrocyte co-culture. Biochem Biophys Res Commun 443(4):1176–1181
15. Heikkila TJ, Yla-Outinen L, Tanskanen JM, Lappalainen RS, Skottman H, Suuronen R et al (2009) Human embryonic stem cell-derived neuronal cells form spontaneously active neuronal networks in vitro. Exp Neurol 218(1):109–116
16. Arnold FJ, Hofmann F, Bengtson CP, Wittmann M, Vanhoutte P, Bading H (2005) Microelectrode array recordings of cultured hippocampal networks reveal a simple model for transcription and protein synthesis-dependent plasticity. J Physiol 564(Pt 1):3–19
17. Pegoraro S, Broccard FD, Ruaro ME, Bianchini D, Avossa D, Pastore G et al (2010) Sequential steps underlying neuronal plasticity induced by a transient exposure to gabazine. J Cell Physiol 222(3):713–728
18. Niedringhaus M, Chen X, Dzakpasu R, Conant K (2012) MMPs and soluble ICAM-5 increase neuronal excitability within in vitro networks of hippocampal neurons. PLoS One 7(8):e42631
19. Otmakhov N, Khibnik L, Otmakhova N, Carpenter S, Riahi S, Asrican B et al (2004)

Forskolin-induced LTP in the CA1 hippocampal region is NMDA receptor dependent. J Neurophysiol 91(5):1955–1962
20. Pre D, Wooten AT, Biesmans S, Hinckley S, Zhou H, Sherman SP et al (2022) Development of a platform to investigate long-term potentiation in human iPSC-derived neuronal networks. Stem Cell Rep 17(9):2141–2155
21. Bardy C, van den Hurk M, Eames T, Marchand C, Hernandez RV, Kellogg M et al (2015) Neuronal medium that supports basic synaptic functions and activity of human neurons in vitro. Proc Natl Acad Sci U S A 112(20): E2725–E2734
22. Shi Y, Kirwan P, Livesey FJ (2012) Directed differentiation of human pluripotent stem cells to cerebral cortex neurons and neural networks. Nat Protoc 7(10):1836–1846
23. Beers J, Gulbranson DR, George N, Siniscalchi LI, Jones J, Thomson JA et al (2012) Passaging and colony expansion of human pluripotent stem cells by enzyme-free dissociation in chemically defined culture conditions. Nat Protoc 7(11):2029–2040

Index

Yu-Wen Alvin Huang and ChangHui Pak (eds.), *Stem Cell-Based Neural Model Systems for Brain Disorders*, Methods in Molecular Biology, vol. 2683, https://doi.org/10.1007/978-1-0716-3287-1,

MIX
Papier aus verantwortungsvollen Quellen
Paper from responsible sources
FSC® C105338

If you have any concerns about our products,
you can contact us on
ProductSafety@springernature.com

In case Publisher is established outside the EU,
the EU authorized representative is:
Springer Nature Customer Service Center GmbH
Europaplatz 3, 69115 Heidelberg, Germany

Printed by Libri Plureos GmbH
in Hamburg, Germany